Applied Plant Science Experimental Design and Statistical Analysis Using SAS® OnDemand for Academics

Applied Plant Science Experimental Design and Statistical Analysis Using SAS® OnDemand for Academics

Edward F. Durner, Ph.D.
Associate Research Professor
Department of Plant Biology
Rutgers, The State University of New Jersey
59 Dudley Road
New Brunswick, New Jersey 08901-8520

 CABI

CABI is a trading name of CAB International

CABI
Nosworthy Way
Wallingford
Oxfordshire OX10 8DE
UK

CABI
WeWork
One Lincoln St
24th Floor
Boston, MA 02111
USA

Tel: +44 (0)1491 832111
Fax: +44 (0)1491 833508
E-mail: info@cabi.org
Website: www.cabi.org

Tel: +1 (617)682-9015
E-mail: cabi-nao@cabi.org

A catalogue record for this book is available from the British Library, London, UK.

Library of Congress Cataloging-in-Publication Data

Names: Durner, Edward Francis, author.
Title: Applied plant science experimental design and statistical analysis using the SAS® OnDemand for Academics / Edward F. Durner, Ph.D., Associate Research Professor, Department of Plant Biology, Rutgers, The State University of New Jersey.
Description: Wallingford, Oxfordshire ; Boston, MA : CAB International, [2021] | Includes bibliographical references and index. | Summary: "Presenting the most frequently used statistical methods for plant science in a non-intimidating fashion and teaching the appropriate use of SAS® within the context of plant science research, this book includes numerous examples and is a key resource for students and scientists undertaking statistical analysis and experimental design"-- Provided by publisher.
Identifiers: LCCN 2020052562 (print) | LCCN 2020052563 (ebook) | ISBN 9781789249927 ISBN 9781789245981 (paperback) | ISBN 9781789245998 (ebook) | ISBN 9781789246001 (epub)
Subjects: LCSH: Botany--Statistical methods. | Experimental design. | SAS (Computer file)
Classification: LCC QK46.5.S7 D87 2021 (print) | LCC QK46.5.S7 (ebook) | DDC 580.7/27--dc23
LC record available at https://lccn.loc.gov/2020052562
LC ebook record available at https://lccn.loc.gov/2020052563

References to Internet websites (URLs) were accurate at the time of writing.

ISBN-13: 9781789249927 (hardback) 9781789245981 (paperback)
        9781789245998 (ePDF)
        9781789246001 (ePub)

DOI: 10.1079/9781789249927.0000

Commissioning Editor: Rachael Russell
Editorial Assistant: Emma McCann
Production Editor: James Bishop

Typeset by SPi, Pondicherry, India
Printed and bound in the UK by CPI Group (UK) Ltd, Croydon, CR0 4YY

Contents

Online Resources

Available at http://www.cabi.org/openresources/45981

Introduction to Plant Science Research and Experimentation

Plant Science Research

Plant science research is an organized investigation using the scientific method to increase our knowledge regarding some aspect of plant biology. It requires the collection, analysis and interpretation of data, which are obtained from one of three sources: observational studies, sample surveys and comparative experiments.

Sources of Data

Observational studies

In observational studies, large amounts of data are collected and the analysis is usually difficult. Results can often be interpreted in many different ways, resulting in controversy. There is no real control as to what is going on with the experimental units, but rather, one simply observes them. For example, the experimental units might be peach trees in an orchard and the researcher might simply observe the state of each tree (living or dead), trunk circumference, canopy height and width, etc. No treatments are applied and variability among the trees due to location in the field is not considered in the analysis.

Sample surveys

With sample surveys, individuals from a population are randomly selected in order to make an inference about characteristics of the general population. Systematic sampling techniques have been developed which lead to much easier analysis and interpretation compared with observational studies.

Comparative experiments

In comparative experiments, the researcher imparts some 'treatment' on the experimental units and measures a response or lack thereof to the treatment. Observed variability among the experimental units prior to treatment application is accounted for by using statistical techniques such as blocking or covariance analysis. Blocking accounts for variation due to gradients in the environment such as temperature, moisture or soil type. Differences in some characteristic of the experimental units such as height, trunk circumference, or vigor can also be used as criteria for blocking. An analysis of covariance can be used to account for any observed pre-treatment variability. The analysis of covariance is covered in Chapter 17.

DOI: 10.1079/9781789249927.0001

In comparative experiments randomization is crucial in order to prevent bias. Randomization ensures that each experimental unit has an equally likely chance of receiving any particular treatment as any other experimental unit. Replication is also necessary in order to estimate variation, a key component of statistical hypothesis testing.

Steps in the Research Process

Finding ideas for research

Research requires an orderly investigation using the scientific method. Ideas for research are generated from problems seen 'in the field', and investigations may address applied and/or basic aspects of the problem. Problems in the field are often solved with applied research. The solution often leads to research into the basic mechanisms of how the applied solution worked. There may also be interest in a particular problem for academic reasons (knowledge for knowledge's sake).

In either case, the problem must be clearly and concisely defined. A broad general question prevents a focused evaluation of the problem at hand and often leads to general rather than specific conclusions. A carefully defined objective often leads to a specific conclusion and a focused answer to the problem. In order to carefully define the problem, a thorough literature review of the current knowledge regarding the problem must be conducted. There is no need to research a problem that has already been solved.

Determine a hypothesis and establish goals

Once these steps have been completed, a hypothesis must be formulated which clearly identifies the research objectives. The clear and concise definition of the problem along with clear research goals are the foundation of good research. There is no point in conducting an extensive research project and ending up with few answers because the problem was never clearly defined and the goals were never clearly identified. Once the hypothesis is established and the goals are clearly defined, appropriate experiments can be constructed.

Key elements of experimentation

Planning

There are a number of key elements to good experimentation. The most important component of research is planning. The time spent reviewing the literature and thinking about the problem is time well spent. With good planning, maximum information will be obtained and problems often encountered in experimental implementation and data analysis are avoided.

Treatment selection

Appropriate treatments that will answer specific questions established by the hypothesis must be selected. Suppose a fertility experiment designed to optimize tree growth was needed and current knowledge indicates that the general recommendation for nitrogen (N) is from 5 to 20 kg of actual N per tree per year. Does it make sense for the treatments to be 0, 50, and 150 kg of N per tree per year? Probably not. Zero N is not a good choice, since all plants need some N, and 50 or 150 kg N are way above the current recommendations. Perhaps 15, 20, 25 and 30 kg might be wiser choices. It all depends on what other researchers have evaluated and what makes sense economically and from a grower's viewpoint. The selection of treatments is often guided by what is discovered in the review phase of planning. It is best to keep experiments as simple as possible with a reasonable number of treatments.

Determining the number of replications

Replication will be covered in more detail in Chapter 3. For now, a replication is an exact duplication of a treatment on more than one experimental unit. The amount of replication is often governed by monetary and time resources. An experiment should: (1) detect differences on the order of magnitude desired (for example, kg or g?); and (2) provide confidence in the results. A general rule of thumb for the number of replicates used in plant science research is somewhere between five and 15, even though there are many examples of experiments which utilized fewer or more replicates.

Identification of experimental units

An experimental unit is that entity to which a treatment is applied. A treatment can also be some inherent property of the experimental unit, such as cultivar. In the plant sciences, the experimental unit is usually either a single plant or a group of plants called a plot. When working with perennial crops, especially trees, the experimental unit may be a single branch. In any case one must be able to clearly identify the experimental unit.

Determining the experimental design

The selection of an appropriate experimental design is just as important as treatment selection. An appropriate experimental design holds treatments efficiently and minimizes variability. This topic will be covered extensively throughout this book.

Experiment implementation

Once planned, an experiment must be implemented. This is the careful integration of the treatments with the design. Careful observation and collection of pertinent data must also be performed. If it is unclear as to whether or not particular data are needed, even after careful thought, they should be recorded anyway. Data can never be retrieved from a terminated experiment.

Be aware of systematic errors, especially in applying treatments or taking data. There must be no bias in treatment application or data collection, whether conscious or unconscious. For example, suppose in an experiment 30 plant apices were observed to count the

---

number of cells in the outer layer of the meristem and the experiment was repeated over 14 days. There are six treatments and five apices per treatment; samples 1 to 5 are treatment 1, 6 to 10 are treatment 2, etc. The treatment cell counts might be inadvertently biased if samples are systematically evaluated from samples 1 to 30 each day. Early in the day counting may be performed meticulously and counts would be very precise. Later in the day the person counting cells is likely tired and may count too quickly, so cell counts might not be very reliable.

The solution could be to impose a design on the counting process and each day randomly assign each sample a number from 1 to 30 and count in that order. This would be a completely random design. Time of day could be considered and blocked off such that one sample from each treatment would be counted during each block of time during the day. This would be a randomized complete block design. Similarly, if there is concern that over the 14 days there may be a day effect, one could consider the days as a main plot and the time of the day as a sub-plot. This would be a split-plot design. Designs will be covered in Chapters 6 and 12–14.

Experiments should be repeated over time and space if possible.

Data analysis and interpretation

The thorough and appropriate analysis of data based on the hypothesis, treatment structure and experimental design is crucial to the scientific process. A step-by-step analysis ensures that key components of the analysis are not skipped. The analysis must be appropriate for the type of data collected, the goals of the work, and the experimental design that 'holds' the treatments. Both of these components will be addressed at length in the chapters which follow.

Interpretation of the results is also important and statistical significance versus practical significance must be examined. For example, a difference in yield of a peach tree of 1 kg per tree may be statistically significant, but may be of little practical significance. Consider that 1 kg is the equivalent of four or five peaches and a mature peach tree may produce 500 to 1000 fruits. One of the responsibilities of the researcher is to determine which statistical differences have practical significance and to discuss them when interpreting results.

Condensation and presentation of results

One of the final steps in research is to condense and present the results in an enlightening and meaningful manner. This includes deciding whether tables, graphs or charts are appropriate and what key components should be included in each of these options. In addition, the work will likely be written as an article or report for publication, thus the researcher must know the specific requirements for publication in different outlets. These topics will be covered in Chapter 21.

Re-evaluating the problem and future work

Finally, work should be evaluated to determine if further experimentation is needed to clarify any issues that may have not been addressed with the present research. In addition, experiments often produce results which lead to questions, thus the process begins all over. It is always good to think of research as a cycle.

Special problems and characteristics of plant science research

There are a number of factors that are somewhat unique to plant science research. Plants are annuals, biennials or perennials and their life cycle and stage of development are often important in an experiment. Plant science research may be conducted in the field, a greenhouse, high or low tunnels, growth chambers or a lab. Each location has unique characteristics which should be considered when planning, implementing and analyzing experiments. In addition, experiments are often performed over a long time and subjected to many environmental and biotic stresses. These stresses often lead to plant injury or death, which leads to missing data. All of these factors will be considered in later chapters.

Goals of this book

The purpose of this book is to provide researchers with appropriate tools for implementing, managing and analyzing their experiments. This is accomplished by teaching SAS® and illustrating why specific procedures fit specific research situations. SAS® is available on different platforms and this text will use SAS® OnDemand for Academics. Setup and use of SAS® OnDemand for Academics will be covered in the next chapter.

Basic statistics will be reviewed, thus previous statistical training, while helpful, is not a prerequisite for using this book. The analysis of variance (ANOVA) and experimental design principles for field research will be covered as well as many experimental designs, including completely random, randomized complete block, Latin square, split-plot, and split-split plot. The estimation of treatment effects using contrasts, regression (with diagnostics) and mean separation techniques will provide powerful tools for the researcher.

Other topics include: analyzing a series of experiments conducted over seasons, years and locations; what to do when data doesn't behave 'normally'; the analysis of covariance; non-parametric techniques; sampling; and presentation of results.

2 An Introduction to SAS® OnDemand for Academics

SAS® (*Statistical Analysis System*) is a software system for data management, analysis and presentation. SAS® is available for different platforms and all versions basically perform the same procedures. The main differences among them are how the user interacts with the software via the operating system. Data set size used to be an issue, often limited by the computer's memory, but that problem hardly exists anymore with today's modern machines.

For this book, SAS® OnDemand for Academics will be used. It is an online version of the software and it replaces the SAS®UniversityEdition implementation of SAS®. SAS® OnDemand for Academics is free and requires no software installation. Once you register with SAS® and set up your free account, you simply need broadband internet access to use the software.

SAS® software is composed of many modules, each of which perform a specific function for the system. There are a number of major components to SAS®, including SAS/Base®, SAS/Stat®, SAS/Graph®, etc. We will be using SAS®OnDemand for Academics thus there is no need to digress into the different modules available. Everything needed for this book is included as part of SAS®OnDemand for Academics. There are many different options in SAS®, thus there may be more than one way to perform specific tasks. In general, the simplest method is presented here.

Registering with SAS®

In order to use SAS®OnDemand for Academics you must register with SAS®. Simply go to the following web address and follow the instructions:

https://www.sas.com/en_us/software/on-demand-for-academics.html#928a7e06-1416-4d88-b966-311df1bdfea6

At this address you will see:

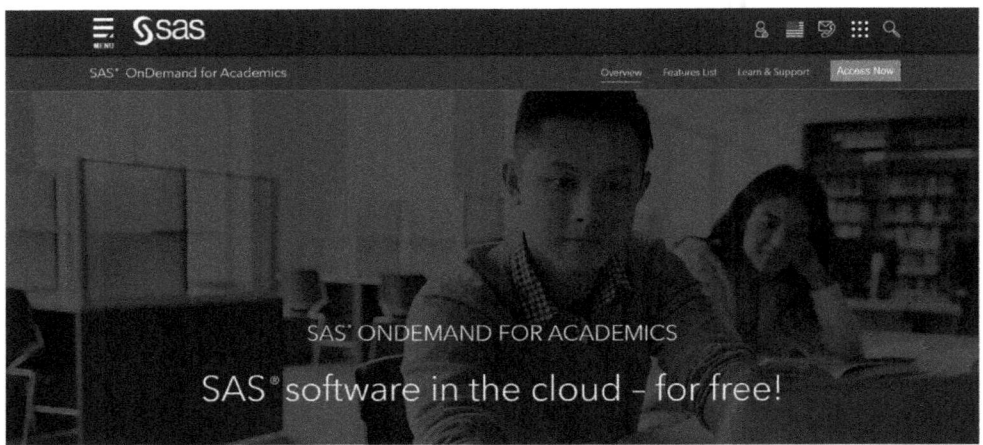

DOI: 10.1079/9781789249927.0002

From this page you can register with SAS® and register to use SAS® OnDemand for Academics which is simple and takes only a few minutes. You are then ready to use SAS®. That's it! Just click the 'Access Now' button in the upper right corner of the

Once you have registered for access and clicked the small box 'Get Access' in the upper right hand corner of the screen. You will then see the sign in window:

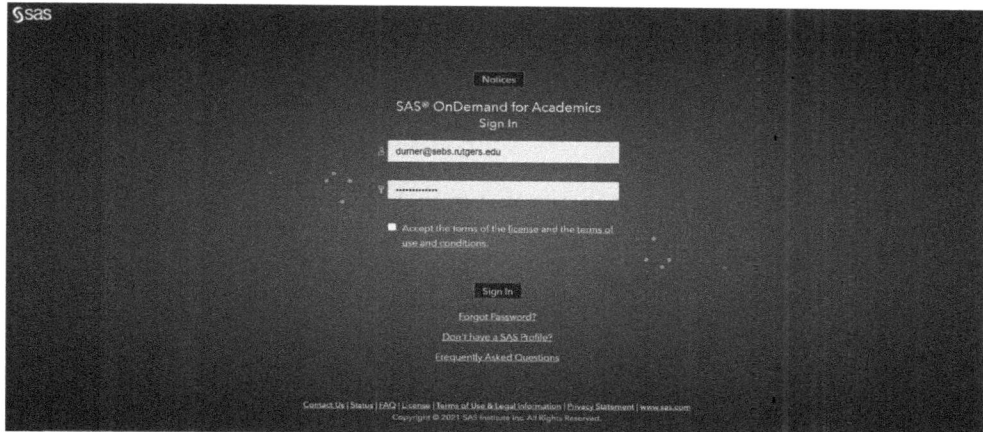

Once you Accept the terms of use and enter your login information, you will be taken to the following screen:

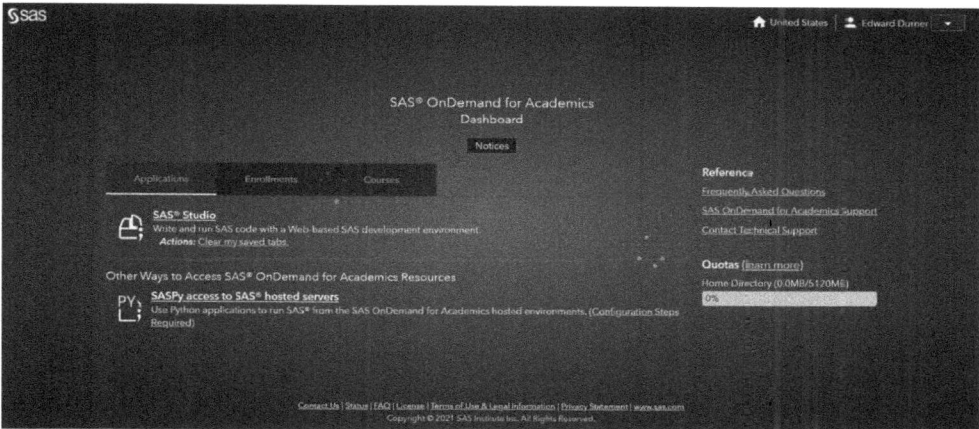

To use SAS® with this book, simply click the SAS® Studio button which will take you to the SAS® Studio which looks like this:

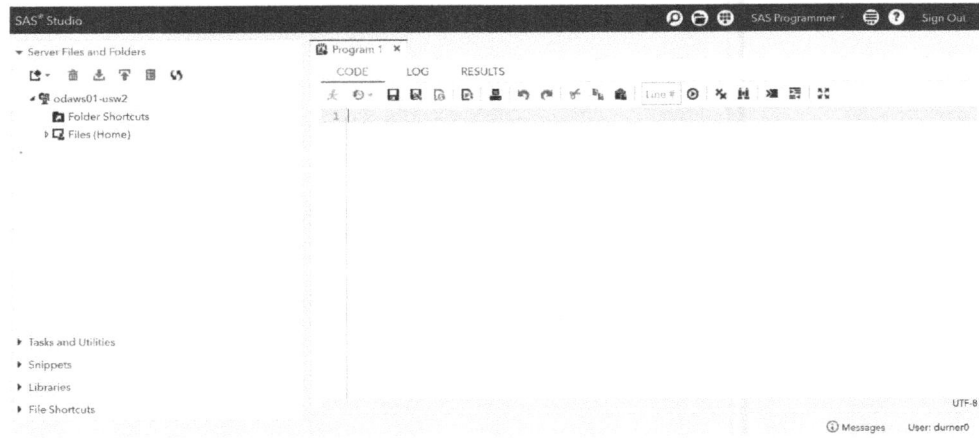

The right window is where all SAS® work is performed. To the left notice a directory tree of SAS® folders. These folders are created by SAS®. If you want to create separate folders in which to save files and data, you can follow directions for creating new folders using SAS® help (the '?' in the upper right corner of the SAS® Studio page).

To create your own folder in which to store files, highlight the 'Files (Home)' label on the menu tree underneath the 'Server Files and Folders' label on the upper left side of the SAS® Studio window above. Click the 'New' icon (the first icon on the left just below the heading 'Server Files and Folders' in the SAS® Studio window above. Do this while the 'Files (Home)' is still highlighted. Select 'Folder' and the following will appear:

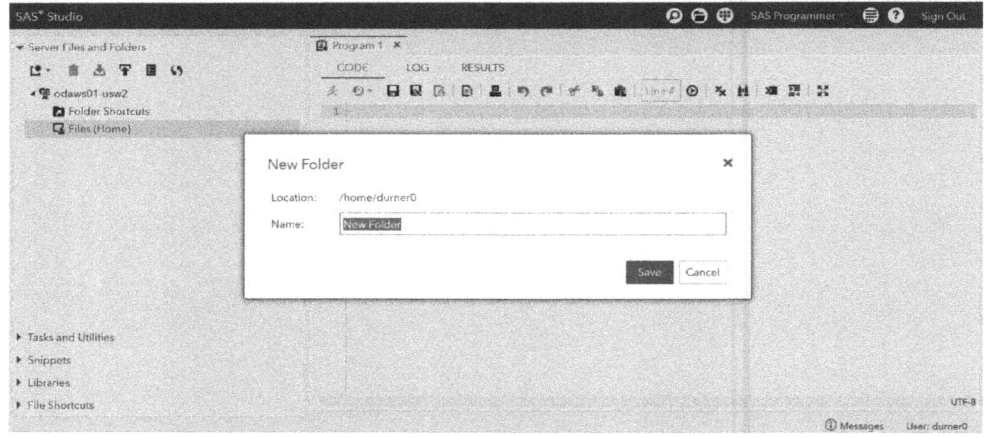

Enter a name for your folder in the 'Name' box. I'll use 'Edward Durner'. After clicking 'Save', your screen should look like this:

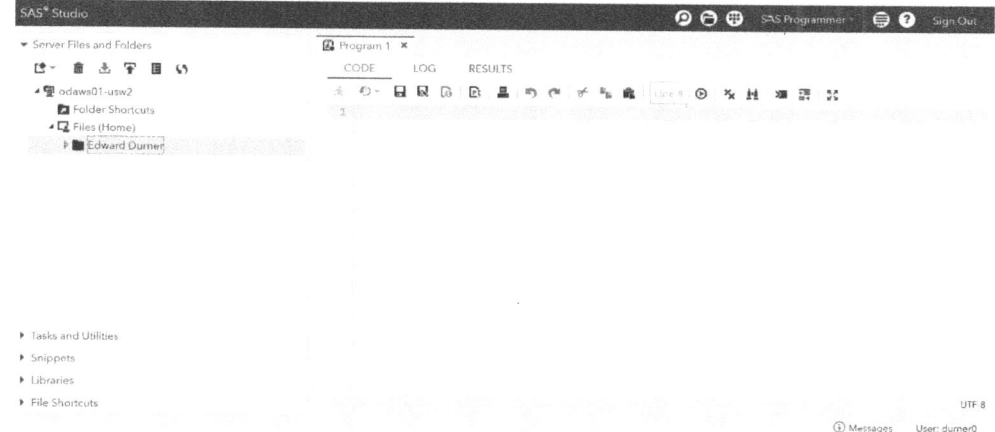

Notice my new folder (Edward Durner) is located in the 'Files (Home)' folder. Now I can save files to this folder.

A first SAS® program

Now let's write our first SAS® program. Notice the right window is named 'Program 1' and 'CODE' is highlighted, with a line numbered one also highlighted. CODE is where you enter SAS® instructions to create a SAS® program which will produce output that will be presented in its own window. The LOG window will list what SAS® has done with respect to the program (errors will be shown there if there are any in the program).

This first simple SAS® program will create a SAS® dataset and create a printout of it. A SAS® dataset is the information SAS® works with when analyzing data and is created within the program. Enter the following code on line 1 (don't forget the semi-colon, omitting it will be the source of errors, especially later on with more complicated programs):

```
DATA one;
```

and hit enter. SAS® will go to a second line in the 'Program 1' window :

Now enter the following lines:

```
INPUT block $ treat yld;
CARDS;
a 2 34
b 3 21
RUN;
PROC PRINT DATA = one;
RUN;
```

Note the semicolon (;) at the end of some of the lines but not the others. The screen should now look like:

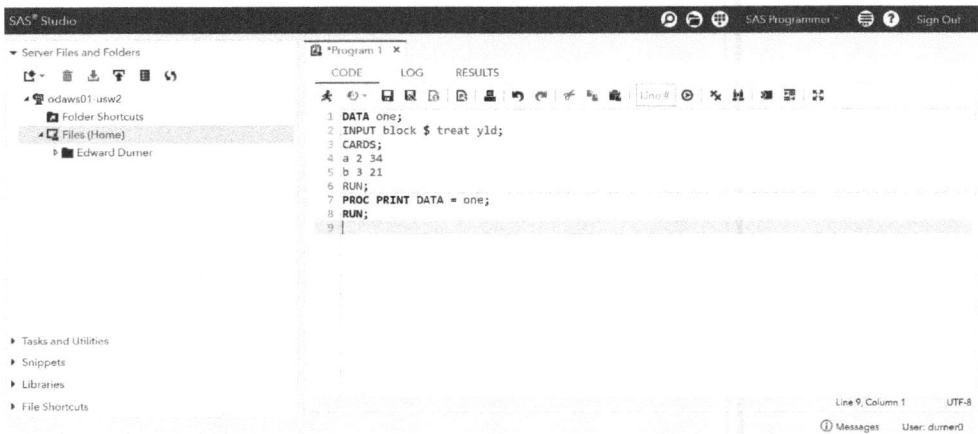

Save the file using the Save As icon, the third icon from the left just below the 'Program 1' tab. SAS® will automatically add the '.SAS' to the filename. You simply need to identify where you want to save the file. Highlight the folder you just created as the location to save your file. For example:

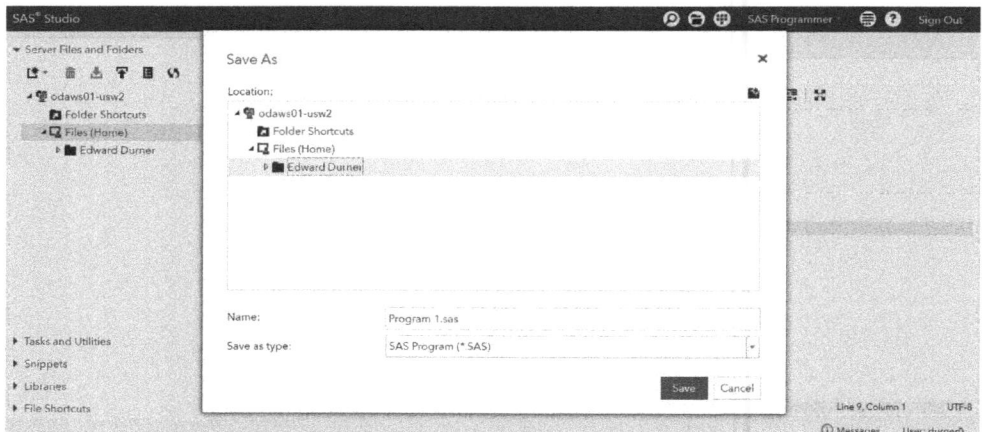

Will save 'Program 1.SAS' in the 'Edward Durner' folder. Now notice, the file just saved is listed in the 'Edward Durner' folder:

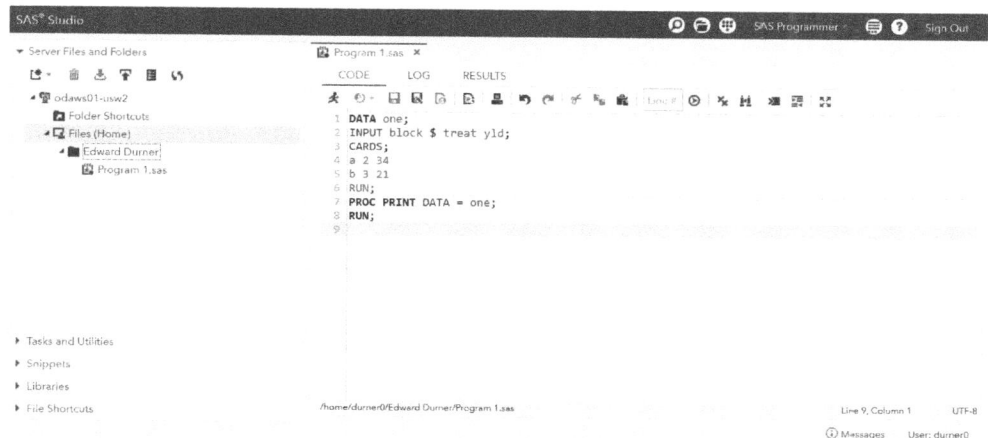

If you don't specify a folder, SAS will save files in the 'Files (Home)' folder.

Submit the program for SAS® processing by pressing the icon that looks like a person running located just under the 'Program 1' tab of the right window on the screen or by pressing the 'F3' key.

After submitting the program for processing, the 'Results' window will be presented with a printout of your dataset:

Obs	block	treat	yld
1	A	2	34
2	B	3	21

When SAS runs a program, it generates a log of activity which is presented in the LOG window. Errors, warnings and notes are provided for documenting SAS activity. We are most interested in the 'Error' section since this is where we will look if our program fails to run correctly. This is particularly important with more complex programs.

If your program fails to run, check the LOG under 'Errors'. Errors will be listed in red. The most common error is forgetting to put a semicolon at the end of each program statement. Other errors are more difficult to find. When you need to correct an error, simply click on 'CODE' and fix the error, then resubmit the program for processing.

You can print a copy of nearly any window or sub-window by clicking the printer icon located on each window's toolbar. To exit SAS®, select 'Sign Out' in the upper right hand corner of the screen.

The Mechanics of SAS®

The SAS® system is a set of related, integrated computer programs which provide a powerful system for data management, analysis and presentation. We won't need to be concerned with the particulars of the system since their implementation and integration are managed by the main program.

Some basic concepts help in learning SAS®. The *SAS® language* is the computer vocabulary and syntax with which you tell SAS® what you want it to do in a language that it "understands". You use this language to develop *SAS® statements*, which are sets of SAS® keywords, SAS® names or special characters that request an operation or gives SAS® information. All SAS® statements must end in a semicolon. Repeat this to yourself five times! Missing semicolons are probably the number one reason for the dreaded error message. Several (or many) SAS® statements are put together to perform some task and the group of statements is called a *SAS® program*. A word which identifies the type of statement a SAS® statement is, is called a *SAS® keyword*. Keywords include DATA, INFILE, INPUT, etc. For this book, a keyword will always be in all caps, but SAS® doesn't require all caps. In SAS® programs we provide SAS® information using *SAS® names*. SAS® names can be up to eight characters long, no blanks, $, @, or # are allowed and the first character must be a letter or an underscore. Certain names both begin and end with an underscore, _N_ or _ERROR_, and can't be used. We will get more specific about SAS® names shortly.

SAS® Datasets

SAS® has a special way of storing information via SAS® files. SAS® files cannot be accessed by other programs. The most important SAS® file is the *dataset*. Think of a SAS® dataset as a rectangular grid of rows and columns, much like a spreadsheet. The rows represent individual observations while the columns represent characteristics about each observation. We normally call these characteristics *variables*.

SAS® Variables

SAS® variables are either numeric or character.

SAS® Programs

The statements we provide SAS® in order to get SAS® to do what we want are collectively called a SAS® program. SAS® programs can be envisioned as a series of steps, each of which accomplish a specific task. The sum of the individual tasks equals the final desired result. Two main types of steps in a SAS® program are the DATA and PROC steps. There may be many DATA and PROC steps in a single SAS® program. The DATA step(s) is(are) used to get data into the form SAS® utilizes. The PROC step(s) is (are) used to perform the analysis we want.

SAS® Statements

A SAS® statement is an instruction to SAS®. It may request an action, request some modification of data or of another statement, or it may direct the execution of the entire SAS® program. All SAS® statements end with the semicolon (;) and most of them begin with a

SAS® keyword. SAS® is not case sensitive and is not affected by spacing within or among statements. However, neat and consistently written programs are easier to follow and edit than sloppy and haphazardly written ones. A good rule of thumb is to put one statement per line, with reasonable spacing within the line. For example, the following statements, all entered onto one line are valid in SAS®:

```
DATA one;SET two;IF a>b;PROC PRINT DATA=one;RUN;
```

The program is much easier to read and follow if entered as:

```
DATA one;
SET two;
IF a > b;
PROC PRINT DATA = one;
RUN;
```

SAS® statements usually contain additional elements such as *statement modifiers, names, format identifiers, operators, function terms,* and *expressions.* Statement modifiers are keywords which tell SAS® to make special refinements to the way it carries out the instructions in the statement. We will see many of these when we use the statistical procedures. SAS® names are names we create to identify information for the SAS® system, such as files or variables. As previously mentioned, SAS® names may be up to 8 characters long and the first character of the name MUST BE a letter or an underscore (no numbers as a first character). The remaining characters can be a combination of letters, numerals and underscores, no spaces are allowed within a name, and no non-alphanumeric characters other than the underscore are allowed. In other words, you can't use *&^%#@/[]{} etc., in a SAS® name. It's a good idea to never name a variable or a dataset using both beginning and ending underscores, since SAS® uses this convention for special variables, such as _N_ or _ERROR_.

SAS® statements often have lists of variables which may be abbreviated in certain instances. If the list consists of a constant character portion followed by an incrementing digit portion, the entire list can be abbreviated by listing the first and last variable separated by a hyphen. For example the variable list:

```
VAR DIR1 DIR2 DIR3 DIR4;
```

can be abbreviated as:

```
VAR DIR1-DIR4;
```

This type of abbreviation can be used even if the variables haven't been created yet. They will be created when listed. (This will make more sense later).

If a set of variables that have already been created follow one another in the SAS® dataset, you can often use the double hyphen to abbreviate the list. For example:

```
KEEP weight height age sex race score detail;
```

can be shortened to:

```
KEEP weight--detail;
```

KEEP is a SAS® keyword. You may also indicate that you are referring to only numeric or character variables within a list using a variation of the double hyphen method. To keep only numeric variables in the previous KEEP statement, you would write:

```
KEEP weight - numeric - detail;
```
To keep only character variables in the previous KEEP statement, you would write:
```
KEEP weight - character - detail;
```

Sometimes you might want SAS® to output variables in a certain format such as date or monetary, or with a certain number of decimal places. Format identifiers are used for this and they will be covered when appropriate. Operators are used for comparing and/ or combining items, data manipulation, mathematical calculations or value assignment. SAS® also has many built in functions such as SQRT( ), which would find the square root of the number in parenthesis. A SAS® expression consists of a set of operators and operands which is resolved into one single value. It is important to know the order of expression evaluation in SAS®. Sub-expressions within parentheses are evaluated first. After parentheses, the order is as follows:

1. functions
2. right to left : **, +(prefix), -(prefix), NOT, <>, ><
3. left to right : *, /, +(infix), -(infix), ||, comparisons, AND, OR

It is often desirable to add comments to a SAS® program to provide insight and to document work flow. Comments can be included in a SAS® program by starting a new line with an asterisk (*) followed by the comment and ending with a semicolon. For example:
```
*This is a SAS® comment;
```

Dataset names are used extensively in SAS® programming. SAS® automatically works with the most recently created dataset if one is not specified. This can cause major headaches in certain situations (like regression analysis with diagnostics), thus always specify the dataset explicitly in program statements. Another good practice is to end all programs with the RUN; statement.

SAS® Datasets

In order to process data with SAS®, the data must be in a form that SAS® recognizes: a SAS® dataset. This is accomplished using a series of statements collectively called the data step. The first statement in the data step is the DATA statement which assigns a name to the data set about to be created. The format for the DATA statement is:
```
DATA data-set-name;
```
where *data-set-name* is an eight character or less name you assign to this data set. There are also two-level dataset names to indicate whether the dataset is permanent or is just being used during the current session. Permanent datasets are stored via the SAS® library facility. In the lower left corner of the left window, locate the 'Libraries' indicator and click the triangle to open the 'Libraries' subfolder. When opened, a toolbar with five icons will appear with a 'My Libraries' subfolder underneath it. Subfolders within 'My Folders' include 'MAPS', MAPSGFK', 'MAPSSAS', 'SASDATA', 'SASHELP', 'SASUSER', 'STPSAMP', 'WEBWORK' and 'WORK'.

The 'WORK' subfolder is where temporary SAS® datasets are stored until a session is terminated. Once the current SAS® session is terminated, these datasets are deleted. Only permanent datasets remain on the hard drive after session termination. The other subfolders have functions which will be discussed as required later in the text.

Creating Permanent SAS® datasets.

To create a permanent SAS® dataset, first create a library. On the Libraries toolbar, click the leftmost icon which looks like a filing cabinet. The 'New Library' window will appear. Enter a name for your library such as 'testlib'. SAS® will ask for the name of a folder (in the PATH box of the pop-up window which appears when you're creating the new library) where you want to save your library. By clicking the 'Browse' button for the PATH box, the 'Files (Home)' folder will be selected by default.

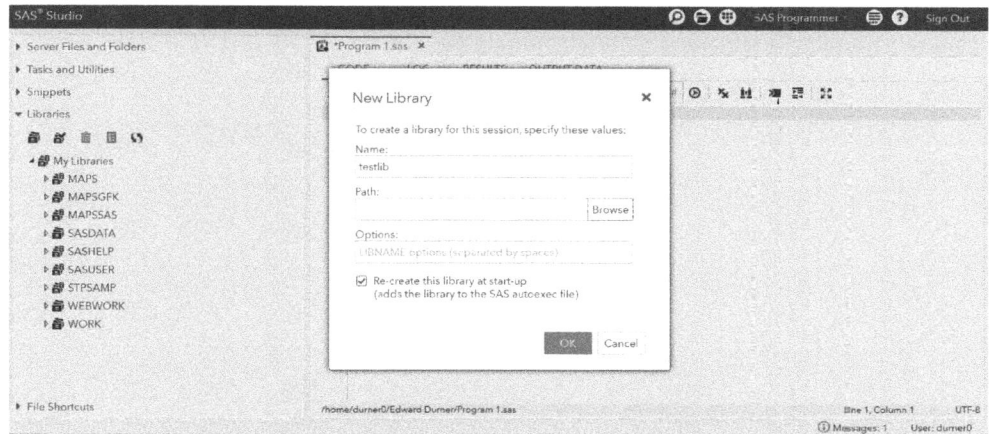

You can change this if you wish.

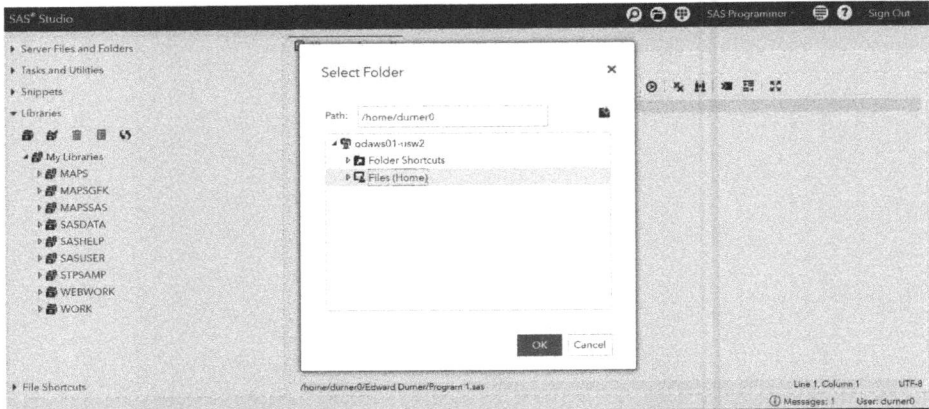

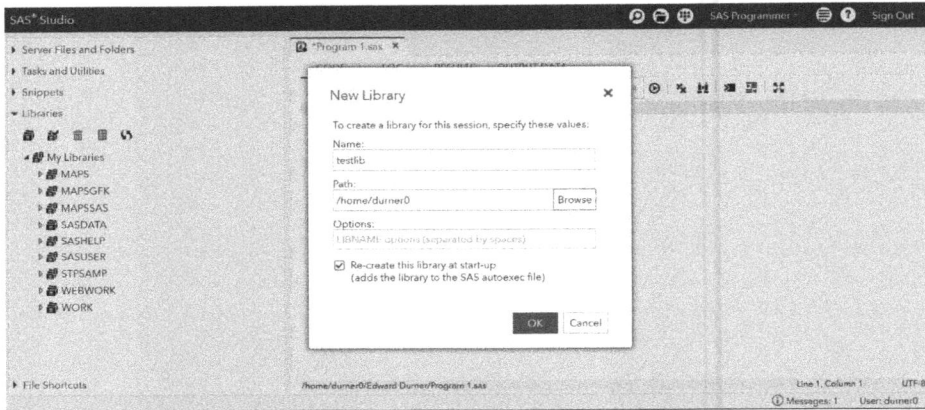

Since you want your new library for use the next time you use SAS®, check the box labeled
'Re-create this library at start-up (adds the library to the SAS autoexec file)'. Click the 'OK' box
and you're done. Now notice the 'TESTLIB' library within your 'My Libraries' folder.

When you create your new library, it will be empty. Try this short program to create a permanent dataset in your newly created library.

Type the following program into the 'Program Editor' window and submit it for processing.

```
DATA testlib.one;
INPUT a b;
CARDS;
1 2
5 6
RUN;
```

Check the 'Log' window to make sure there were no errors. If there were errors, look in the 'Program Editor' window and make sure the program in the 'Program Editor' was typed in exactly as it appears above. Correct any mistakes and resubmit the program. Hopefully, there will be no errors. We will cover the parts of the program shortly.

When the program has finished running (less than a second) the 'OUTPUT DATA' window will appear providing all the information about the dataset you just created and stored in your 'TESTLIB' library. Most of the time you won't need the information displayed in the 'OUTPUT DATA' window, but rather would want to use the dataset for some sort of analysis. It is presented here, so that you will know what it is when it pops up when you create this test permanent dataset in your 'TESTLIB' library.

Look in the left window under 'Libraries'. In the 'My Libraries' folder, you should see your 'testlib' library listed. If you don't see any list under 'Libraries', make sure you created 'testlib' as instructed above. Within the 'testlib' library subfolder, you should see a spreadsheet-like icon with a dot and its name should be 'one'. That's the dataset named one you just created. If you double-click the icon, a new window will open on the right side of your screen to show you the contents of this dataset. If you would like a printout of the dataset, submit the following program and results will appear in the 'Results' window. You can then hit the printer icon (fifth from the left on the 'Results' window toolbar) to produce a hardcopy of the results.

```
PROC PRINT data = testlib.one;
RUN;
```

Datasets in libraries are called upon to be used in various ways in SAS® by using their two-level name, *library.dataset*. When you don't specify a library, SAS® automatically creates a temporary one for you called WORK, which is deleted when you end the SAS® session. Since SAS® already knows this is the default library, you don't even have to be concerned with it. SAS® sees the statement:

```
DATA one;
```

as

```
DATA work.one;
```

Now with a named dataset, we will get real world data into it. There are a number of ways in which you can do this in SAS®. One way is to enter the data directly into the program editor window. We did this earlier to test our library creation. This is called in-stream data. Before you actually enter the data, you need a few statements in the data step to make it work.

The first necessary statement in the data step is the DATA statement which assigns a name to the dataset we are about to create. Create a dataset called 'testdata' in the 'testlib' library by using the statement:

```
DATA testlib.testdata;
```

The next statement in the data step is the INPUT statement. The INPUT statement immediately follows the DATA statement. The INPUT statement tells SAS® the variable names, whether the variables are numeric or character and how the variables are arranged on each line of data. There may also be some informat (additional formatting) information included in the INPUT statement. Character variable names must always be followed by a $ on the INPUT line, as this is the SAS® indicator that the variable is a character variable. Variables without the $ are by default assumed to be numeric. An example:

```
INPUT var1 var2 $ var3;
```

where var1 and var3 are numeric and var2 is character.

The final statement required in the data step is CARDS; which tells SAS® that the lines following the CARDS; statement are actual data values. *Data lines DO NOT end in a semicolon.* Following the last line of data, the final command RUN; should be entered on a separate line.

Here is the complete set of statements for the example of a SAS® dataset called testdata created with in-stream data:

```
DATA testlib.testdata;
INPUT varone vartwo $ varthree;
CARDS;
1 yes 2
2 no 1
RUN;
```

This creates a dataset named testdata with two observations in it. Note the dollar sign to indicate that vartwo is character, and that all lines except the actual data lines end with a semicolon.

Suppose that you wanted to develop a library of SAS® datasets related to a certain study we'll call "Nitrogen" and the dataset we just created above is the first dataset we want to include in this library. Create a 'Nitrogen' library as per the preceding instructions, then create the dataset via:

```
DATA nitrogen.one;
INPUT varone vartwo $ varthree;
CARDS;
1 yes 2
2 no 1
RUN;
```

One advantage to creating permanent datasets is that they let you access the data without having to recreate a new data set for new SAS® sessions. This is particularly useful for large datasets. Many analyses are performed over time, with many revisits to the dataset. Once you create a dataset for a given set of data, you easily access it for processing with a simple statement such as:

```
PROC PRINT DATA=nitrogen.one;
RUN;
```

This two-line program simply creates a printout of the data. You don't have to recreate the dataset each time you want to use it.

The actual data in the preceding examples were entered in what is known as 'list' input. Each observation consists of a list of the values for the variables identified in the INPUT statement in the same order as the INPUT statement, each separated by one blank space. If a data value is missing, it is indicated via a period. Do not use 0 or leave it blank. (Be sure it is missing. There may be times when you measure 0. In that case the observed value is a 0, thus you would input it as such). There may be times when you have a character variable which has an embedded blank space in its' actual values. For example, if one of your variables is 'name', you might encounter a value such as 'John Doe'. In this case you must inform SAS® in the INPUT statement that the name variable has an embedded blank in it. You must also have two blank spaces after each name value. You indicate the embedded blank to SAS® by inserting the & symbol after the $ (indicating a character variable) in the INPUT statement. For example:

```
DATA one;
INPUT name $ & age;
CARDS;
john doe 21
RUN;
```

(There are 2 spaces between doe and 21 even though it may not be evident on the printout.) Character values must be eight or fewer characters in length unless a longer length is specified via formatted input. If in the previous example we wanted to ensure that names up to 20 characters would be read rather than truncated to 8 characters, we would modify the INPUT statement to read :

```
INPUT name $ & 20. age;
```

Notice the period following the 20. This informs SAS® that it is an informat (formatted input). You can also informat numeric values.

Besides list input, SAS® also supports column input. This type of input is useful when data are entered in specific columns. In column input you must be sure of where each variable value lies, columnwise, and it must be consistent for the entire data set. Column input is useful for coded forms. Column input requires planning data input before it occurs. Suppose name data were entered such that the name value for each observation was in columns 1 to 20. Valid name data will be located within the column range of 1 - 20. Leading blanks will be discarded. For example, if the name john doe actually began in column 5, SAS® would discard the blank spaces in columns 1, 2, 3 and 4. Additionally, the space between john and doe would be preserved without including any notification to SAS® that it exists, as was required for list input. Column input is indicated on the INPUT line, for example:

```
INPUT name $ 1-20 age 21-23;
```

This indicates that the character variable 'name' is located in columns 1 - 20 and the numeric variable 'age' is located in columns 21-23.

When numeric values are entered via column input, decimals can be implied via a modification to the INPUT statement. For example, suppose you enter yield data as grams, but now wish to consider it in kilograms (granted, this is a simple example). You could either convert grams to kilograms by entering an expression after the INPUT statement as follows:

```
DATA one;
INPUT yield 1-5;
kg = yield/1000;
CARDS;
1243
2645
RUN;
```

You could save a line of code by entering the following instead:

```
DATA one;
INPUT yield 1-5 3;
CARDS;
1243
2645
23
RUN;
```

The number 3 following the column specification for yield indicates to SAS® that there should be 3 digits to the right of a decimal point, thus 1243 grams is 1.243 kilograms. What about the value of 23? SAS® will interpret it correctly and produce the value of 0.023 ! Try it to prove to yourself it works.

Column input also works well if you want to 'reuse' data in each observation for several different variables. As a simple example suppose you had the following data observation: 23547. This code stands for a specific plant in a breeding program, plant number 23547. In addition, the 54 stands for the year of the original cross, and the 7 stands for the male parent. You need a dataset containing the plant number, the year of the cross and the male parent. There are several ways you could create the dataset. One might be to recode the data and enter it as list input. This would take some time for a large data set. If you used column input, it would be very easy:

```
DATA one;
INPUT plant 1-5 year 3-4 male 5;
CARDS;
23547
RUN;
```

This was a simple example, but it illustrates the point.

Sometimes data is entered on more than one physical line. SAS® has an internal variable called the line pointer to keep track of which line of input it is on for a particular observation. In addition, it maintains a column pointer. Thus you have the power to specifically tell SAS® where to go in your input line of data to get specific information. You reset the line and column pointers either absolutely or relatively. To move to a column 2 columns ahead of the present column you would use a + sign. To move to an absolute column number, you would use the @ sign. Similarly, the relative line pointer movement is accomplished with the / symbol while the absolute line pointer movement is accomplished with the # symbol. An example should help illustrate these points. Suppose you had performed a taste test on new versus old strawberry cultivars and had individuals fill out lengthy questionnaires. You input the data as columns into SAS® and found that you needed 3 physical data lines to hold all the information for one individual. The following data (rightmost columns 20 to 80 truncated for brevity) is entered into SAS® for one individual

```
john doe
123159
25344
```

These three items represent name, birth date, annual salary. You would access the data as follows:

```
DATA one;
INPUT name $ 1-20 / bdate 1-6 / salary 1-7;
john doe
123159
25344
RUN;
```

The '/' tells SAS® to go to the next physical line to get the following data. When SAS® goes to a new line, the column pointer is reset to 1 automatically. What if you wanted only the year of birth? You could access it by using the absolute column pointer

```
DATA one;
INPUT name $ 1-20 / @5 yob 2 / salary 1-7;
john doe
123159
25344
RUN;
```

The 2 following 'yob' informs SAS® that 'yob' is in two columns, while the @5 tells it which column to begin reading the value in. What if you wanted to include the month of birth too, but wanted to read it in after you read the year of birth? You could access it using either the absolute column pointer:

```
DATA one;
INPUT name $ 1-20 / @5 yob 2 @1 mob 2 / salary 1-7;
john doe
```

```
123159
25344
RUN;
```

The 2 following 'mob' indicates that the value is in two columns. The @1 informs SAS®
which column to start with in reading the mob. You could also get the same results using
the relative column pointer:

```
DATA one;
INPUT name $ 1-20 / @5 yob 2 +(-5) mob 2 / salary 1-7;
john doe
123159
25344
RUN;
```

Notice that you move the column pointer with the + symbol then the number, in this
case -5 tells it which direction and how far to move. You might use the line pointer if you
wanted to read the data in this order: birth date, name, salary:

```
DATA one;
INPUT #2 birth date 1-6 #1 name $ 1-20 #3 salary 1-7;
john doe
123159
25344
RUN;
```

You could also use line and column pointers at the same time. Much of the time data will
fit on one physical line thus the line pointers aren't necessary. Similarly, you might not
ever use the column pointers, but it's nice to know that they are there if you need them.
SAS® allows you to mix column and list input, for example:

```
DATA one;
INPUT block $ trtmnt $ yield 33-50 @65 year;
CARDS;
(data goes here)
RUN;
```

The INPUT line informs SAS® that the data should be read as block and trtmnt as
character variables in list input, yield as numeric in columns 33-50, and year as numeric
in list input at or after column 65.

We have seen how SAS® creates a dataset with in-stream data, but what if the date is
stored as a file on a disk? Data can be read into the program editor, but with large files
that is quite cumbersome. There is an easier way. To read data into a SAS® dataset from
a file, the following procedure is used. Immediately after the DATA statement, place an
INFILE statement to inform SAS® of the data file name. The INFILE statement is followed
by an INPUT statement to assign variable names to the data in the file. If the data presented in the example above where stored in the file 'data.dat' on a data stick assigned
the drive letter 'b:', the following statements would be used to create a data set:

```
DATA one;
INFILE 'b:data.dat';
INPUT block $ trtmnt $ yield 33-50 @65 year;
CARDS;
RUN;
```

Besides in-stream and infile methods of creating SAS® datasets, they can also be created from existing SAS® datasets. You use this method to subset a data set or to create new variables in an existing data set. For example, suppose you had a data set named 'one' which contained weight and number of harvested apples and you needed to analyze average fruit weight. You could create this new variable from the original data set:

```
DATA one;
INPUT rep cult $ harwt harnum;
CARDS;
(data lines)
RUN;
```

using the commands:

```
DATA two;
SET one;
avesze = harwt/harnum;
RUN;
```

You could also use this method to drop variables from a data set:

```
DATA three;
SET two;
DROP harwt harnum;
RUN;
```

Actually, you could achieve all of this in one run, but it is presented here in three steps to illustrate programming methods. The one step version might be:

```
DATA one;
INPUT rep cult $ harwt harnum;
avesze = harwt/harnum;
DROP harwt harnum;
CARDS;
(data lines)
RUN;
```

There are many ways of getting data into SAS®, however, to minimize confusion, a single method will be illustrated here and used throughout the rest of this book. I prefer to enter data in a spreadsheet, save the file as a comma delimited (csv) file, then access the data via statements illustrated in the previous paragraphs.

You need to upload the data to the SAS® cloud by clicking the fourth icon from the left (arrow pointing upward into a box) just under the 'Server Files and Folder' heading. First I highlight the 'Edward Durner' folder, since that's where I want the file placed when it is uploaded. I then click the upload icon and a popup window appears letting me select the file from my computer that I want to upload. In this example, I would upload the file 'Chapter Two Data Example 1.csv' from my hard drive. Once it is uploaded, the file appears in my 'Edward Durner' folder.

In this example, data was collected from a strawberry experiment. Each observation occupies one row of a spreadsheet and indicates the row in the field, plot, treatment, number of surviving plants per plot, total number of fruit harvested per plant and total weight (g) of fruit harvested per plant. The following SAS® statements retrieve the data, stored as 'Chapter Two Data Example 1.csv' (comma delimited file), and produces a printout of the dataset created. Explanatory notes follow the SAS® statements.

```
0001  TITLE 'Chapter Two Data Example 1';
0002  DATA one;
0003  INFILE '/home/durner0/Edward Durner/Chapter Two Data Example
1.csv' dlm =',' firstobs=2;
0005  INPUT row plot trt $ plants totnum totwt;
0006  CARDS;
0007  PROC PRINT DATA = one;
0008  RUN;
0001  TITLE 'Chapter Two Data Example 1';
```

Give the program and pages of output a title. Enclose the text of the title in single quotes following the keyword TITLE. You can have as many as 10 titles per program, consecutively indicated as TITLEn where n = 1 to 10. You do not need to indicate that a single title is TITLE1, using the keyword TITLE without the 1 is sufficient. You may also have as many as 10 FOOTNOTEs, using the same format as the TITLE statement, for example:

```
000n  FOOTNOTE1 'This is footnote 1';
000n  FOOTNOTE2 'This is footnote 2';
```

You do not need to indicate that a single footnote is FOOTNOTE1, using the keyword FOOTNOTE without the 1 is sufficient.

```
0002  DATA one;
```

Create a dataset (not permanent) and give it the name 'one'.

```
0003  INFILE '/home/durner0/Edward Durner/Chapter Two Data Example
```

The file with data values is located in the file called 'Chapter Two Data Example 1.csv' located in the subfolder '/home/durner0/Edward Durner'.

To determine where your file is stored, highlight the filename you are interested in, usually a file in one of your libraries since that's where you should upload data to, and right click 'Properties'. The files location will be indicated in the 'Location' box of the properties popup window.

```
0004  1.csv' dlm =',' firstobs=2;
```

The 'dlm' option indicates that the delimited used to separate variable values within each observation. In this case we use the comma, thus we indicate that with the 'dlm = ',' portion of the statement. Note that the delimiter is enclosed in single quotes. The firstobs option indicates to SAS® that the first observation begins on line number 2. Note that 2 is not enclosed in single quotes.. By using this option we avoid having to remove the header row (the row indicating row, trt, plants, etc) from our csv file before saving it on the hard drive. It's a good idea to keep the header row as part of the data file so that you easily recall what each column of the spreadsheet identifies.

```
0005  INPUT row plot trt $ plants totnum totwt;
```

Data is read from the input file as plot, trt, plants, totnum and totwt where plot = plot number, trt = treatment, plants = number of surviving plants per plot, totnum = total number of fruit harvested per plant and totwt = total weight (g) of fruit harvested per plant. The '$' following trt indicates that it is a character value. All other values are numeric.

```
0006  CARDS;
```

The CARDS keyword is used to indicate the end of the DATA step in SAS® processing. Lines 1 through 5 are collectively called the DATA step, a term used to describe the process of creating a SAS® dataset.

```
0007 PROC PRINT DATA = one;
```

Calls the SAS® procedure (hence the keyword PROC) PRINT, which instructs SAS® to create a printout of the dataset indicated by the 'DATA = one' portion of the statement. If no dataset is indicated explicitly, SAS® will use the most recently accessed DATA set. In our example we only have one data set. In later examples we may have several data sets involved in an analysis. For this reason, it is a good practice to explicitly inform SAS® which dataset to use for any procedure or analysis when (almost always) it is an option to do so.

```
0008 RUN;
```

Instructs SAS® to perform the preceding statements.

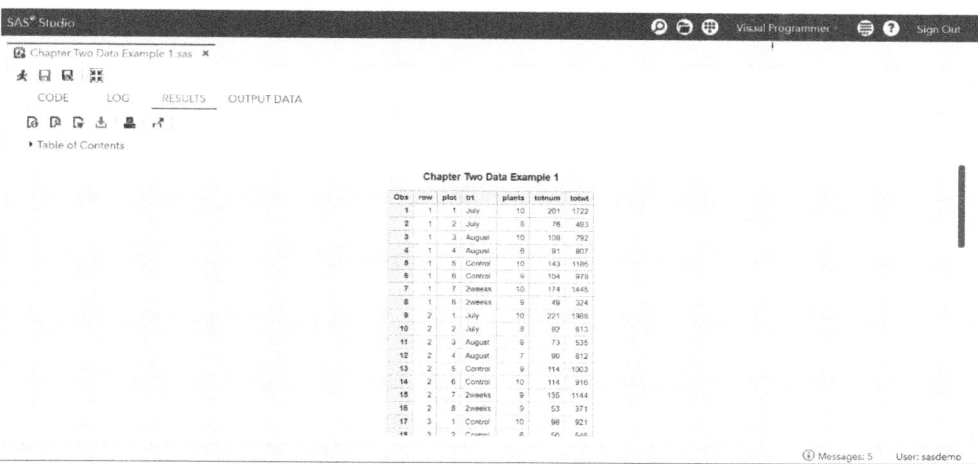

Besides the DATA step, SAS® utilizes PROC steps. We used a PROC step in the previous example of data retrieval when we requested a printout of the dataset we created. PROC statements request that SAS® utilize a procedure from its library on the hard disk to process the data in a SAS® dataset created by a DATA statement. Some PROC statements include: PROC PRINT, PROC ANOVA, PROC MEANS, PROC GLM, PROC REG, etc. Each PROC statement has sub-statements. We will cover these as we progress.

This brief overview should get you started with SAS®. There is much more to learn about SAS®, and we will cover details as they are appropriate to the statistical methods we are covering at any time. Two important things to remember: SAS® will do almost anything you tell it to do so it's up to you to tell it the right thing to do. Secondly, SAS® is just a tool, since you could do the procedures by hand, so you must learn to use the tool correctly. If you understand the statistical theory behind what SAS® is doing, you will be able to interpret SAS® effectively. One last comment to remind you that there are often several ways to accomplish the same goal in SAS®. Your friends or colleagues may suggest approaches different than those presented in this text. It is often a matter of preference and you should use what you find most comfortable.

3 A Review of Basic Statistics and Terms

Some General Terms

In order to effectively communicate ideas and concepts, a basic framework of terms and general statistics is needed. For many using this text, this is a review, while for others, it is not. As long as you understand the general concepts presented within this chapter, you should have no problems with the rest of the text. Terms and statistics are presented in just enough detail to make things clear but not too much detail as to overwhelm you.

Population, sample, statistic, parameter, estimate

A *population* is a large group of individuals. A *sample* is a sub-group of a population. We take samples to estimate something about a population, for example, average yield. The values we derive from samples are called *statistics*. Statistics estimate the corresponding values of the population called *parameters*. For example, we may be interested in the average yield of peaches in New Jersey. The population would include all peach trees in New Jersey. Even if we precisely define our population, we may never be able to measure every individual in the population. We therefore consider the population in its theoretical context. To *estimate* the average yield, we measure the yield of a certain number of trees, say 50, and then estimate the population yield with a statistic calculated from the sample. Any time you take a measurement, you are taking a sample to calculate an estimate of the population value (a statistic) of what you have measured. The most common statistics include the mean, variance and standard deviation. Remember, we never know population parameters, we always estimate them with sample statistics.

Random

Random means that something is not predictable, such as the assignment of treatments to experimental units. We use randomness to ensure that our work is not biased.

Deduction and induction

Deduction is the process of using known facts to predict a result. For example, if we put a glass of water into a freezer that is set to a temperature at or below 0°C, we deduce that the water will turn to ice. Why? Because we know the fact that water freezes at 0°C. *Induction* is the process of using observations to guess what caused some result. Suppose we didn't know that the freezing point of water was 0°C. We could carry out some experiments to determine when water froze. Experimentation using induction is called *inference*.

DOI: 10.1079/9781789249927.0003

Estimates

Estimates are our best guess (educated, of course) of a value. For example, our best guess of the value for a population parameter is a sample statistic. Estimates are often indicated with a carat (^) over the symbol of what we are estimating. Suppose we did several freezing experiments to determine the freezing point of some unknown substance and had five measurements of the freezing point. The five measurements are not exactly the same, because of errors (more later on error). We could estimate the freezing point by calculating the mean for the sample, noted $\bar{Y}$, which is estimating the population mean (μ). In this case, since $\bar{Y}$ is an estimate of μ, we can say $\bar{Y}$ is $\hat{\mu}$.

Accuracy and precision

Instruments are *accurate* and measurements made with them are accurate if they are not biased too high or too low. If a 2 kg weight is put on a scale and the scale reads 3 kg, it is not accurate. *Precise* is a relative term. One instrument may be more precise than another. One balance may measure milligrams while another may measure nanograms. Precision without accuracy is useless. A scale reading nanograms must really be measuring nanograms (it must be accurate). If it's not measuring nanograms, then it is not very useful.

Effect versus affect

Effect is noun and is the change brought about by some action, while *affect* is a verb which means to have an effect on. *Between* is used for comparing two items, while *among* is used for comparing more than two items. I've included the terms here since they are sometimes misused in the scientific literature.

Discrete and continuous variables

A variable is some measured characteristic such as weight, height, yield, etc. *Discrete variables* are variables capable of assuming only certain values, usually whole numbers, such as the number of leaves per plant, i.e. it cannot be 1.3. *Continuous variables* are variables which can have any value, depending on the precision of the instrument used for measuring.

Experimental units, treatments and plots

An *experimental unit* is that entity to which a treatment is applied. A *treatment* can also be some inherent property of the experimental unit, such as cultivar. In the plant sciences, the experimental unit is usually either a single plant or a group of plants we call a *plot*. When we are working with perennial crops, especially trees, the experimental unit may be a single branch. You must be able to recognize your experimental unit.

Replication versus repetition

A *replication* is an exact duplication of a treatment on more than one experimental unit. A *repetition* is simply another item. For example, suppose the experimental unit was a plot of ten tomato plants. When harvested, yield per plant is measured. The individual plants are not replications but rather repetitions (sub-samples). The plot is the replication. Researchers often confuse replication and repetition.

Replicates and blocks

This brings us to the concept of replicates versus blocks. All blocks are replicates but not all replicates are blocks. Individuals in a block have something in common, while individuals in a replication do not necessarily have something in common. Blocking is a form of replicating. This will become clearer after reviewing completely randomized and randomized complete block designs.

Variation

When there are differences among individuals, there is variation. Variation that can be controlled is called *non-random variation* and is often caused by sloppy technique. On the other hand, *random variation* is variation that cannot be controlled and is inherent to a system. Patterns of variation are predictable and one of the main goals of using statistical methods is to account for that variation. Different experimental designs allow us to systematically account for variation. That is why good experimental design is crucial for good research.

We are concerned with three major attributes of variation:

1. What is the magnitude of the variation?
2. How confident are we in our estimate of the variation?
3. Are the differences among treatments due to the treatments or to random variation?

The unaccounted-for variation in an experiment is called the *experimental error*. We can estimate this error; and the better our estimate, the higher is our confidence in our results.

Probability

A random experiment is any investigation in which the outcome is uncertain (most experiments). The flip of a coin is a random experiment. The yield response of wheat to plant growth regulators is a random experiment. The outcome of any experiment always has some degree of uncertainty associated with it. This uncertainty is measured by using the theories of probability. A mathematical description of a physical phenomenon is often used to describe an experiment. This mathematical description, or model, of a random experiment is called a probability model.

There are three elements to a probability model: (i) the sample space which is the set of all possible outcomes; (ii) an event which is one of the possible outcomes; and (iii) the probability for each event, P(E), such that $0 < P(E) < 1$. Note that the probability is always between 0 and 1. The probability model for the coin toss experiment has three elements: (i) the sample space, which is heads or tails; (ii) an event, which could be either heads or tails; and (iii) the probability for each event, which is 0.5 for heads and 0.5 for tails (assuming a fair coin!). (Note that you should always put a leading zero as part of a decimal number. That way, a reader will not overlook the decimal point.)

We won't explore probability models in this text, but they are important. When we experiment and the outcomes are random variables, such as yield or height, the probability model is called the probability distribution of the random variable. It is important to know the probability distribution for the random variables in our experiments, since many of the statistical tests we use are only valid if the data we are analyzing come from a particular distribution. Many of the procedures we will use in this class assume that the data are from a normally distributed population. Much research data is normally distributed; however, there are times when data are not normally distributed. In these cases, we use non-parametric statistical procedures. Fortunately, many of the procedures we use that are designed for analyzing normal data are robust. This means that the procedures will 'work' even if the data are not 'exactly' normal. There is no exact test to determine if data are normal or not, but there is a procedure in SAS® that performs an approximate test for normality.

PROC UNIVARIATE Normal

We now will learn one of the many procedures (hence the SAS® keyword 'PROC') available in SAS®. We can use the univariate procedure with the 'Normal' option to perform an approximate test of normality. We first need a dataset to use, in this case a very simple one. The following statements will generate a dataset with 25 observations for the variable 'yld', and will then invoke the univariate procedure to test those 25 observations for normality.

```
DATA one;
INPUT yld @@;
CARDS;
1 2 5 3 4
7 5 6 2 9
5 6 7 8 3
4 7 2 6 5
7 6 4 6 7
RUN;
PROC UNIVARIATE normal;
VAR yld;
RUN;
```

Note the '@@' symbol at the end of the INPUT line. This is the 'double trailing at' sign. It is instructing SAS® to stay on the current line of data to read the next observation for the preceding specified variable(s), rather than reading one observation and moving to the next line of data. When SAS® reaches the end of the line of data, it knows to go to the next line. This option is useful in situations where you may have more than one observation per line. Actually, the preceding program could be written as:

```
DATA one;
INPUT yld @@;
CARDS;
1 2 5 3 4 7 5 6 2 9 5 6 7 8 3 4 7 2 6 5 7 6 4 6 7
RUN;
PROC UNIVARIATE normal;
VAR yld;
RUN;
```

You are not limited to one variable when using this option. For example:

```
DATA one;
INPUT height yld @@;
CARDS;
1 2 5 3 4 7 5 6 2 9 5 6 7 8 3 4 7 2 6 5 7 6 4 6 7 5
RUN;
```

would generate a dataset with 13 observations of 'height' and 'yld'.

Back to our normality question. Try it yourself by typing in the program as shown below, submitting the program to SAS® for processing and comparing the output with that shown in the second figure:

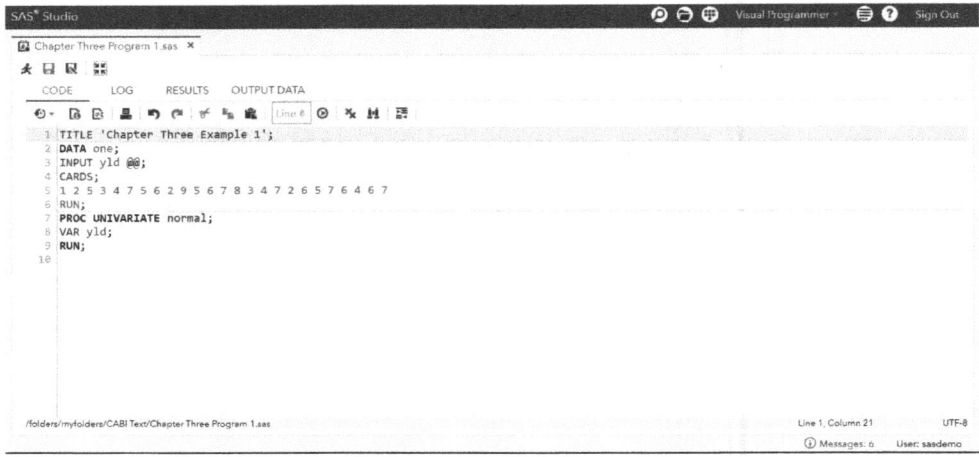

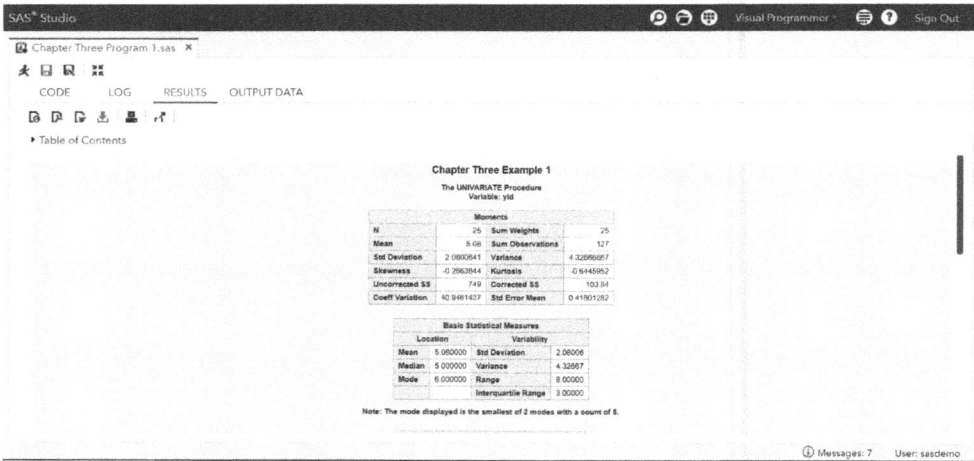

There is a great deal of output with PROC UNIVARIATE, but at the moment we are only interested in whether or not our data come from a normal distribution. The test statistic which SAS® uses (W) is the Shapiro–Wilks test if $N \leq 2000$, or otherwise the

Kolmogorov D test. This brings up the concept of a test statistic and a hypothesis test. We'll get back to our normality test momentarily.

Hypothesis testing

The null hypothesis (H₀)

A hypothesis is a question stated in statistical terms. For example, one may be studying the yield of two cultivars and want to know: does yield differ between cultivar a and b?

We develop what is called the 'null' hypothesis, that is no difference, and statistically state it as:

H_0: Yield of cultivar a = yield of cultivar b.

The alternate hypothesis (Hₐ)

If you have enough evidence to reject this idea of a = b, then some alternate must be true. We call this hypothesis, quite appropriately, the alternate hypothesis. The alternate hypothesis can take one of three forms and it is important to know which form you are interested in:

1. The two cultivars do not yield equally:
   H_a: Yield of cultivar a ≠ yield of cultivar b.
2. Cultivar a yields more than cultivar b:
   H_a: yield of a > yield of b.
3. Cultivar a yields less than cultivar b:
   H_a: yield of a < yield of b.

When testing a hypothesis, we need to know what the alternate hypothesis is because that will determine if you are using a one- or two-sided test. Always know the hypothesis you are testing when performing a statistical test and what the alternate is.

When you reject a null hypothesis you then accept that the alternate must be true. You *never* accept the null hypothesis, but rather, *fail to reject it.* You may simply not have enough proof that the null hypothesis is not true, thus you can't say that it is true. If you collected more data, you might obtain better estimates and then have enough information to reject H_0. When you obtain enough evidence that H_o is false and therefore reject it, that only leaves the alternate. One can accept the alternate since the null has been proven to be false.

Level of a test

The *level* of a statistical test is the probability of rejecting the null hypothesis when the null is in fact true. One usually wants this probability to be small. It is designated as α.

Power of a test

The *power* of the test (β) is the probability of rejecting the null hypothesis when a specific alternate is true. One generally wants this to be high since one wants to reject the null if the alternate is actually true. To ask for the power of the test you must have a specific alternate in mind (i.e. some numerical value). Most of the time, we specify α and speak in general terms about the power of the test being high or low. Since much of the time we don't have a specific numerical value in mind for the alternate, we can't generate a value for the power.

Different test statistics are used for testing different types of hypotheses. Some common test statistics include: F, t and chi-squared (χ^2). Regardless of the test statistic, the determination of whether to reject or fail to reject the null hypothesis is similar. The test statistic is calculated (thankfully by SAS®) and the probability of obtaining a test statistic of the magnitude or greater (or sometimes smaller, depending on the test statistic) than was calculated, simply by chance, given that the null hypothesis is true, is also provided. For example, suppose the probability of a greater value for the test statistic 't' (statistic which tests for equality of two means) was reported as 0.25. This indicates that there is a 25% chance of getting a t statistic as large as or larger than the one calculated, if our null hypothesis that our two means do not differ is true. We would fail to reject the null hypothesis since we don't have much evidence that it isn't true. However, what if the probability level were reported as 0.02? This indicates that there is only a 2% chance of getting a test statistic larger than the one calculated if the null hypothesis is true. This indicates that it is likely that the null hypothesis is not true, thus we would reject it in favor of the alternate. The level of the test gives us a clue as to whether we should reject or fail to reject. In general, if the probability reported is $P \leq 0.05$, we reject the null hypothesis in favor of the alternate. There is nothing wrong with stating that a null hypothesis was rejected at an α other than 0.05. In fact, I prefer to present the exact α value since it provides more information than simply saying a test was significant.

Now back to our original question: are our data normally distributed? Since our sample size is < 2000, we are looking at the Shapiro–Wilks test statistic:

Tests for Normality				
Test	Statistic		p Value	
Shapiro-Wilk	W	0.958383	Pr < W	0.3832
Kolmogorov-Smirnov	D	0.150862	Pr > D	0.1461
Cramer-von Mises	W-Sq	0.07824	Pr > W-Sq	0.2159
Anderson-Darling	A-Sq	0.468422	Pr > A-Sq	0.2344

With this test statistic, we reject the null hypothesis of normally distributed data if W is sufficiently smaller than 1, indicated by a low α value (generally < 0.05) for $P < W$. For our example, $\alpha = 0.3832$, thus we fail to reject the null hypothesis of normality. Pay attention: we are not saying that we accept that our data is normal, we are saying that we fail to reject the assumption that it is normal. This is a bit picky, but it is an important concept to keep in mind. In other words, it looks like we're dealing with normal data, we don't have any proof that it isn't, but we realize that a different or larger sample from the same population may indicate otherwise. This concept is often difficult to grasp, but the point is that we are reasonably confident that the data are normal, we just cannot say emphatically, with 100% certainty, that it is.

Other distributions

Data are not always normally distributed. Some other distributions often encountered in plant science research include:

- *Binomial* – when outcomes of an experiment can be considered success or failure, yes or no, like categorizing a plant as diseased or not diseased.
- *Poisson* – when the outcomes of an experiment are events that occur over time, such as the number of insects visiting a flower in an hour.
- *Exponential* – when the outcomes of an experiment are the times for some event to occur, such as the time between bee visits to flower x on tree 12.

There are many other distributions. Realize that all data are not distributed normally and to make such an assumption is not always valid. Again, many of the procedures we use will give reasonably valid results for many situations where the data is not 'exactly' normally distributed. Unless otherwise stated, the procedures set forth from here on are for normally distributed data.

The normal distribution

Any normal distribution can be completely defined by two parameters: the *mean* (μ) and *standard deviation* (σ). Since we usually cannot measure all of the individuals in a population, we take a sample and estimate the mean ($\overline{Y}$, $\hat{\mu}$) and standard deviation (s, $\hat{\sigma}$). When we have these estimates, we can perform a number of statistical procedures to answer questions about the population from which we took the sample.

Some basic statistics

We collect data to make inferences about the population from which our sample is drawn. From our sample we calculate some descriptive statistics which gives us important information about our sample. Usually we are interested in the arithmetic mean of the sample, $\overline{Y}$. The mean is what is called a measure of central tendency, that is, where the data are centered. There are other measures of central tendency such as the median, mode and midrange. We are concerned with the sample mean.

Sample mean, or average

An *average* (henceforth called the mean unless otherwise stated) is an estimate of the mean of a population. Think of population parameters as 'unknowable', thus we will always estimate them with samples. Suppose you measured the yield of n potato plants. The mean would be calculated as:

$$\overline{Y} = \frac{\sum_{i=1}^{n} y_i}{n}$$

This equation is read as: the mean of a sample of y's is equal to the sum from $i = 1$ to n of the individual y's, divided by n, where n is the number of observations. In other words, add up all the y's and divide the result by n.

Besides measures of central location, we also are interested in the variability of our data. In other words: how do our observations differ from each other? We will need estimates of variability to perform many of our statistical tests.

Variance and standard deviation

The spread of the data in a sample gives an idea of variability and is measured by the sample standard deviation or sample variance. The *sample standard deviation* (henceforth, standard deviation) is abbreviated as *s* and the *sample variance* (henceforth, variance) is the square of the standard deviation, s^2, and these are estimates of the *population standard deviation*, σ, and the *population variance*, σ^2.

The sample variance, s^2, is:

$$\left(\sum_{i=1}^{n} y_i^2\right) - \frac{\left(\sum_{i=1}^{n} y_i\right)^2}{n} \bigg/ n-1$$

The standard deviation, *s*, is:

$$\sqrt{s^2}$$

Estimates of the mean, variance and standard deviation are easily obtained from several different procedures in SAS®. We will hold off on this for a bit.

Standard error, or standard error of the mean

The *standard error* (sometimes called standard error of the mean) is a very important value. The concept behind it is quite simple. The standard error is actually the standard deviation of a sample of means. Visualize this: suppose you took a sample of five plants and measured their yield to estimate the yield of a certain treatment. From this sample you could calculate the mean, standard deviation and variance. Now suppose you were a bit nervous that the sample might not represent the population you are trying to estimate. So you take another sample of size 5 and calculate the mean, standard deviation and variance for this sample number two. Now imagine you do this say, ten times. Now you actually have a sample of ten means, each with their own standard deviation and variance. You could calculate a grand mean (a mean of the ten means) and its standard deviation and variance.

Theory shows that you could really estimate the standard deviation of the grand mean as the standard error of a single sample mean (say, the first sample you derived) divided by the square root of *n*, the sample size. Your single sample estimate of the mean is still a good estimate of the grand mean. Now instead of having to take nine more samples to estimate the standard deviation of the grand mean (the standard deviation of the sample of ten means), you can estimate it from the first sample's standard deviation. Why is this important? This concept is built into statistics formulas for *t*-tests and confidence intervals. But sometimes, researchers like to give an estimate of the mean plus or minus the standard error (I guess they do this because theory shows that about 67% of the sample observations will lie within plus or minus 1 standard deviation, about 95% of the observations will lie within plus or minus 2 standard deviations). Many people confuse standard deviation and standard error and thus report the standard deviation rather than the standard error that they say they are reporting. The standard error is always smaller than the standard deviation, because the standard error is simply the standard deviation divided by the square root of n. For example, given the mean is 10, $s = 8$ and $n = 16$, consider that the mean ± the standard deviation is 10 ± 8 while the mean ± the standard error is 10 ± 2. That's a big difference. The first option (± standard deviation) is interpreted as meaning that 67% of the observations lie between 2 and 18, while the second observation (± standard error) is interpreted as meaning that 67% of a sample of mean estimates would lie between 8 and 12. Think about it. The first option concerns individual observations. Do you really care what the possible value of one single plant's yield is? Probably not. You are more likely to be interested in the *average* yield per plant (the second option).

In defense of those who report means this way, you can calculate a confidence interval around the mean (we'll get to these in a moment) if you are given the mean, standard error (or standard deviation) and the sample size.

z Distribution, t distribution and degrees of freedom (df)

There are many different random variables one might measure in research. Since each random variable Y has its own true unknowable population mean and standard deviation, which are estimated with samples, one would need an infinite number of tables of probabilities to derive confidence intervals or test hypotheses, one for each random variable. To get around this, random variables can be standardized by subtracting the mean and dividing by the standard deviation. For a population, a *standardized value* (z) is:

$$z = \frac{Y - \mu}{\sigma}$$

The standardized value is a random variable. Its distribution, derived theoretically, is a standard normal random variable (z) with $\mu = 0$, variance $= \sigma^2$. Now one table of probabilities (the z table) can be used for all standardized variables from any normally distributed population.

But this works for a population and we are usually (always?) dealing with samples. Statistical theory shows that the mean of a population is distributed with mean, μ, variance, σ^2/n. From this it is derived that the sample mean is distributed with mean, $\bar{Y}$, and variance, s^2/n. Taking the square root of the sample variance of the mean gives the standard error, s. This is important (I just told you that a couple of paragraphs ago). Many times, we are interested in hypothesis about the true population mean. When we take a sample we can calculate the sample mean and standard error. If we have a hypothesis concerning μ, we can calculate a statistic called t which is analogous to the z statistic of a population. Instead of subtracting the population mean from an individual observation and dividing by the population standard error, we subtract our fixed value for μ (based on our hypothesis) from our sample mean, and divide by the sample standard error. Note that often our hypothesis leads to testing whether or not $\mu = 0$. For a sample:

$$t = \frac{\bar{Y} - \mu}{s_{\bar{Y}}}$$

The t statistic is defined as the difference between a sample mean and the population mean in standard error units. This standardized value from a sample is a random variable distributed as a Student's t random variable with $(n-1)$ degrees of freedom (df). We can now use one t-table to look up probabilities associated with hypothesis tests for samples.

What are degrees of freedom? Since μ and σ^2 are unknown, they are estimated from a sample. A 'cost' ensues. These costs are measured in degrees of freedom. Think of degrees of freedom as bits of information. If there are n observations in total, there are n bits of information or n degrees of freedom. If μ were known, then σ^2 would also be known. Thus if one estimates μ, one can determine s (review the formula for s to verify this). The estimate costs 1 df, and the standardized value is based on $(n-1)$ df (bits of information), rather than n df. You have to pay for the estimate. (There is a much more complicated mathematical explanation of df, but I think this one suffices.)

A Real World Example

We have a small set of data and we want to calculate the mean and standard error. With this information, we can summarize our data nicely, as we shall see. We collected yield data (grams per plant) from ten strawberry plants growing in a field. The field was full of thousands of strawberry plants, but I only sampled ten to estimate the yield of all of them. Take my word that I randomly sampled the ten plants (more on random sampling later). The following ten yields were recorded: 235, 312, 294, 123, 333, 438, 310, 299, 267, 341.

Means

Let's get the data into a SAS® dataset and use SAS® to calculate some basic statistics for us using the PROC MEANS statement:

The output from PROC MEANS is:

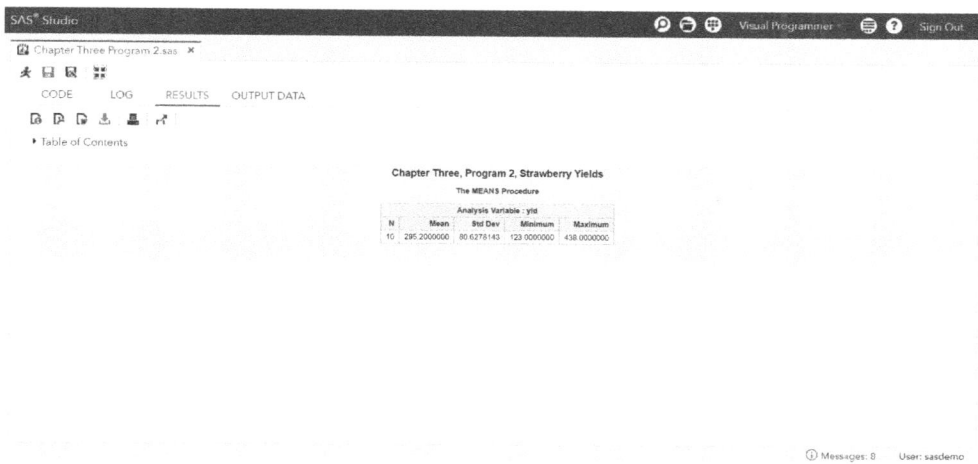

The standard error could be calculated as $80.63/\sqrt{10}$. The variance is $(80.63)^2$. We will use this information for the next sections.

Confidence intervals for μ

When we estimate μ we can report our point estimate, $\bar{Y}$. However, this provides no information about the variability of our sample. Thus we often choose to present a range in which our estimate likely lies, otherwise called a ***confidence interval*** or confidence limit (CL) around our estimate of μ. The formula to calculate the confidence limits around μ is:

$$CL = \bar{Y} \pm ts_{\bar{Y}}$$

where t is a tabular value with $(n-1)$ df at the $1 - (\alpha/2)$ significance level. For example, if you wanted an α level of 0.05, and your sample had ten observations, you would need to look up the t-value in a t-table for t 9,0.975. The subscript is shorthand to indicate df, significance level. You look at $(\alpha/2)$ because you are, in a sense, using half the α for the upper limit and half the α for the lower limit. If you had a one-sided confidence interval, you would not use $(\alpha/2)$ but rather, α.

Let's calculate a 95% confidence interval for our estimate of the mean yield. We already have our estimate of the mean and standard deviation from our PROC MEANS run. We obtain a t-value with the following code:

```
DATA tvalue;
t = tinv(0.975,9);
PROC PRINT DATA = tvalue;
TITLE 'Value of t-statistic for 95% CI with 9 df';
RUN;
```

Note that for the function 'tinv( )', you first supply the value of $1- (\alpha/2)$ followed by a comma, followed by the df (sample size minus 1). This code produces the following output:

'Value of t-statistic for 95% CI with 9 df'

Obs	t
1	2.26216

We can then use the output from the PROC MEANS run and the *t*-value run to calculate a confidence interval by hand, or, better yet, let's put all the above together into one SAS® program and let SAS® do the work:

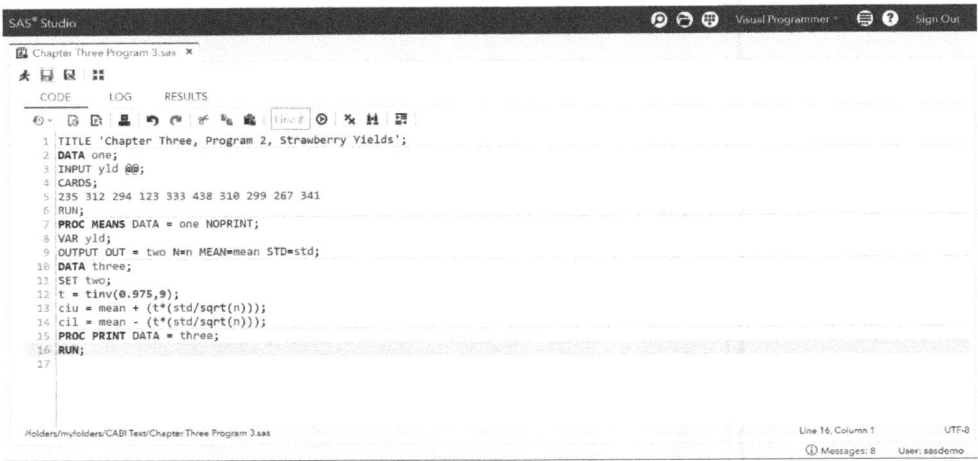

Output produced:

Chapter Three, Program 2, Strawberry Yields

Obs	_TYPE_	_FREQ_	n	mean	std	t	ciu	cil
1	0	10	10	295.2	80.6278	2.26216	352.878	237.522

Let's examine parts of this SAS® program more closely.

`PROC MEANS data = one NOPRINT;`

The 'NOPRINT' option of PROC MEANS instructs SAS that you do not want a printout of the results. Rather, you are going to create a dataset from the output. This is accomplished with the next statement:

`OUTPUT OUT = two N=n MEAN=mean STD=std;`

This statement instructs SAS to create a dataset called 'two' from the output of the previous procedure (PROC MEANS). In this dataset, we want to use three of the estimates from the procedure, namely N, MEAN and STD. We also want SAS to rename these 'n', 'mean' and 'std', respectively. That was done with the keyword = name.

```
DATA three;
SET two;
t = tinv(0.975,9);
ciu = mean + (t * (std/sqrt(n)));
cil = mean (t * (std/sqrt(n)));
PROC PRINT DATA = three;
RUN;
```

These statements instruct SAS to create a dataset called 'three' from dataset 'two' and include some new variables, 't', 'ciu' and 'cil' in the new dataset. These values are calculated via the formulas entered. Finally, we get SAS to print out dataset three.

You can also request confidence intervals from the PROC MEANS, but the above program allows tremendous flexibility as far as α levels go. It also teaches you some new SAS® tricks. But if you're interested, here's the easy way out:

```
TITLE1 'Strawberry Yields';
DATA one;
INPUT yield @@;
CARDS;
235 312 294 123 333 438 310 299 267 341
RUN;
PROC MEANS DATA = one CLM;
VAR yield;
RUN;
```

which produces the following output:

Strawberry Yields

The MEANS Procedure

Analysis Variable : yield	
Lower 95% CL for Mean	Upper 95% CL for Mean
237.5223362	352.8776638

Notice how when you request an option in PROC MEANS, you only get that option as your output. If you want other outputs, such as the mean, n, etc., you would have to explicitly request them.

The t-test

The t-test is used for testing hypotheses about the population mean. Usually the null hypothesis is that $\mu = 0$ versus the alternate that $\mu \neq 0$. The t-test is also often used to test that the difference between paired samples is 0. Paired samples are most easily comprehended with the example of a before-and-after test. This might involve estimating the number of insects on a plant before an insecticide is applied, then estimating insect numbers after spraying. Did the insecticide have an effect on insect numbers? Another good example is number of flowers before and after application of a chemical thinning agent. Even though both examples involve counts, it is reasonable to assume that the populations are normally distributed. In either case, you would calculate the differences of the paired observations, then calculate an average difference and variance and perform a t-test on the 'difference' data.

Confidence interval for a variance

Confidence intervals can be constructed for variance estimates or standard deviations also. Theory indicates that for a normal distribution:

$$\frac{(n-1)s^2}{\sigma^2} \approx \chi^2_{(n-1)}$$

with $(n-1)$ df. One can look up probabilities in a χ^2 table, and construct confidence intervals. One might also write a SAS® program to perform the calculations. The equation above reads 'the variance multiplied by $(n-1)$ df, all divided by the population variance, is distributed as a chi-squared random variable with $(n-1)$ degrees of freedom'.

A final comment concerning confidence intervals and probabilities. Suppose one has 95% confidence limits about the mean. This indicates that one is 95% confident that the confidence interval would contain the population mean if one could sample an infinite number of times. It does not mean that the population mean lies within the confidence interval with the probability of 95%.

Most of the time we deal with two-sided confidence limits and thus will not consider one-sided confidence limits here.

F-distribution

If one obtains two independent estimates of a sample variance, say s^2_1 and s^2_2 which have (n_1-1) and (n_2-1) df, respectively, theory shows that:

$$\frac{\dfrac{(n_1-1)s^2_1}{\sigma^2_1}}{\dfrac{(n_2-1)s^2_2}{\sigma^2_2}} \approx F(n_1-1, n_2-1)$$

This is an F random variable with (n_1-1), (n_2-1) df. We will use the F statistic throughout the rest of this book.

Summary

In this chapter, a number of basic terms have been defined and some of the most commonly referenced test statistics have been presented. An in-depth consideration of these statistics is not warranted at this point as these statistics will be referred to again and again throughout the rest of this book. Each time they are referenced, you will become a little more familiar with their importance and use in statistical analysis.

4 Some Basic Experimental Design Principles

Planning

Planning is the most critical phase of research, since decisions made during planning will impact the entire course of any experiment. When experiments are planned well, implementation, analysis and interpretation flow freely. The obvious first stage of planning always includes a thorough review of the literature. With the internet, this review process has become relatively easy.

Hypotheses Development

Once the review is complete, a series of well-thought-out hypotheses must be developed. Keep in mind that a greater number of hypotheses will require larger and potentially more complicated experiments. A good rule of thumb is to keep any experiment as simple as possible while obtaining sufficient information to intelligently and confidently perform hypotheses tests. The easiest way to develop hypotheses is to write down specific questions and then convert these questions into statistical shorthand.

With hypotheses in hand, you can now determine the treatments necessary to effectively answer your questions. Most experiments usually involve 'applying' some treatment to experimental units, followed by the collection of data from each of them. Treatments are not always 'applied' in the sense of fertilizer application, insecticides, etc., but may be inherent to the initial layout of an experiment. These inherent treatments include things such as cultivar, rootstock, location or time of planting, to name a few. The data we collect should be associated with the hypothesis of interest, and usually is some measured characteristic such as height or weight.

Fixed versus Random Effects

In statistics there are two kinds of effects: fixed and random; and the type of effect is important when determining correct denominators in an F-test calculation. In many instances, the correct denominator is 'built in' to the experimental design. In a few others, particularly those involving tests over time or location, denominators are not 'built in' and must be determined based on equations derived knowing whether the effects being tested are fixed or random. This will be clearer in Chapters 15 and 16.

The selection of treatments determines whether your effects are fixed or random. Effects are fixed if the k-categories of interest exhaust all possible categories of interest, and are not merely samples of possible categories taken from a larger population. Effects are random if the k-categories of interest constitute a sample from a larger population of interest, and inferences about the population are made based on the results obtained from the experiment with the sample.

Suppose one wished to study the productivity of snap beans, of which there are 100 cultivars (the hypothetical population of cultivars). If ten cultivars are chosen specifically because they are adapted to a particular climate, then cultivar would be a fixed effect.

DOI: 10.1079/9781789249927.0004

Inferences would not and could not be made regarding the other 90 cultivars and their productivity. On the other hand, if ten cultivars were randomly chosen from the possible 100, say, to make a general observation on productivity of snap bean in a given climate, then cultivar would be random. Inferences about the entire population of snap bean cultivars could be made. However, since cultivars are quite climate specific, it would not be a good idea to make such inferences even though they would be 'statistically legal'.

Experimental Design

Once treatments have been selected, an appropriate experimental design is chosen to hold the treatments. The selection and use of specific designs will be covered in future chapters. Just remember to try to keep your experiments as simple as possible. I repeat this since in my experience I have observed folks conducting overly complicated experiments to answer rather simple questions. And I reiterate, experiments should always be well planned so that the information collected is relevant to the problem being investigated. Data collection is expensive and many times unnecessary data is collected just for the sake of having data to analyze. Consider the cost of data collection when deciding whether or not to gather it. When in doubt about whether or not to collect specific data, collect it since it can never be gathered once the experiment is terminated. However, if your experiments are well thought out and properly planned, there will be no doubt. Good planning eases many of the difficulties encountered at the analysis stage.

Replication versus Repetition

One basic principle of experimental design is the concept of replication. Simply put, a replication is a single repeat of the basic experiment. Replication provides an estimate of the experimental error used in hypothesis testing and confidence interval estimation. Under certain conditions (such as when there is likely to be a lack of higher order interactions (but you don't know of these yet)), experimental error can be estimated without replication. These instances are generally rare and should be avoided if possible. Multiple readings are not replicates. For example: suppose there was an experiment where the interest was in daylength effect. Two growth chambers were available, one was set to long days and one was set to short days and there were ten plants in each chamber. Each plant is not a replicate, since the experiment is the daylength effect. Each chamber is a single replicate of either short day or long day. The plants are simply multiple readings. In this case, each chamber, not each plant, is an experimental unit. An experimental unit is that entity to which a single treatment is applied in one replication of the basic experiment. In this case the treatment is daylength. If you want replication in this experiment, you would need more growth chambers or you would need to repeat the experiment over time, randomly assigning which chamber is set to short days and which one is set to long days.

Experimental Error

We have mentioned experimental error several times so far. Let's clarify the issue a bit. Experimental error is the failure of two identically treated experimental units to respond similarly. We can estimate it numerically. A single experiment can have several experimental errors, such as in split plots, and use of the correct experimental error for testing specific hypothesis is critical. The experimental error actually consists of errors in experimentation such as treatment application (faulty techniques), errors in observations or measurements (again, faulty technique), and the natural variability among experimental units. Always strive to minimize the experimental error in an experiment by:

1. Having good treatment application and measuring techniques.
2. Using efficient experimental designs.
3. Using homogeneous experimental units if possible or block if necessary.

Confounding

Another important concept is confounding. This is the inability to determine which of two or more treatments caused some observed difference. For example, suppose there were two tomato plants, one in the shade and given extra nitrogen (poor yield) and the other in the sun and given no extra nitrogen (great yield). Did the one in shade yield poorly due to excess nitrogen or did the one in the sun yield more because of the sunshine?

Randomization

Randomization is also important. Randomization guarantees that each experimental unit has an equally likely chance of receiving any treatment as any other experimental unit. We can assume that experimental units that were randomly assigned treatments are independent of each other (meaning that one experimental unit has no effect on another experimental unit). Most of the statistical analyses of normal data that we do will assume independence, thus we ensure independence with randomization. Randomization also eliminates bias.

Controlling Experimental Error with Grouping or Blocking

We can do certain things called 'local control' to minimize experimental error. One is grouping. Grouping is the random assignment of experimental units together to receive a treatment as a group. The entire population of units are homogeneous (nearly identical experimental units). This is not always the case. Thus we may choose blocking. Blocking is the assignment of experimental units such that maximum homogeneity occurs within a block while maximum heterogeneity occurs between blocks. The blocking is usually based on some prior knowledge of the researcher, such as knowledge of vigor, prior productivity in the case of perennials, water or nutrient gradients in the field, light gradients in the greenhouse, etc.

Balanced Experiments

One final thought is to try as much as possible to make experiments balanced. The concept of a balanced experiment is easiest to explain with an example. Suppose we are interested in the effect of source (NO_3 versus NH_3) and rate of nitrogen fertilizer on crop yield. If we use two rates of nitrate and three rates of ammonia, the treatment assignment is unbalanced. The statistical analysis is made more complicated. Even if we weren't really interested in a third rate for nitrate, it might be wise in the long run to add a reasonably sensible one, if only to simplify the analysis.

---

5 Variation and the Analysis of Variance (ANOVA)

The Analysis of Variance (ANOVA)

The goal of applied statistics is to account for the variability observed in an experiment and determine whether it is random or due to a treatment effect. Responses to treatment(s) are evaluated by a systematic accounting of variation called the analysis of variance, or ANOVA, where variability is attributed to various sources. The sources of variability and complexity of an ANOVA depend on an experiment's design and treatment structure. When an experimental design effectively and efficiently holds the treatments of an experiment, the ANOVA is simple and quite elegant. When the experimental design and/or the treatment structure is or are not appropriate for the experiment or each other, the analysis can be cumbersome and difficult. A good researcher learns to combine appropriate treatment structures with efficient experimental design. This book will help guide you in this direction.

ANOVAs and the F-test

The simplest ANOVA compares two groups (say, two different cultivars or two different treatments) testing the hypothesis that the means of the two groups are not different, versus the alternate that they are different. In most ANOVAs, regardless of the complexity, the hypothesis tested is that of equality of treatment means (that there is no treatment effect). Estimates of the variation *within* each group (within each cultivar or treatment) are obtained and pooled to obtain experimental error, our best estimate of the naturally occurring random variation in our dataset. Estimates of the variation *between* the two groups are also obtained. If the variability between the two groups is significantly greater than the variability within the groups, we then have statistical evidence that the two groups differ, and that differences in group means are real and not due to chance. We determine if the two estimates of variability are significantly different from each other by performing an *F*-test. If the *F* statistic is large enough, we reject the null hypothesis that the two means are the same. If the *F* statistic is not large enough, we fail to reject the null hypothesis. We determine if the *F* statistic is large enough by looking at a table of *F* values, or by looking at the Pr>F on a SAS® printout. We will see how this is done when we look at our first design, the completely random design.

Fixed and Random Effects: Expected Mean Squares

In the previous chapter, we learned about fixed and random effects and that most plant science experiments involve fixed effects. Knowing whether effects are fixed or random is important in constructing equations called 'expected mean squares' (EMS), which guide

DOI:10.1079/9781789249927.0005

us in selecting the proper denominator for an *F*-test. Expected mean squares are equations, not numbers. Mean squares are actual numbers we calculate in our ANOVA and represent our estimates of variation (variance estimates). In a complicated ANOVA, there may be many mean squares. One must always use the correct numerator and denominator in an *F*-test for results to be valid. Correct *F*-testing is often 'built-in' to a design. However, sometimes it can be confusing as to which mean squares should be used in *F*-tests. When expected mean squares are determined for a particular ANOVA, correct tests are always performed since appropriate test denominators are based on EMS. Expected mean squares are particularly useful with large factorial experiments where some effects are fixed and some are random. Derivation of EMS will be covered in Chapter 16. Once you know how to calculate EMS, you can figure out appropriate tests for any ANOVA.

The ANOVA Table

The ANOVA table is a convenient way to present the statistical results of an experiment. ANOVA tables should be constructed before an experiment is performed. Below is the simplest ANOVA table possible. The experimental design is completely random, but this won't mean anything to you until we cover this design.

Source	df	SS	MS	F	Pr>F	EMS
Total						
Correction (μ)						
Treatment						
Error						

Note the different columns in the ANOVA table. The 'source' column indicates the sources of variation in an experiment. The other columns ('df' (degrees of freedom), 'SS' (sums of squares), 'MS (mean square), 'F', 'Pr>F' (probability of a greater *F* statistic) and the 'EMS' (expected mean square)) each contain information needed to complete the table. Degrees of freedom are numbers representing the 'bits of information' in each source, kind of an accounting system. The theoretical derivation of df is beyond the scope of this book. Sums of squares are calculated values that, when divided by their corresponding df, produce a mean square, an estimate of variance. The calculated *F* statistic allows us to assign a probability (Pr>F) to the chances of obtaining an *F* statistic as large or larger than the calculated one with repeated experimentation. A relatively low Pr>F indicates that obtaining an *F* statistic as large or larger than the calculated one simply by chance is highly unlikely. This suggests that the 'large' *F* value is attributable to something other than chance, namely treatment. The EMS are equations used to determine the appropriate numerators and denominators for *F*-tests. In many situations there is no need to include this column, since the appropriate *F*-test denominators are built into many designs. The EMS column is often only included in more complex designs to ensure correct *F*-statistic calculation.

The sources of variation in the ANOVA table are important. The 'total' is the total variation observed within our experiment. We partition this total variation into various components depending on the experimental design and the treatment structure. The treatment structure is a statistical description of the treatments and how they are or are not related to each other. The 'correction' source of variation is included in all ANOVAs and is that amount of variation that is accounted for by estimating the grand mean of our experiment. The 'correction' source always has 1 df. The 'treatment' source is that variation that is accounted for by the treatments applied. It has $(t-1)$ df. In many experiments we partition this treatment variation out into additional sources. This usually can make specific statistical tests more powerful. The 'error' source is that variation left over, or that variation which cannot be accounted for by any other source of variation that we know of. We assume that this variation estimates the natural variability in our experiment and we use this error source for performing various tests. Some experiments have more than one error term.

We will construct and use a real ANOVA table in the next chapter where we cover our first experimental design, the completely random design.

The Completely Random Design

6

The completely random design (CRD) is the simplest experimental design available. The treatments are applied to the experimental units randomly and every experimental unit has an equal chance of receiving any treatment assignment. If there is any outcome of the random treatment assignment to the experimental units that you are not willing to accept, then the CRD is not the design you need. A restriction on the possible randomization indicates that you need to perform some sort of blocking, which is discussed in Chapter 12 (The Randomized Complete Block Design).

The CRD has the major advantage of being very flexible in terms of the number of treatments and replications that can be included in the experiment. Some designs, such as the Latin square, restrict the number of treatments that can be included in an experiment. While it is a good idea to keep the number of replications the same for each treatment, having different numbers of observations in each treatment does not cause excessive calculation problems, even with missing data.

A word or two regarding missing data. When recording data, it is important to record a zero if no response was observed. Do not leave the observation blank, since that implies missing data. If you measured nothing, then report nothing. Missing data implies a missing or dead experimental unit and should be indicated on data sheets via a period ('.'). When data is 'lost' or 'missing', we lose statistical information. With the CRD, the relative amount of information lost with missing or lost data is minimal compared with other designs. Thus if you think you might lose a considerable number of experimental units during the course of the experiment, a CRD might be appropriate.

The major objection to using a CRD is that of accuracy in the experiment. Since variation from experimental unit to experimental unit is not accounted for by any sort of blocking, any inherent variation among experimental units will be pooled in with the true experimental error. Recall that experimental error is the failure of any two identically treated experimental units to respond similarly. The lack of similarity in response may be due to differences in vigor, growth stage, pre-experiment handling, or any number of other factors. Thus a CRD should only be used when the experimental units are extremely homogeneous. If you cannot group experimental units by some characteristic, then the CRD is acceptable.

With a set number of replications and treatments, the CRD provides the greatest number of degrees of freedom possible for estimating experimental error. We want as many df for estimating error as possible.

Once a CRD is deemed appropriate, the treatments must be randomly assigned to the experimental units. SAS® provides a convenient procedure, PROC PLAN, with which to do this.

There are two statements used in PROC PLAN: PROC PLAN and FACTORS. The PROC PLAN statement has only one option and that is for you to give the computer a random number with which to start generating random numbers. It is called the seed. The option is: 'PROC PLAN SEED = n;', where 'n' is a 5, 6 or 7-digit odd integer. If you don't specify a seed (and most users don't) the computer uses the time of day from your computer's clock as the seed.

The second statement, FACTORS, indicates whether the factors are ordered or random. The FACTORS request takes the form 'name = n [ordered]', where 'name' is the factor name, such as cultivar, and 'n' is the number of cultivars being randomized. The '[ordered]' is a special command indicating that the factor preceding it is ordered, not random. If '[ordered]' is omitted, then the factor is assumed by default to be random. The ordered option will be used with the randomized complete block where we require separate randomization of *n* factors for each of *m* blocks and we want SAS® to list the randomizations in order from block 1 through *m*.

The following example is for the treatment assignment of a completely random design with 36 experimental units to which six replications of six treatments will be assigned. The process is as follows:

1. Assign numbers to the experimental units, in order, from 1 to 36. This assignment is not random.

2. Run the following SAS® program:

```
TITLE1 'Treatment Assignment For Completely Randomized
    Design';
TITLE2 '6 treatments, 6 replications';
TITLE3 '36 Experimental Units';
PROC PLAN;
FACTORS trt = 36;
RUN;
```

3. The results of this program look like this:

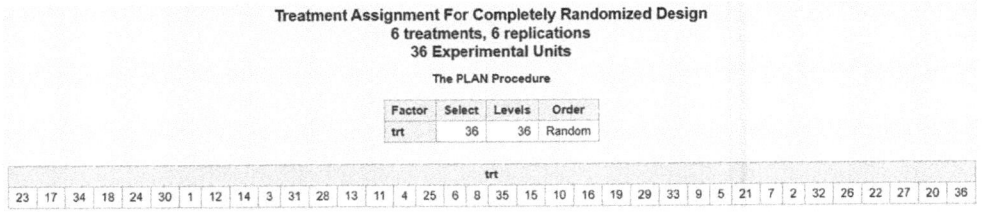

4. The first treatment is assigned to the first six plants listed in the PROC PLAN output, the second treatment to the next six plants, etc., until all six treatments have been assigned to six experimental units each for the total of 36 units.

The first treatment would go on plants numbered 23, 17, 34, 18, 24 and 30. The second treatment would go on 1, 12, 14, 3, 31 and 28. Simply continue the process until you have assigned the six treatments to the 36 units.

Once you have assigned the treatments to the experimental units, you would conduct the experiment according to the protocol you have developed and collect relevant data along the way. Once data has been collected, you would proceed with the statistical analysis.

Regardless of design, a typical analysis proceeds in a stepwise manner. Data is entered, then verified to make sure no mistakes were made during data entry. A test for normality is run to ensure our assumptions regarding normality are justified. If the data are not normal, some sort of transformation should be used to get the data into a form where the transformed data is normally distributed, thus parametric testing (i.e. an ANOVA) is justified. Treatment of non-normal data will be covered in Chapters 18 and 20. Finally, we get to the actual analysis.

The analysis of data in a completely random design is simple and straightforward. In this example, the experiment evaluated the effect of fall nitrogen application on the yield in spring-bearing strawberries. A total of 32 experimental units consisting of 3 m plots were available for this study. There were eight replications of four nitrogen treatments applied to the test plots. The four treatments were 0, 30, 60 and 90 kg of nitrogen per hectare applied before planting in the fall. The field map below (Table 6.1) shows that there are four rows of eight 3 m plots available for the experiment.

To assign the four treatments to the 32 plots randomly, the following steps would be carried out:

1. Number the plots in order from 1 to 32 if they are not already labeled (in Table 6.1, the plots are already labeled).

2. Use SAS® to generate a random treatment assignment to the 32 plots:

```
TITLE1 'Treatment Assignment For Completely Randomized
    Design';
TITLE2 '4 treatments, 8 replications';
TITLE3 '32 Experimental Units';
PROC PLAN;
FACTORS trt = 32;
RUN
```

Note that you can only use the completely random design if you are willing to accept *any* outcome to the randomization. For example, there is the chance, slight as it is, that the eight replications of any one of the four treatments could end up in adjacent plots. If you are unwilling to accept this treatment assignment, for example, you should not use a completely random design. For the sake of illustrating the completely random design, and since we will use this example later in the text, we will accept any randomization. In reality, though, you probably would not be willing to accept such a randomization and would have to use a different design (such as the randomized complete block design).

SAS® PROC PLAN results:

Treatment Assignment For Completely Randomized Design
4 treatments, 8 replications
32 Experimental Units

The PLAN Procedure

Factor	Select	Levels	Order
trt	32	32	Random

trt																															
27	21	12	16	11	24	29	4	1	17	32	15	18	10	8	22	23	2	25	20	30	7	31	9	28	19	6	5	3	14	13	26

Treatment 1 (0 kg N per ha) applied to plots 27, 21, 12, 16, 11, 24, 29 and 4.
Treatment 2 (30 kg N per ha) applied to plots 4, 1, 17, 32, 15, 18, 10 and 8.
Treatment 3 (60 kg N per ha) applied to plots 22, 23, 2, 25, 20, 30, 7 and 31.
Treatment 4 (90 kg N per ha) applied to plots 9, 28, 19, 6, 5, 3, 14, 13 and 26.

3. Apply the treatments to the appropriately assigned plots and collect data as indicated in your experimental protocol.

For this experiment, the dependent variable, Y, was yield per plot. The field plan and subsequent data are presented below in Table 6.2. Note the arrangement of treatments in the field is completely random.

Table 6.1. Field plot map for strawberry example of completely random design with eight replicate 3 m plots for each of four nitrogen treatments.

Row 1	Row 2	Row 3	Row 4
1	9	17	25
2	10	18	26
3	11	19	27
4	12	20	28
5	13	21	29
6	14	22	30
7	15	23	31
8	16	24	32

Table 6.2. Field plot map with treatment assignments for strawberry example of completely random design with eight replicate 3 m plots for each of four nitrogen treatments.

Row 1	Row 2	Row 3	Row 4
Plot 1, Trt 30	Plot 9, Trt 90	Plot 17, Trt 30	Plot 25, Trt 60
Plot 2, Trt 60	Plot 10, Trt 30	Plot 18, Trt 30	Plot 26, Trt 90
Plot 3, Trt 90	Plot 11, Trt 0	Plot 19, Trt 0	Plot 27, Trt 0
Plot 4, Trt 30	Plot 12, Trt 0	Plot 20, Trt 60	Plot 28, Trt 90
Plot 5, Trt 90	Plot 13, Trt 90	Plot 21, Trt 0	Plot 29, Trt 0
Plot 6, Trt 90	Plot 14, Trt 90	Plot 22, Trt 60	Plot 30, Trt 60
Plot 7, Trt 60	Plot 15, Trt 30	Plot 23, Trt 60	Plot 31, Trt 60
Plot 8, Trt 30	Plot 16, Trt 0	Plot 24, Trt 0	Plot 32, Trt 30

Yield (Yld) per plot (kg)											
Plot	Trt	Yld	Plot	Trt	Yld	Plot	Trt	Yld	Plot	Trt	Yld
1	30	7.1	9	90	7.7	17	30	7.2	25	60	8.4
2	60	8.2	10	30	7.5	18	30	6.9	26	90	7.1
3	90	7.1	11	0	6.2	19	0	7.8	27	0	6.2
4	30	7.6	12	0	7.0	20	60	8.4	28	90	7.4
5	90	7.2	13	90	7.3	21	0	7.0	29	0	6.4
6	90	6.9	14	90	7.5	22	60	8.5	30	60	8.3
7	60	8.0	15	30	7.2	23	60	8.2	31	60	8.7
8	30	7.7	16	0	6.3	24	0	6.7	32	30	7.1

A summary table of the data is included here since we will need some of the values in it for our calculations later in this chapter.

Treatment (kg N/ha)			
0	30	60	90
Yield (kg/plot)			
6.2	7.1	8.2	7.1
7.0	7.6	8.0	7.2

Continued

Continued.

	Yield (kg/plot)			
	6.3	7.7	8.4	6.9
	7.8	7.5	8.5	7.7
	7.0	7.2	8.2	7.3
	6.7	7.2	8.4	7.5
	6.2	6.9	8.3	7.1
	6.4	7.1	8.7	7.4
Total	53.8	58.6	66.9	58.4
n	8	8	8	8
Means	6.7	7.3	8.4	7.3

In any statistical analysis we want to account for as much variability as possible. In this analysis, we can account for variation due to treatment. The variability that remains after accounting for that due to treatment is our best estimate of the natural random variability among our experimental units. Understand at this point that the variability we estimate may in fact contain variability to some other source we have not accounted for, such as variability in water or fertility across the field we used. If we know of some inherent variability among our test plots and can account for it, then we would use a different design. For now, we will assume that we do not know of any other source of variability, thus the estimate of natural random variability we obtain is a reasonable one. This estimate is called experimental error. We are always trying to obtain the best estimate of error possible by using good technique and appropriate experimental designs.

We keep track of our variability estimates by constructing an ANOVA table. In the ANOVA for a completely random design there are two main sources of variation: the variability we can account for due to treatment (treatment); and the variability that we cannot account for (error). The complete ANOVA table will ultimately have the following columns: (i) source; (ii) degrees of freedom (df); (iii) sums of squares (SS); (iv) mean squares (MS); (v) F-test statistic (F); and (vi) probability associated with obtaining an F as large or larger than that calculated simply by chance (Pr>F). Sometimes an additional column for expected mean square equations (EMS) will be included, but not now.

The basic ANOVA table for a completely random experiment is:

Source	df	SS	MS	=	Pr>F
Total	N				
μ	1				
Treatment	$t-1$				
Error	$N-(t-1)-1$				

N = number of replications × number of treatments; t = number of treatments.

For our specific example:

Source	df	SS	MS	F	Pr>F
Total	32				
μ	1				
Treatment	3				
Error	28				

Note that the df for μ (the population mean) is always 1. This is associated with the 'bit of information' used to estimate the grand mean. Some books present an ANOVA table with a corrected total source row rather than the two rows which include the total and μ. I prefer the expanded table as shown.

Notice that there are blanks in all columns to the right of the df column. We will calculate the values that should be entered into these spots shortly. Not all spaces will require an entry. The table below shows which spaces need values.

Source	df	SS	MS	F	Pr>F
Total	32		X	X	X
μ	1		X	X	X
Treatment	3				
Error	28			X	X

We really only have to calculate the SS column, since the MS column is simply SS/df. We'll see where we get the other values in a bit. Normally we will use SAS® to calculate all values needed for our ANOVA table. We are only performing the hand calculations for this example so that you see where these values come from.

We begin with the total sums of squares. This value is simply the sum of the squared observations:

$$SS\,Total = \sum_{i=1}^{p}\sum_{j=1}^{n} Y_{ij}^2$$

where n is the number of observations in each of p treatments.

We also need to calculate the correction factor for the mean. This is the SSμ. Most statistics texts call it the correction factor for the mean, but for this text it is referred to as SSμ. It is calculated as:

$$SS\mu = \frac{\left(\sum_{i=1}^{p}\sum_{j=1}^{n} Y_{ij}\right)^2}{N}$$

which is the sum of the observations, squared, divided by N (the total number of observations).

Thus the calculated values for our example are:

$$SS\,Total = 1766.96$$

$$SS\mu = 1752.3$$

We must calculate one more SS, that attributed to treatment. The formula for $SS_{\text{Treatment}}$ is as follows:

$$SS\,Treatment = \sum_{i=1}^{p}\left(\frac{T_i^2}{n_i}\right) - SS\mu$$

where T_i is the total for treatment i, and n_i is the number of observations in each treatment. Our calculated value is:

$$SS\,Treatment = \frac{\left(53.6^2 + 58.3^2 + 66.7^2 + 58.2^2\right)}{8} - 1752.3 = 11.2$$

We obtain the SS due to error (SS *Error*) by subtraction:

$$SS\,Error = SS\,Total - SS\,\mu - SS\,Treatment$$

$$=$$

$$SS\,Error = 1767 - 1752.3 - 11.2 = 3.5$$

We now complete the ANOVA table with our calculated values. The MS column is SS/df for each source:

Source	df	SS	MS	F	Pr>F
Total	32	1767.0	X	X	X
μ	1	1752.3	X	X	X
Treatment	3	11.2	3.7		
Error	28	3.5	0.125	X	X

Note that we don't calculate a total MS or an MS for μ. The null hypothesis we are testing is that there are no differences among treatments with respect to yield. This is the usual null hypothesis for any ANOVA, that is, equality of treatment means. The alternate is that there is at least one inequality. We test the null hypothesis by comparing the estimate of variance among treatments (MS treatment) to our best estimate of the naturally occurring variability among experimental units (MS Error). This comparison is MS Treatment / MS Error, which yields an F-value. This F is a statistic with associated probabilities. For us, we are interested in the probability that we could get an F-value larger than our calculated value simply by chance (Pr>F). If this probability is low, usually < 0.05, we have evidence that there is little chance of getting the F we got simply by chance, and our 'large' F must be due to some treatment effect. We would therefore then reject the null hypothesis that the means are equal in favor of at least one inequality.

The F-value for our example is 29.6 (MS Treatment / MS Error) and we place it in our ANOVA table. Notice we don't have a value for Pr>F yet.

Source	df	SS	MS	F	Pr>F
Total	32	1767.0	X	X	X
μ	1	1752.3	X	X	X
Treatment	3	11.2	3.7	29.6	
Error	28	3.5	0.125	X	X

Since we are using SAS® we don't need to look up probabilities for F in a table. We will see how to run this analysis in SAS® and wait to see what the Pr>F is from SAS®. We will hold off on interpretations and significance of F-values until after using SAS® to verify that these hand calculations are correct. For the overall ANOVA, we first need to generate a SAS® dataset, then use the procedure ANOVA to obtain the analysis of variance for the data in the set.

To generate the SAS® dataset and perform the ANOVA we would use the following program. Recall keywords are all caps, information you supply is in lowercase. Also, this is an example of an ANOVA with a minimum number of statements.

```
TITLE1 'Strawberry Experiment';
TITLE2 'Analysis of Variance';
DATA one;
INPUT trt yld @@;
CARDS;
30 7.1 90 7.7 30 7.2 60  8.4 60 8.2 30 7.5 30 6.9 90 7.1
90 7.1 0   6.2 0   7.8 0   6.2 30 7.6 0   7.0 60 8.4 90 7.4 90 7.2
90 7.3 0   7.0 0   6.4 90 6.9 90 7.5 60 8.5 60 8.3 60 8.0
30 7.2 60 8.2 60 8.7 30   7.7 0   6.3 0   6.7 30 7.1
RUN;
PROC ANOVA DATA = one PLOTS = NONE;
CLASSES trt;
MODEL yld = trt;
MEANS trt;
RUN;
```

Results from the SAS® run:

Strawberry Experiment
Analysis of Variance

The ANOVA Procedure

Class Level Information		
Class	Levels	Values
trt	4	0 30 60 90

Number of Observations Read	32
Number of Observations Used	32

Strawberry Experiment
Analysis of Variance

The ANOVA Procedure

Dependent Variable: yld

Source	DF	Sum of Squares	Mean Square	F Value	Pr > F
Model	3	11.17750000	3.72583333	30.13	<.0001
Error	28	3.46250000	0.12366071		
Corrected Total	31	14.64000000			

R-Square	Coeff Var	Root MSE	yld Mean
0.763490	4.752085	0.351654	7.400000

Source	DF	Anova SS	Mean Square	F Value	Pr > F
trt	3	11.17750000	3.72583333	30.13	<.0001

Let's look at the statements used in this program:

PROC ANOVA;	This calls the ANOVA procedure from the SAS® library of procedures. There are options you can have with the ANOVA procedure, but we don't need any of them at this time.
CLASSES trt;	The **CLASSES** statement identifies variables that are main factor or block sources of variation in the ANOVA. These terms will be clearer after we cover factors and the randomized complete block design.
MODEL yld = trt;	The **MODEL** statement is a critical statement in the ANOVA. Even with the wrong model statement, SAS will perform an analysis, albeit the wrong one. With other incorrect statements, SAS usually won't perform the analysis. After the keyword **MODEL**, the dependent variable(s) are listed followed by the '=' sign. All of the sources of variation in the ANOVA table except μ, total and error are included on the right-hand side of the '=' sign.
MEANS trt;	This statement tells SAS to calculate the means for the treatments.
RUN;	Perform the analysis.

The first thing SAS prints out with the ANOVA is the Class Level Information, which is helpful in finding errors. You should look at this to make sure that all main effects of interest are listed and that SAS has identified the correct number of levels of each factor. In addition, make sure the correct total number of observations has been read.

The next thing printed out is the ANOVA table. It looks a little bit different than ours, but the same information is there. Actually SAS® gives more information than we need at this time. The 'Model' source is not of concern at this time. The 'Error' source is the same Error as in our table, just in a different place. The 'Corrected Total' is the same as our [Total – Correction factor]. We are really interested in the 'trt' source. The F-value is 30.13. The probability of getting a larger F simply by chance is <0.0001, indicating that we should reject the null hypothesis in favor of the alternate. The difference in F values between the hand and SAS calculations is due to rounding error.

In the next chapter, we will look at factors, levels and factorial treatment structures. Then we'll see what to do after the F-test.

7 Factorials

Independent variables evaluated for their influence on some characteristic of our experimental units are called *factors*. Many things can be considered factors, including cultivar, rootstock, fertilizer or growth regulator treatment, etc. The different variations within a single factor are called *factor levels*. For example: 10, 20 and 30 kg of N per tree are levels of the factor 'N fertilization' in an experiment studying the effect of N and P nutrition on tree productivity. In many experiments, different levels of a single factor are studied for their effect on the experimental units. In other cases, we are also often interested in the effects which several factors might have, and if one factor in any way influences the response to another factor. Experiments involving two or more factors, each at two or more levels, are called *factorial experiments*. There is no such thing as a factorial design. Factorial refers to treatment structure. Factorial treatment structures are combined in an efficient experimental design.

When we consider two or more factors, we introduce the concepts of main effects and interactive effects (interactions). The **main effect** of a factor is a measure of the change in the response variable to changes in the level of the factor, averaged over all levels of all other factors in an experiment. An **interaction** occurs when the response to various levels of one factor changes as the levels of another factor change.

Factorial Notation

Factorial treatment structures are often described using exponential notation, such as M^n, where M is the number of levels of the factors and n is the number of factors. So 6^3 would be read as three factors, each at six levels (a rather large experiment with $6 \times 6 \times 6 = 216$ treatments). This is sufficient if all the factors in the experiment are at the same number of levels. When several factors are under consideration in an experiment, and each is at a different number of levels, a different notation is employed. If we had one factor (A) at two levels, another factor (B) at three levels and a third factor (C) at six levels, we would note this as a $2 \times 3 \times 6$ factorial treatment structure.

Now let's return to our original experiment, a single-factor experiment, with the factor 'nitrogen application rate' at four levels and modify it to now include the concept of a factorial treatment structure in a completely random design. The same field is available for us to use (32 plots). The difference now is that we are interested in two factors: (i) the time of application (fall or spring); and (ii) the rate of application (0, 30, 60 and 90 kg perhectare). This would be written as a 2×4 factorial. With two times of application and four rates of application, we have a total of eight treatments that must be randomly assigned to the field plots. With 32 plots and eight treatments, we can have four replications per treatment. We used PROC PLAN in SAS® to assign treatments. The field map below (Table 7.1) shows that there are four rows of eight 3 m plots available for the experiment.

DOI: 10.1079/9781789249927.0007

Table 7.1. Field plot map for strawberry example of completely random design with four replicate 3 m plots for each of eight nitrogen treatments.

Row 1	Row 2	Row 3	Row 4
1	9	17	25
2	10	18	26
3	11	19	27
4	12	20	28
5	13	21	29
6	14	22	30
7	15	23	31
8	16	24	32

To assign the eight treatments to the 32 plots randomly, the following steps would be carried out:

1. Number the plots in order from 1 to 32 if they are not already labeled (in Table 7.1, the plots are already labeled).

2. Use SAS® to generate a random treatment assignment to the 32 plots.

```
TITLE1 'Treatment Assignment For Completely Randomized
    Design';
TITLE2 '8 treatments, 4 replications';
TITLE3 '32 Experimental Units';
PROC PLAN;
FACTORS trt = 32;
RUN;
```

Note again that you can only use the completely random design if you are willing to accept *any* outcome to the randomization.

SAS® PROC PLAN results:

Treatment Assignment For Completely Randomized Design
8 treatments, 4 replications
32 Experimental Units

The PLAN Procedure

Factor	Select	Levels	Order
trt	32	32	Random

																trt															
22	8	17	30	12	15	25	20	27	4	29	23	9	28	31	18	21	2	19	16	3	13	11	14	6	1	10	24	26	5	7	32

Treatment 1 Fall (0 kg N per ha) applied to plots 22, 8, 17 and 30.
Treatment 2 Fall (30 kg N per ha) applied to plots 12, 15, 25 and 20.
Treatment 3 Fall (60 kg N per ha) applied to plots 27, 4, 29 and 23.
Treatment 4 Fall (90 kg N per ha) applied to plots 9, 28, 31 and 18.
Treatment 5 Spring (0 kg N per ha) applied to plots 21, 2, 19 and 16.
Treatment 6 Spring (30 kg N per ha) applied to plots 3, 13, 11 and 14.
Treatment 7 Spring (60 kg N per ha) applied to plots 6, 1, 10 and 24.
Treatment 8 Spring (90 kg N per ha) applied to plots 26, 5, 7 and 32.

3. Apply the treatments to the appropriately assigned plots and collect data as indicated in your experimental protocol.

Table 7.2. Field plot map with treatment assignments for strawberry example of completely random design with four replicate 3 m plots for each of eight nitrogen treatments.

Row 1	Row 2	Row 3	Row 4
Plot 1, Trt S-60	Plot 9, Trt F-90	Plot 17, Trt F-0	Plot 25, Trt F-30
Plot 2, Trt S-0	Plot 10, Trt S-60	Plot 18, Trt F-90	Plot 26, Trt S-90
Plot 3, Trt S-30	Plot 11, Trt S-30	Plot 19, Trt S-0	Plot 27, Trt F-60
Plot 4, Trt F-60	Plot 12, Trt F-30	Plot 20, Trt F-30	Plot 28, Trt F-90
Plot 5, Trt S-90	Plot 13, Trt S-30	Plot 21, Trt S-0	Plot 29, Trt F-60
Plot 6, Trt S-60	Plot 14, Trt S-30	Plot 22, Trt F-0	Plot 30, Trt F-0
Plot 7, Trt S-90	Plot 15, Trt F-30	Plot 23, Trt F-60	Plot 31, Trt F-90
Plot 8, Trt F-0	Plot 16, Trt S-0	Plot 24, Trt S-60	Plot 32, Trt S-90

Yield (Yld) per plot (kg)

Plot	Trt	Yld	Plot	Trt	Yld	Plot	Trt	Yld	Plot	Trt	Yld
1	S-60	7.1	9	F-90	7.7	17	F-0	7.2	25	F-30	8.4
2	S-0	8.2	10	S-60	7.5	18	F-90	6.9	26	S-90	7.1
3	S-30	7.1	11	S-30	6.2	19	S-0	7.8	27	F-60	6.2
4	F-60	7.6	12	F-30	7.0	20	F-30	8.4	28	F-90	7.4
5	S-90	7.2	13	S-30	7.3	21	S-0	7.0	29	F-60	6.4
6	S-60	6.9	14	S-30	7.5	22	F-0	8.5	30	F-0	8.3
7	S-90	8.0	15	F-30	7.2	23	F-60	8.2	31	F-90	8.7
8	F-0	7.7	16	S-0	6.3	24	S-60	6.7	32	S-90	7.1

For this experiment, the dependent variable, Y, was yield per plot. The field plan and subsequent data are presented above in Table 7.2. Note that the arrangement of treatments in the field is completely random.

The data is more easily analyzed if it is summarized in a table.

	Treatment							
	F-0	F-30	F-60	F-90	S-0	S-30	S-60	S-90
	7.7	7.0	7.6	7.7	8.2	7.1	7.1	7.2
	7.2	7.2	8.2	6.9	6.3	6.2	6.9	8.0
	8.5	8.4	6.2	7.4	7.8	7.3	7.5	7.1
	8.3	8.4	6.4	8.7	7.0	7.5	6.7	7.1
Total	31.7	31.0	28.4	30.7	29.3	28.1	28.2	29.4
n	4	4	4	4	4	4	4	4
Mean	7.9	7.8	7.1	7.7	7.3	7.0	7.1	7.4

$N = 32$. Grand total = 236.8

As in the first example, we will perform an analysis of variance to account for variation due to treatment and that due to error to determine if there are any differences in yield due to treatment.

We first calculate the total sums of squares and correction factor as before. The calculated values for our example are:

$$SS_{Total} = 1766.96$$

$$SS\mu = 1752.32$$

We calculate $SS_{Treatment}$ as follows:

$$SS\,Treatment = \left(\frac{31.7^2 + 31^2 + 28.4^2 + 30.7^2 + 29.3^2 + 28.1^2 + 28.2^2 + 29.4^2}{4} \right)$$
$$- 1752.32 = 3.36$$

We obtain the SS due to error (SS_{Err}) by subtraction:

$$SS_{Error} = SS_{Total} - SS_{\mu} - SS_{Treatment}$$

$$1766.96 - 1752.32 - 3.36 = 11.28$$

We now need to construct an ANOVA table:

Source	df	SS	MS	F	Pr >F
Total	32	1766.96			
μ	1	1752.32			
Treatment	7	3.36	0.48	1.01	0.4468
Error	24	11.28	0.47		

Verify the results with SAS® (We won't calculate values by hand for most of the examples. Most of the time we will simply rely on SAS®, but we perform hand calculations here to illustrate where the values SAS® generates are coming from.)

```
TITLE1 'Strawberry Nitrogen Study 8 Treatments';
TITLE2 'Analysis of Variance';
DATA one;
INPUT trt $ yld @@;
CARDS;
S-60 7.1 F-90 7.7 F-0 7.2 F-30 8.4
S-0 8.2 S-60 7.5 F-90 6.9 S-90 7.1
S-30 7.1 S-30 6.2 S-0 7.8 F-60 6.2
F-60 7.6 F-30 7.0 F-30 8.4 F-90 7.4
S-90 7.2 S-30 7.3 S-0 7.0 F-60 6.4
S-60 6.9 S-30 7.5 F-0 8.5 F-0 8.3
S-90 8.0 F-30 7.2 F-60 8.2 F-90 8.7
F-0 7.7 S-0 6.3 S-60 6.7 S-90 7.1
RUN;
PROC ANOVA DATA = one PLOTS = none;
CLASSES trt;
MODEL yld = trt;
MEANS trt;
RUN;
```
Results from the SAS® run:

Strawberry Nitrogen Study 8 Treatments
Analysis of Variance

The ANOVA Procedure

Class Level Information		
Class	Levels	Values
trt	8	F-0 F-30 F-60 F-90 S-0 S-30 S-60 S-90

Number of Observations Read	32
Number of Observations Used	32

Strawberry Nitrogen Study 8 Treatments
Analysis of Variance

The ANOVA Procedure

Dependent Variable: yld

Source	DF	Sum of Squares	Mean Square	F Value	Pr > F
Model	7	3.34000000	0.47714286	1.01	0.4468
Error	24	11.30000000	0.47083333		
Corrected Total	31	14.64000000			

R-Square	Coeff Var	Root MSE	yld Mean
0.228142	9.272608	0.686173	7.400000

Source	DF	Anova SS	Mean Square	F Value	Pr > F
trt	7	3.34000000	0.47714286	1.01	0.4468

Strawberry Nitrogen Study 8 Treatments
Analysis of Variance

The ANOVA Procedure

Level of trt	N	yld	
		Mean	Std Dev
F-0	4	7.92500000	0.59090326
F-30	4	7.75000000	0.75498344
F-60	4	7.10000000	0.95916630
F-90	4	7.67500000	0.75883683
S-0	4	7.32500000	0.84606934
S-30	4	7.02500000	0.57373048
S-60	4	7.05000000	0.34156503
S-90	4	7.35000000	0.43588989

The *F*-value for treatment is 1.01 and is significant at the 0.4468 level. Notice that, instead of simply rejecting or failing to reject a null hypothesis at some prescribed level such as 0.05 or 0.01, I've chosen to present the exact α level provided by SAS® since it conveys much more information than a simple reject or fail to reject. Since the *F*-value is significant at $\alpha = 0.4468$, we would indicate that there does not appear to be a treatment effect and we fail to reject the null hypothesis of equal treatment means.

By analyzing the data as eight treatments, we are ignoring a tremendous amount of information inherent to a factorial treatment structure. Using the eight-treatment approach we can answer one question: is there a treatment effect? Using the analysis that is suggested by the factorial treatment structure we can quickly answer three questions:

1. Is there an effect of nitrogen on yield?
2. Is there an effect of the time of year at which nitrogen is applied?
3. Is there an interaction of rate with time of application?

Statistically, there are two ways to approach answering these questions. We will start with the traditional method. We will then add a twist to the experiment and introduce the concept of contrasts. Contrasts are extremely useful but many researchers do not use them even when they are appropriate.

Consider the factorial treatment structure. We have two main factors of interest: time of application and rate of application. Each factor is a source of variation in our ANOVA table. Part of the power in the factorial treatment structure is the ability to detect interactions between factors. Thus we must include a source of variation in our ANOVA table to account for this interaction between time and rate. Thus our three new lines in the ANOVA table (which will replace the 'treatment' line) are: Time, Rate and Time*Rate. Time will have $(t-1)$ df, where t = levels of the factor 'time'. Rate will have $(r-1)$ df where r = levels of the factor 'rate'; and Rate*Time will have $(t-1)*(r-1)$ df. This could be expanded for more factors. Note that in the ANOVA table, the df for rate, time and rate*time will sum to what the treatment df would have been if we had looked at this experiment as having eight treatments (2 × 4 rates).

The sources of variation and corresponding df for larger factorial experiments could be determined by expanding the ideas in the previous paragraph. With more than two factors, there will be more than main effects and two-way interactions (interactions involving two main effects). For example, suppose we had three main factors, A (a = 4 levels), B (b = 3 levels) and C (c = 5 levels). This is a total of 60 treatments (4 × 3 × 5). There would be 59 df assigned to this source of variation in an ANOVA table if we looked at it this way. However, we should look at it as a factorial. Our sources of variation and df would be:

Source	df	df (formula)
A	3	(a–1)
B	2	(b–1)
C	4	(c–1)
A * B	6	(a–1) * (b–1)
A * C	12	(a–1) * (c–1)
B * C	8	(b–1) * (c–1)
A * B * C	24	(a–1) * (b–1) * (c–1)

Note that the df for the factorial treatment structure ANOVA will sum to the treatment df considering the experiment without the factorial treatment structure. How do you know how many different interactions you should have for different factorial structures? Given a set number of factors in an experiment, the following table provides the numbers of main effect and interaction sources of variation in the ANOVA table. We limit the number of factors to five because it is doubtful that you will ever have an experiment with more than five factors.

Number of factors	Main effects	2-way int.	3-way int.	4-way int.	5-way int.
1	1				
2	2	1			
3	3	3	1		
4	4	6	4	1	
5	5	10	10	5	1

Back to the example. We need to construct the ANOVA table:

Source	df (formula)	df
Total	N	32
μ	1	1
Time of application	(t–1)	1
Rate of application	(r–1)	3
Time * rate of application	(t–1)*(r–1)	3
Error	N–(t–1)–(r–1)–(t–1)*(r–1)–1	24

Now we proceed with the analysis of variance. Again we will calculate the SS by hand to illustrate where these numbers are coming from. We calculate the SS_{Total} and the $SS\mu$ as we did before. Now instead of calculating an $SS_{Treatment}$, we need to calculate SS_{Time}, SS_{Rate}, and $SS_{Rate*Time}$.

The SS_{Time} is calculated as:

$$SS\,Time = \sum_{i=1}^{p}\left(\frac{T_i^2}{n_i}\right) - SS\mu$$

where T_i's are the time of application totals. You would get these from the data table below.

	Treatment							
	F-0	F-30	F-60	F-90	S-0	S-30	S-60	S-90
	7.7	7.0	7.6	7.7	8.2	7.1	7.1	7.2
	7.2	7.2	8.2	6.9	6.3	6.2	6.9	8.0
	8.5	8.4	6.2	7.4	7.8	7.3	7.5	7.1
	8.3	8.4	6.4	8.7	7.0	7.5	6.7	7.1
Total	31.7	31.0	28.4	30.7	29.3	28.1	28.2	29.4
n	4	4	4	4	4	4	4	4
Mean	7.9	7.8	7.1	7.7	7.3	7.0	7.1	7.4

Continued

Continued.

	F-0	F-30	F-60	F-90	S-0	S-30	S-60	S-90
				Treatment				
			Fall total				Spring total	
			121.8				115	
n			16				16	
	Nit = 0		Nit = 30		Nit = 60		Nit =90	
Total	61		59.1		56.6		60.1	
n	8		8		8		8	

$N = 32$. Grand total = 236.8

Thus:

$$SS\,Time = \left(\frac{121.8^2 + 115^2}{16}\right) - 1752.32 = 1753.765 - 1752.32 = 1.445$$

SS_{Rate} would be calculated as:

$$SS\,Rate = \sum_{i=1}^{p}\left(\frac{R_i^2}{n_i}\right) - SS\mu$$

where R_i's are rate totals, thus:

$$SS\,Rate = \left(\frac{61^2 + 59.1^2 + 56.6^2 + 60.1^2}{8}\right) - 1752.32 = 1753.673 - 1752.32 = 1.353$$

$SS_{Rate*Time}$ is calculated as:

$$SS\,Rate*Time = \sum_{j=1}^{t}\sum_{i=1}^{r}\left(\frac{\left(T_j R_i\right)^2}{n_i}\right) - SS\mu - SS\,Time - SS\,Rate$$

where $TjRi$'s are time*rate totals; in this case, treatment totals thus:

$$SS\,Rate*Time = \left(\frac{\left(31.7^2 + 31^2 + 28.4^2 + 30.7^2 + 29.3^2 + 28.1^2 + 28.2^2 + 29.4^2\right)}{4}\right)$$
$$- 1752.32 - 1.445 - 1.353 = 0.542$$

We obtain the SS due to error (SS_{Err}) by subtraction, and construct an ANOVA table:

Source	df	SS	MS	F	Pr>F
Total	32	1766.96			
μ	1	1752.32			
Time of application	1	1.445	1.445	3.068	
Rate of application	3	1.353	0.451	0.958	
Time * Rate	3	0.542	0.181	0.384	
Error	24	11.3	0.471		

Notice there are no Pr>F-values in the ANOVA table. We will let SAS® calculate them, and then we will proceed with the *F*-tests and determine if there are significant main effects or interactions. We will then interpret and discuss the results. The SAS® program is:

```
TITLE1 'Strawberry Nitrogen Study Factorial Treatments';
TITLE2 'Analysis of Variance';
DATA one;
INPUT time $ rate yld @@;
CARDS;
S 60 7.1 F 90 7.7 F 0 7.2 F 30 8.4
S 0 8.2 S 60 7.5 F 90 6.9 S 90 7.1
S 30 7.1 S 30 6.2 S 0 7.8 F 60 6.2
F 60 7.6 F 30 7.0 F 30 8.4 F 90 7.4
S 90 7.2 S 30 7.3 S 0 7.0 F 60 6.4
S 60 6.9 S 30 7.5 F 0 8.5 F 0 8.3
S 90 8.0 F 30 7.2 F 60 8.2 F 90 8.7
F 0 7.7 S 0 6.3 S 60 6.7 S 90 7.1
RUN;
PROC ANOVA DATA = one PLOTS = none;
CLASSES time rate;
MODEL yld = time rate time*rate;
MEANS time rate time*rate;
RUN;
```

The results:

Strawberry Nitrogen Study Factorial Treatments
Analysis of Variance

The ANOVA Procedure

Class Level Information		
Class	Levels	Values
time	2	F S
rate	4	0 30 60 90

Number of Observations Read	32
Number of Observations Used	32

Strawberry Nitrogen Study Factorial Treatments
Analysis of Variance

The ANOVA Procedure

Dependent Variable: yld

Source	DF	Sum of Squares	Mean Square	F Value	Pr > F
Model	7	3.34000000	0.47714286	1.01	0.4468
Error	24	11.30000000	0.47083333		
Corrected Total	31	14.64000000			

R-Square	Coeff Var	Root MSE	yld Mean
0.228142	9.272608	0.686173	7.400000

Source	DF	Anova SS	Mean Square	F Value	Pr > F
time	1	1.44500000	1.44500000	3.07	0.0926
rate	3	1.35250000	0.45083333	0.96	0.4287
time*rate	3	0.54250000	0.18083333	0.38	0.7654

Strawberry Nitrogen Study Factorial Treatments
Analysis of Variance

The ANOVA Procedure

Level of time	N	yld	
		Mean	Std Dev
F	16	7.61250000	0.76496187
S	16	7.18750000	0.54267854

Level of rate	N	yld	
		Mean	Std Dev
0	8	7.62500000	0.74785407
30	8	7.38750000	0.73180306
60	8	7.07500000	0.66708320
90	8	7.51250000	0.59865922

Level of time	Level of rate	N	yld	
			Mean	Std Dev
F	0	4	7.92500000	0.59090326
F	30	4	7.75000000	0.75498344
F	60	4	7.10000000	0.95916630
F	90	4	7.67500000	0.75883683
S	0	4	7.32500000	0.84606934
S	30	4	7.02500000	0.57373048
S	60	4	7.05000000	0.34156503
S	90	4	7.35000000	0.43588989

Our hand-calculated SS agree with those of SAS®. Interpreting the results is straightforward. The first thing you should look at is the SS interaction. Whether or not it is significant will determine how to proceed. This does not mean that we look at the results and then determine how we should separate means. Before you perform the experiment, you should figure out how you will analyze the data. In this case we have two scenarios: (i) no interaction, and thus we would concern ourselves with main effects; and (ii) a significant interaction, which, if present, would suggest determining the effect of rate within the time of application. When you get a significant interaction, it is telling you that you need to look at the effect of one of the factors within each level of the other factor. It makes more sense in this case to look at the rate effect in the spring versus the rate effect in the fall, rather than the difference between fall and spring within rate 0, within rate 30, etc. In fact, since rate is a quantitative factor, identifying the rate effect within the time of application is the only logical approach.

The significance level (α) for the interaction of Time with Rate is 0.7654. This indicates that there is no significant interaction between Time and Rate, thus we can look at main effects. If you were so inclined, you could say that there was a significant interaction between Time and Rate at the 0.7654 level. However, most researchers generally stick with significances at α levels of 0.05 or less. Some may go as high as 0.10, but not usually. If there were a significant interaction, one would concentrate on the interaction, not the main effects. But since there was no interaction, we will look at the main effects.

We look at the significance levels for Time (0.0926) and Rate (0.4287) and see that neither is really significant. Thus it doesn't seem to matter when you apply nitrogen or not; there is no effect of nitrogen application on strawberry yield.

Let's suppose that we did detect a rate effect or a time*rate interaction. The appropriate analysis of the rate effect would be to determine a regression line describing the yield response to rate. For the significant main effect, there would be one regression analysis and resulting equation of the line describing the response. For the rate effect analysis of the significant interaction, there would be two analyses, one for spring and one for fall. We know this because 'Rate' is a quantitative effect and the appropriate analysis for quantitative effects is almost always a regression of some sort. We will visit regression in Chapters 9 and 10. If we had a significant main effect for a qualitative factor, we would use either a mean separation technique or a set of contrasts. It might seem logical to present regression here; however, I want to throw a wrench into our analysis.

When considering the 0 treatment level of the factor 'Rate', isn't nothing applied in the fall the same treatment as nothing applied in the spring? One could argue this and insist that you really don't have a factorial treatment structure, but rather, have seven total treatments: six nitrogen treatments replicated four times with a control (no N) replicated eight times. What can you do? That is the subject of our next chapter: Contrasts.

8 Contrasts

Take a look at the treatment SS from our example in the previous chapter when we considered the experiment as one with eight treatments. Compare this SS with the sum of the SS for rate, time and rate*time. They are identical! The treatment SS was partitioned into component pieces: the pieces suggested by the factorial treatment structure. Check the df too. Notice that they are partitioned out as well. But what about the experiments where you don't have such a structure? This is where contrasts (also called planned comparisons, single df contrasts, planned F-tests) come in.

In any experiment we start with hypotheses that can often be stated in very specific terms. For example, in our experiment in Chapter 7 our questions were:

1. Is there an effect of nitrogen on yield?
2. Is there an effect of the time of year at which nitrogen is applied?
3. Is there an interaction of rate with time of application?

We answered these questions using a factorial analysis.

These questions were posed prior to running the experiment, and this is crucial to the validity of the following analysis. Questions posed before the experiment are called *a priori* questions. Planned contrasts, or simply contrasts, are used to powerfully answer *a priori* questions. We cover planned contrasts before mean separation techniques because contrasts are more powerful than mean separations. Power is the ability of a method to detect differences when they in fact really exist. We want powerful techniques.

Recall that the treatment SS represents the amount of variation within our experiment that can be attributed to treatment effects. This SS treatment can be further partitioned into specific components of variation due to specific treatment effects, and we did this with the factorial analysis of the last chapter. We can also partition treatment SS using contrasts. You could perform the factorial analysis using contrasts (some would be multiple df contrasts, which we won't cover; we will only cover single df contrasts). However, there is no need to learn a factorial analysis using contrasts. We will address contrasts in a somewhat more useful context: how to answer specific questions when a factorial analysis is not an option. We do this with single df contrasts.

Now return to the problem at the end of Chapter 6 where the argument was made that we did not have a 2 × 4 factorial, but rather an experiment with seven treatments. (You could also argue that you have a 2 × 3 factorial (two times and three rates) with eight replicate control plots. Our partitioning of the treatment SS with contrasts performs the analysis supporting this 2 × 3 factorial argument.)

When partitioning any treatment SS into single df contrasts, you may partition the SS into *n* contrasts where *n* is the df for treatment. In the example from Chapter 7, we could partition the treatment SS into seven single df contrasts since we had 7 df for treatment. If we consider that we only really have seven treatments (six nitrogen treatments and a control), we could partition the treatment SS into six contrasts, since there would be 6 df for treatment. It should be emphasized that contrasts are not performed to test effects that

DOI: 10.1079/9781789249927.0008

are suggested by the data. In other words, you don't decide that you 'planned' to do specific contrasts after you have seen treatment means.

Again, contrasts are mathematical calculations designed to partition a treatment SS into meaningful pieces. A contrast (C) is defined as a linear combination of treatment totals; the linear combination you use is determined by your set of treatments and your set of questions:

$$C = c_1 T_1 + c_2 T_2 + \ldots + c_k T_k$$

where $T_1, \ldots T_k$ are treatment totals and $c_1, \ldots c_k$ are coefficients which you will learn how to figure out shortly.

In order for a contrast to be valid, two conditions must be met:

1. The sum of the coefficients (c_i's) must equal zero.

In other words:

$$\sum n_i c_i = 0$$

with equal n_i:

$$\sum c_i = 0$$

2. Two single df contrasts are orthogonal (which is what they must be to be valid) if the sum of the cross-products of their coefficients is equal to zero.

The rule of orthogonality ensures that the partitioned SS will add up to the original treatment SS. In other words, given:

$$L_1 = c_{11} T_1 + c_{12} T_2$$

$$L_2 = c_{21} T_1 + c_{22} T_2$$

L_1 and L_2 are orthogonal if:

$$(n_{11} c_{11} * n_{21} c_{21}) + (n_{12} c_{12} * n_{22} c_{22}) = 0$$

with equal replication (equal n), given:

$$L_1 = c_{11} T_1 + c_{12} T_2$$

$$L_2 = c_{21} T_1 + c_{22} T_2$$

L_1 and L_2 are orthogonal if:

$$(c_{11} * c_{21}) + (c_{12} * c_{22}) = 0$$

Note that there will likely be k treatments (rather than only two) in a contrast, thus you would have to extend the formula to include the cross-products of all k pairs of coefficients. If the number of contrasts is greater than two, you must check all possible pairs for orthogonality. Any SS with n df can be partitioned into n orthogonal single df contrasts.

The SS for any given contrast is calculated as:

$$SS(C) = \frac{C^2}{n\left(\sum c_i^2\right)}$$

where n is the number of observations (replications) that went into calculating each treatment total, T_i in the contrast of interest. If you have a balanced, replicated experiment with no missing data, all of the n's will be equal. The previous formula expanded gives:

$$SS(C) = \frac{[c_1 T_1 + c_2 T_2 + \ldots + c_k T_k]^2}{n[c_1^2 + c_2^2 + \ldots + c_k^2]}$$

When single df contrasts are to be used for partitioning treatment SS, a contrast table must be constructed. Note the arrangement of the contrast table below. On the left-hand side we have the contrasts listed. In addition, you might include a description of the contrast in a column too. The treatments are listed in order across the top row. For each contrast line, the coefficients, c_{ik}'s, are listed. The right-most column is the sum of the c_{ik}'s (which, remember, in order to be a valid contrast, must sum to 0 for each row, assuming equal n for each treatment; when the n's are different, the $n_k c_{ik}$'s must sum to 0). The bottom row is usually not included in the table but is presented to remind you that you must test for orthogonality. When there are more than two contrasts, all possible combinations of two contrasts must be tested for orthogonality. For example, with three contrasts, this means testing contrasts 1 and 2, contrasts 1 and 3, and contrasts 2 and 3.

This table is for an experiment looking at nitrogen fertilization:

Contrast	Control	urea lo	urea hi	nitrate lo	nitrate hi	Σc_{ik}
Control vs nitrogen	4	−1	−1	−1	−1	0
Urea vs nitrate	0	1	1	−1	−1	0
Orthogonal?	0	−1	−1	1	1	0

These coefficients are used in the calculation of the partitioned sums of squares for treatment. Since the contrasts are orthogonal, all partitioned sums of squares derived through the contrasts should sum to equal the non-partitioned, original SS treatment. This is a good check on your math. You need coefficients for SAS® also, so tables such as the one above will be needed often.

But how do you derive the coefficients for a set of contrasts? The procedure can be put into four basic steps. Let's use those steps to come up with a contrast table for our example with the strawberry nitrogen experiment of the previous chapter. Our first contrast is one to determine the SS attributable to nitrogen application, i.e. 'Control vs Nitrogen'.

1. Consider that the contrast has two sides, left of the 'vs' and right of the 'vs'. For example, in the contrast 'Control vs Nitrogen', the left-hand side is Control and the right-hand side is Nitrogen. For our example, we have:
   Treatments: 0   30F   30S   60F   60S   90F   90S
   which can be split into the contrast's two 'sides':
   Control   0
   vs
   Nitrogen   30F   30S   60F   60S   90F   90S

2. Consider the number of observations each side of the 'vs':

Control   8

vs

Nitrogen   4   4   4   4   4   4

Thus control is based on eight observations vs the 24 observations for nitrogen.

3. Multiply the number of observations on the left-hand side by the number of observations on the right-hand side. For our example we get $8 \times 24 = 192$.

4. Remember that the $n_{ik}c_{ik}$'s must sum to 0. Thus the right-hand side of the equation must sum to a number equal to but opposite in sign to the left-hand side. We know the n_{ik}'s, thus we can figure out the c_{ik}'s:

Control   192

vs

Nitrogen   −192

or

$n_{ik}c_{ik}$	192	vs −192					
Treatments:	0	vs	30F	30S	60F	60S	90F	90S
n_{ik} :	8	vs	4	4	4	4	4	4

Thus for the left-hand side of the equation, you divide 192 by 8 to get the coefficient of 24. For the right-hand side, there are six treatments, each with four observations, and (−192) divided by (6 × 4, or 24) equals (−8). Thus for the right-hand side of the equation, each treatment coefficient is −8. Our set of c_{ik}'s is thus:

Con vs nitrogen   24   −8   −8   −8   −8   −8   −8

The coefficients can be reduced to:

Con vs nitrogen   3   −1   −1   −1   −1   −1   −1

Let's set up a set of contrasts for the strawberry data and partition the SS treatment into sources of variation due to specific contrasts. As mentioned before, simple examples are the exception rather than the rule in agricultural experimentation. This is no exception. Note again that we do not have equal n's for all treatments since the control has eight observations. When the n's are equal, the process is simpler.

Setting up the contrast table:

Contrast	0	30F	30S	60F	60S	90F	90S	
Control vs nitrogen	24	−8	−8	−8	−8	−8	−8	
n_i	8	4	4	4	4	4	4	
n_ic_{ik}	192	−32	−32	−32	−32	−32	−32	$\Sigma nc = 0$
Fall vs spring	0	1	−1	1	−1	1	−1	
n_i	8	4	4	4	4	4	4	
n_ic_{ik}	0	4	−4	4	−4	4	−4	$\Sigma nc = 0$
C_1 orthogonal to C_2 ?	0	−128	128	−128	128	−128	128	$\Sigma = 0$

In this experiment, we want to know if there is an effect of the rate of nitrogen applied. There may be a linear effect and quadratic effect. We will cover linear and quadratic when we cover regression. We include it here for completeness. Suffice to say that a linear response is a straight line while a quadratic response has a curve to it. Coefficients for linear and quadratic effects can be found in a basic statistics text. In reality, you won't have to worry about that because if you were to perform a regression analysis on this data, you would use PROC REG of SAS®. The coefficients you would use are presented in the following table:

Contrast	0	30F	30S	60F	60S	90F	90S
Rate effect: linear	0	−1	−1	0	0	1	1
$n_j c_{3k}$	0	−4	−4	0	0	4	4
Rate effect: quadratic	0	1	1	−2	−2	1	1
$n_j c_{4k}$	0	4	4	−8	−8	4	4

Remember that the c_{jk}'s for the linear and quadratic components are obtained from a statistical textbook table of coefficients for equally spaced quantitative treatment levels. You should verify that contrasts 3 and 4 are orthogonal to each other and also to all other contrasts.

We are also interested in whether or not there is a differential response to rate of nitrogen depending on the time of application. This is the interaction of time of application with rate. Our test for the interaction of time of application and rate uses the remaining unaccounted-for SS treatment after subtracting out the SS due to the stated contrasts. Each contrast uses 1 df, thus we have used 4 df of the possible 6 df in our contrast estimation. The remaining 2 df are attributable to the interaction previously mentioned. Verify that the df work out and that the 'left-over' df = df for the interaction of rate × time of application. There are three rates, thus 2 df for rate (which we subdivided into linear and quadratic components), and there are two times of application, thus 1 df, which we used in our contrast. The interaction should have (2 × 1) df, and it does. The 2 df for the interaction can be further broken down into two single df contrasts: (1) Time of application × Rate$_{\text{linear}}$ and (2) Time of application × Rate$_{\text{quadratic}}$, each with 1 df. The coefficients for estimating these contrasts are determined by cross-multiplying the coefficients for the contrasts in the interaction. For the (Time of application × Rate$_{\text{linear}}$) interaction, the coefficients would be the cross-products of the Time of application contrast and the Rate$_{\text{linear}}$ contrast:

				Treatments			
Contrast	0	30F	30S	60F	60S	90F	90S
Fall vs spring	0	1	−1	1	−1	1	−1
Rate Effect							
Rate$_{\text{Linear}}$	0	−1	−1	0	0	1	1
Rate$_{\text{Quadratic}}$	0	1	1	−2	−2	1	1
Interactions							
Time × Rate$_{\text{Linear}}$	0	−1	1	0	0	1	−1
Time × Rate$_{\text{Quadratic}}$	0	1	−1	−2	2	1	−1

To verify that the last two contrasts are orthogonal:
Cross-products   0   −1   −1   0   0   1   1

which sums to 0. (Since all the 'pieces' in the contrasts we just looked at had equal n, I did not include the row showing $n_i c_{jk}$ in this table.)

You might question why the zero treatment level is not considered a rate of 0 in the contrast. One reason is that the contrast for rate effect would make our set of contrasts non-orthogonal if we include 0 as a rate (which is not necessarily bad, but we want to keep things simple for the time being). Another reason is that we have already determined if there is an effect of no nitrogen vs yes nitrogen in contrast 1. (It is for this reason that we have already used the information concerning the difference in yes vs no nitrogen in contrast 1, which makes a 're-use' of the data in a contrast of rate effect 'unacceptable' and therefore makes the contrasts non-orthogonal. We keep orthogonality to emphasize what the contrasts are doing.)

So far we have only come up with a table of contrasts and have not yet calculated the SS attributable to each. First we will calculate them by hand, and then see how we would use SAS® to perform the ANOVA and subsequent contrasts.

We need the data from Chapter 7 for our calculations, so here it is rearranged so that data are in the same order as the table of contrast coefficients. In addition the S-0 and F-0 treatments have been combined into a singular treatment with eight observations.

Original treatment designation for factorial ANOVA

	F-0	S-0	F-30	S-30	F-60	S-60	F-90	S-90
	7.7	8.2	7.0	7.1	7.6	7.1	7.7	7.2
	7.2	6.3	7.2	6.2	8.2	6.9	6.9	8.0
	8.5	7.8	8.4	7.3	6.2	7.5	7.4	7.1
	8.3	7.0	8.4	7.5	6.4	6.7	8.7	7.1

Original treatment totals, N and means

Total	31.7	29.3	31.0	28.1	28.4	28.2	30.7	29.4
N	4	4	4	4	4	4	4	4
Mean	7.9	7.3	7.8	7.0	7.1	7.1	7.7	7.4

Treatment designation for contrasts

	0	30F	30S	60F	60S	90F	90S

Treatment totals, N and means for contrast calculations

Total	61	31.0	28.1	28.4	28.2	30.7	29.4
N	8	4	4	4	4	4	4
Mean	7.6	7.8	7.0	7.1	7.1	7.7	7.4

N=32. Grand total = 236.8

First, we should perform a simple ANOVA considering seven treatments. From this ANOVA we will develop our complete ANOVA table to include contrasts. The following SAS® statements will generate the dataset and a simple ANOVA:

```
TITLE1 'Strawberry Nitrogen Study Contrasts';
TITLE2 'Analysis of Variance';
DATA one;
INPUT time $ rate trt $ yld @@;
CARDS;
```

```
S  60  60S  7.1  F  90  90F  7.7  F  0    0  7.2  F  30  30F  8.4
S   0   0   8.2  S  60  60S  7.5  F  90  90F  6.9  S  90  90S  7.1
S  30  30S  7.1  S  30  30S  6.2  S  0    0  7.8  F  60  60F  6.2
F  60  60F  7.6  F  30  30F  7.0  F  30  30F  8.4  F  90  90F  7.4
S  90  90S  7.2  S  30  30S  7.3  S  0    0  7.0  F  60  60F  6.4
S  60  60S  6.9  S  30  30S  7.5  F  0    0  8.5  F   0   0  8.3
S  90  90S  8.0  F  30  30F  7.2  F  60  60F  8.2  F  90  90F  8.7
F   0   0   7.7  S  0   0   6.3  S  60  60S  6.7  S  90  90S  7.1
RUN;
PROC PRINT;
PROC ANOVA DATA = one PLOTS = none;
CLASSES trt;
MODEL yld = trt;
MEANS trt;
RUN;
```
The ANOVA results:

Strawberry Nitrogen Study Contrasts
Analysis of Variance

The ANOVA Procedure

Dependent Variable: yld

Source	DF	Sum of Squares	Mean Square	F Value	Pr > F
Model	6	2.62000000	0.43666667	0.91	0.5050
Error	25	12.02000000	0.48080000		
Corrected Total	31	14.64000000			

R-Square	Coeff Var	Root MSE	yld Mean
0.178962	9.370236	0.693397	7.400000

Source	DF	Anova SS	Mean Square	F Value	Pr > F
trt	6	2.62000000	0.43666667	0.91	0.5050

This allows us to complete a simple ANOVA table:

Source	df	SS	MS	F	Pr>F
Total	32	1766.96			
μ	1	1752.32			
Treatment	6	2.62	0.44	0.91	0.50
Error	25	12.02	0.48		

From this simple ANOVA we can state that we fail to reject the null hypothesis of equal treatment means at the 0.50 level of significance. Let's see what happens when we use contrasts to subdivide the Treatment SS into pieces attributable to specific treatment effects. In other words, let's use contrasts to answer some specific, pre-planned questions about treatment effects.

We will calculate the contrast SS by hand to illustrate where these numbers come from. In reality, we will always use SAS® for the calculations.

For Contrast 1 of our example, Control vs Nitrogen:

$SS\,Control\,vs\,Nitrogen\,(C1)$

$$= \frac{\left((24)(61)-(8)(31)-(8)(28.1)-(8)(28.4)-(8)(28.2)-(8)(30.7)-(8)(29.4)\right)^2}{\left((8)(24^2)+(4)(-8^2)+(4)(-8^2)+(4)(-8^2)+(4)(-8^2)+(4)(-8^2)+(4)(-8^2)\right)}$$

$$= \frac{3317.76}{6144} = 0.54$$

The other contrasts are calculated in a similar manner:

$SS\,Fall\,vs\,spring\,(C2)$

$$= \frac{\left((0)(61)+(1)(31)-(1)(28.1)+(1)(28.4)-(1)(28.2)+(1)(30.7)-(1)(29.4)\right)^2}{\left((8)(0^2)+(4)(+1^2)+(4)(-1^2)+(4)(1^2)+(4)(-1^2)+(4)(+1^2)+(4)(-1^2)\right)}$$

$$= \frac{19.36}{24} = 0.81$$

$SS\,Rate_{linear}\,(C3)$

$$= \frac{\left((0)(61)-(1)(31)-(1)(28.1)+(0)(28.4)+(0)(28.2)+(1)(30.7)+(1)(29.4)\right)^2}{\left((8)(0^2)+(4)(-1^2)+(4)(-1^2)+(4)(0^2)+(4)(0^2)+(4)(+1^2)+(4)(+1^2)\right)}$$

$$= \frac{1}{16} = 0.0625$$

$SS\,Rate_{quadratic}\,(C4)$

$$= \frac{\left((0)(61)+(1)(31)+(1)(28.1)-(2)(28.4)-(2)(28.2)+(1)(30.7)+(1)(29.4)\right)^2}{\left((8)(0^2)+(4)(+1^2)+(4)(+1^2)+(4)(-2^2)+(4)(-2^2)+(4)(+1^2)+(4)(+1^2)\right)}$$

$$= \frac{36}{48} = 0.75$$

$SS\,Time*Rate_{linear}$

$$= \frac{\left((0)(61)-(1)(31)+(1)(28.1)+(0)(28.4)+(0)(28.2)+(1)(30.7)-(1)(29.4)\right)^2}{\left((8)(0^2)+(4)(-1^2)+(4)(1^2)+(4)(0^2)+(4)(0^2)+(4)(+1^2)+(4)(-1^2)\right)}$$

$$= \frac{2.56}{16} = 0.16$$

$SS\,Time*Rate_{quadratic}$

$$= \frac{\left((0)(61)+(1)(31)-(1)(28.1)-(2)(28.4)+(2)(28.2)+(1)(30.7)-(1)(29.4)\right)^2}{\left((8)(0^2)+(4)(+1^2)+(4)(-1^2)+(4)(-2^2)+(4)(+2^2)+(4)(+1^2)+(4)(-1^2)\right)}$$

$$= \frac{14.44}{48} = 0.3001$$

Once the SS are calculated, MS values are calculated by dividing the SS by the df for each source. Since the df for each contrast is 1, the MS and SS values are the same. The F-values are calculated by dividing the MS for each source of variation by the MS error. An updated ANOVA table would look like this:

Source	df	SS	MS	F	Pr>F
Total	32	1766.96			
μ	1	1752.32			
Treatment	6	2.6200	0.4367	0.91	0.50
Control vs nitrogen	1	0.5400	0.5400	1.12	
Fall vs Spring	1	0.8067	0.8067	1.68	
Rate linear	1	0.0625	0.0625	0.13	
Rate quadratic	1	0.7500	0.7500	1.56	
Date $\times$ Rate$_{Linear}$	1	0.1600	0.1600	0.33	
Date $\times$ Rate$_{Quadratic}$	1	0.3008	0.3008	0.63	
Error	25	12.0200	0.4808		

Instead of retrieving the significance values for the calculated F-values from a probability table, we will use SAS® to verify our calculations and use the probabilities provided by SAS®. Once we have F-value significance levels, we will interpret them.

To perform the contrasts, we need to use a different procedure in SAS®, called the General Linear Models (GLM) procedure. For a balanced experiment with no missing data, the results of PROC GLM are the same as the results from PROC ANOVA. The statements used for our analysis are given below.

One very important consideration is that SAS® automatically compensates for different n_i's, thus you set up a contrast table as if all treatments had equal numbers of observations even if they don't. This actually saves you time and calculations. Additionally, you need to know how SAS® has sorted your treatments in order to know how to order the top line of the contrast coefficient table. The easiest way to know how SAS® has ordered your treatments is to look at the classes level information printed out with your ANOVA. The order listed is the order for SAS®. For our example, the contrast table *for use with SAS®* (assuming equal numbers of observations in each treatment) would be:

				Treatments			
Contrast	0	30F	30S	60F	60S	90F	90S
Nitrogen effect	6	−1	−1	−1	−1	−1	−1
Fall vs spring	0	1	−1	1	−1	1	−1
Rate$_{Linear}$	0	−1	−1	0	0	1	1
Rate$_{Quadratic}$	0	1	1	−2	−2	1	1
Date $\times$ Rate$_{Linear}$	0	−1	1	0	0	1	−1
Date $\times$ Rate$_{Quadratic}$	0	1	−1	−2	2	1	−1

While this table is no different than the one previously presented for calculating SS by hand, in some cases the two tables (hand calculation vs SAS® calculation) could be quite different. Because it is so important, I repeat: always prepare contrast coefficient tables for use in SAS® assuming equal numbers of observations in each treatment. You will likely never perform contrast calculations by hand once you know how to perform them using SAS®, so remembering to construct tables correctly should not be difficult.

The SAS® program for this analysis is:

```
TITLE1 'Strawberry Nitrogen Study Contrasts';
TITLE2 'Analysis of Variance using PROC GLM';
DATA one;
INPUT time $ rate trt $ yld @@;
CARDS;
S 60 60S 7.1 F 90 90F 7.7 F 0   0 7.2 F 30 30F 8.4
S  0 0   8.2 S 60 60S 7.5 F 90 90F 6.9 S 90 90S 7.1
S 30 30S 7.1 S 30 30S 6.2 S 0   0 7.8 F 60 60F 6.2
F 60 60F 7.6 F 30 30F 7.0 F 30 30F 8.4 F 90 90F 7.4
S 90 90S 7.2 S 30 30S 7.3 S 0   0 7.0 F 60 60F 6.4
S 60 60S 6.9 S 30 30S 7.5 F 0   0 8.5 F  0   0 8.3
S 90 90S 8.0 F 30 30F 7.2 F 60 60F 8.2 F 90 90F 8.7
F  0 0   7.7 S 0  0   6.3 S 60 60S 6.7 S 90 90S 7.1
RUN;
PROC PRINT;
PROC GLM DATA = one PLOTS = none;
CLASSES trt;
MODEL yld = trt;
CONTRAST 'Control vs nitrogen' trt 6 -1 -1 -1 -1 -1 -1;
CONTRAST 'Fall vs Spring'      trt 0  1 -1  1 -1  1 -1;
CONTRAST 'Linear'              trt 0 -1 -1  0  0  1  1;
CONTRAST 'Quadratic'           trt 0  1  1 -2 -2  1  1;
CONTRAST 'Time*Ratelinear'     trt 0 -1  1  0  0  1 -1;
CONTRAST 'Time*Ratequadratic'  trt 0  1 -1 -2  2  1 -1;
MEANS trt;
RUN;
```

Output:

Strawberry Nitrogen Study Contrasts
Analysis of Variance using PROC GLM

The GLM Procedure

Class Level Information		
Class	Levels	Values
trt	7	0 30F 30S 60F 60S 90F 90S

Number of Observations Read	32
Number of Observations Used	32

Strawberry Nitrogen Study Contrasts
Analysis of Variance using PROC GLM

The GLM Procedure

Dependent Variable: yld

Source	DF	Sum of Squares	Mean Square	F Value	Pr > F
Model	6	2.62000000	0.43666667	0.91	0.5050
Error	25	12.02000000	0.48080000		
Corrected Total	31	14.64000000			

R-Square	Coeff Var	Root MSE	yld Mean
0.178962	9.370236	0.693397	7.400000

Source	DF	Type I SS	Mean Square	F Value	Pr > F
trt	6	2.62000000	0.43666667	0.91	0.5050

Source	DF	Type III SS	Mean Square	F Value	Pr > F
trt	6	2.62000000	0.43666667	0.91	0.5050

Contrast	DF	Contrast SS	Mean Square	F Value	Pr > F
Control vs nitrogen	1	0.54000000	0.54000000	1.12	0.2994
Fall vs Spring	1	0.80666667	0.80666667	1.68	0.2071
Linear	1	0.06250000	0.06250000	0.13	0.7215
Quadratic	1	0.75000000	0.75000000	1.56	0.2232
Time*Ratelinear	1	0.16000000	0.16000000	0.33	0.5692
Time*Ratequadratic	1	0.30083333	0.30083333	0.63	0.4364

Strawberry Nitrogen Study Contrasts
Analysis of Variance using PROC GLM

The GLM Procedure

Level of trt	N	yld	
		Mean	Std Dev
0	8	7.62500000	0.74785407
30F	4	7.75000000	0.75498344
30S	4	7.02500000	0.57373048
60F	4	7.10000000	0.95916630
60S	4	7.05000000	0.34156503
90F	4	7.67500000	0.75883683
90S	4	7.35000000	0.43588989

Let's look at the SAS® statements associated with contrasts. You are already familiar with TITLE statements, the DATA statements and PROC PRINT.

```
PROC GLM DATA = one PLOTS = none;
```

This invokes the GLM procedure to perform the analysis on the dataset called 'one' without generating plots we don't need right now.

```
CLASSES trt;
```

This identifies variables that are main effect sources of variation in the ANOVA table.

```
MODEL yld = trt;
```

The dependent variable(s) are listed followed by the '=' sign followed by all sources of variation from the ANOVA table, with the exception of Total, μ, Contrasts and Error.

```
CONTRAST 'Control vs nitrogen' trt 6 -1 -1 -1 -1 -1 -1;
CONTRAST 'Fall vs Spring'      trt 0  1 -1  1 -1  1 -1;
CONTRAST 'Linear'              trt 0 -1 -1  0  0  1  1;
CONTRAST 'Quadratic'           trt 0  1  1 -2 -2  1  1;
CONTRAST 'Time*Ratelinear'     trt 0 -1  1  0  0  1 -1;
CONTRAST 'Time*Ratequadratic'  trt 0  1 -1 -2  2  1 -1;
```

The code performing the contrast begins with the keyword CONTRAST followed by a title for the contrast that you supply, in single quotes. Immediately following the title is the variable name for the independent main effect variable you are performing the contrast on. In our case it is 'trt'. Following this are the coefficients from your contrast table. Note that there is a CONTRAST statement for each contrast you are performing.

```
MEANS trt;
```

```
RUN;
```

This generates treatment means and performs the analysis.

We have now learned about the characteristics of a CRD experiment and performed several different types of analyses of variance by hand and using SAS®.

Our complete discussion of the results from this experiment can center on the contrasts, since they answer all of the questions we posed. The F-value for testing the hypothesis that there was an effect of nitrogen is 1.12 with a $Pr > F = 0.2994$. This means that the probability of generating an F-value as large or larger than the one we generated simply by chance is 0.2994. We would fail to reject the null hypothesis of equal treatment means at the 0.2994 level; thus, overall, nitrogen did not have an effect on yield. However, the beauty of contrasts is that since we posed specific questions prior to conducting our experiment, we can be more specific in terms of whether or not nitrogen had an effect. Our overall hypothesis test simply compares nitrogen with no nitrogen. We were specifically interested in whether the time of application had an effect on yield, whether or not rate had an effect on yield and whether or not there was an interaction between the two factors. We answer these questions using the results from our contrast analysis.

- Does time of application affect yield?
  The F-value for the contrast 'Fall vs Spring' is 1.68 with a $Pr > F = 0.2071$, thus we would fail to reject the null hypothesis that the means for yield were different with fall vs spring nitrogen application.
- Does rate of application affect yield? More specifically, is there a linear response to rate? Is the response quadratic?

The *F*-value for the contrast 'Linear' is 0.13 with a Pr > *F* = 0.7215, thus we would fail to reject the null hypothesis that the means for yield were linearly affected by rate of nitrogen application (straight line response).

The *F* value for the contrast 'Quadratic' is 1.56 with a Pr > *F* = 0.2232, thus we would fail to reject the null hypothesis that the means for yield were affected by rate of nitrogen application in a quadratic manner (curved line response).

- Is there an interaction between the two factors?

There was no 'Time*Rate$_{linear}$' interaction: $F = 0.33$, Pr > $F = 0.5692$.

There was no 'Time*Rate$_{quadratic}$' interaction: $F = 0.63$, Pr > $F = 0.4364$.

This indicates that it didn't matter when the nitrogen was applied. The response to rate was the same for both spring and fall application: no response.

Contrasts are a very powerful method for detecting specific treatment differences based upon questions posed *a priori*. This is our first major method for documenting specific differences among treatment means in an experiment. In the next chapter we will learn a second major tool for examining responses to treatments and that is simple linear regression.

9 Simple Linear Regression

We often want to express the relationship between a dependent variable, such as yield, and an independent variable, such as fertilization rate, with a mathematical equation. With an equation we can illustrate the relationship clearly, using a graph, and can add confidence bands around the line to show the variability inherent in the modeled data. This chapter covers the methods for obtaining and expressing these mathematical equations and their confidence bands. The methodology is linear regression analysis. Four types of regression analysis are presented, including: simple linear regression with no repeated measures or replication; simple linear regression with repeated measures; simple linear regression with replication; and polynomial regression. The approach is a simple, straightforward 'how to' one.

Suppose we treated some plants with several rates of a plant growth regulator and then measured their yield. The treatment with several levels of a growth regulator is a quantitative treatment, rather than a qualitative treatment. Whenever you have quantitative treatments with three or more levels, the usual data analysis involves some form of regression. The treatment variable, the independent variable identified using the letter X, is often called the causal variable because many times we are inferring that X has caused some response in the experimental unit. The response we measure is called the response or dependent variable, identified with the letter Y.

The question we are asking in this type of experiment is: 'Is there a response of Y to a change in the level of X?' In other words, does yield change as the level of the growth regulator changes? If it does, then we want to know how it changes. Does it increase? Decrease? By how much?

We can describe the relationship, if it exists, between X and Y with a simple mathematical equation. The simplest equation is that of a straight line:

$$Y = \beta_0 + \beta_1 X$$

where β_0 is called the intercept and β_1 is called the slope. The intercept is the level of the dependent variable (Y) when the independent variable (X) is at a level of 0. Even if we don't test a rate of $X = 0$, we can still estimate the response in Y at that level. One must be careful in estimating values for Y outside of the range of tested levels of X. This is called extrapolation and it is discouraged, except for estimating the intercept, since we need it to complete the equation of our line. The slope is the amount of change in Y for each unit change in X. The intercept and slope are parameters that we can estimate from our data to describe the unique relationship between X and Y, if one exists in our data. I have noted several times the idea of 'if a relationship exists'. If there is no response in Y to X, then there is no relationship. We can statistically test whether or not this relationship exists.

Now back to our two parameters we want to estimate (β_0 and β_1). How do we estimate these two parameters? First, we must make several assumptions about our data and the relationship between X and Y. We assume that there is a straight-line relationship between X and Y. We also assume that the parameters (β_1, β_0) enter into the equation in a linear manner (that is, they are not an exponent). We can plot our data to visually check

DOI: 10.1079/9781789249927.0009

our assumption of a straight-line relationship using PROC PLOT of SAS®. Even if our plot reveals a slight curve in the relationship we will still use linear regression, just a slight variation (polynomial regression). If we determine that we seem to have a straight-line or polynomial relationship, we can proceed to use SAS® to estimate our parameters using what is called least squares estimation. The nice thing about linear and polynomial regression is that even if the 'true' relationship between X and Y is not linear or polynomial, we can very often approximate the relationship with this method. Just remember that the relationship you uncover from your data is limited to the range of levels of X tested. Also, remember that the whole point of a regression analysis is to try to describe the relationship between the two variables, not to determine the 'true' relationship between the two.

So far we have two variables, X and Y, and we are assuming that there is a simple straight-line relationship of the form $Y = \beta_0 + \beta_1 X$. Since β_0 and β_1 could each theoretically take on any value, we must estimate them for our particular set of data.

But we're not quite ready to estimate them. When we consider a regression analysis we say that there are two different types of errors in regression: fixed and random. **Fixed error** is the error associated with the fact that the model we chose might not fit our data. This is appropriately called 'lack of fit' and we can test for it with a replicated experiment. Note that sometimes we might have observational data and want to perform a regression analysis on it. One problem with this is that we might not have replication to obtain a good estimate of random variability. Thus our tests of parameters might not be too good. We'll see how to deal with this situation later.

The other type of error is called random error. **Random error** is indicated by the symbol 'ε_i'. This error is associated with errors in measurement and natural variability from experimental unit to unit. Each observation in our experiment or dataset can now be described via our model to include the concept of statistical error, thus our model now looks like:

$$Y = \beta_0 + \beta_1 X + \varepsilon_i$$

Each observation can be considered via the equation:

$$Y_i = \beta_0 + \beta_1 X_i + \varepsilon_i$$

for $i = 1$ to n, where n is the number of observations in our dataset. The last piece of the equation, 'ε_i' is our random error and is not observable. However, when we derive β_0 and β_1, we can estimate ε_i, and note it with the symbol $\hat{e}$. This '$\hat{e}$' is called a **residual**. We estimate the residual by subtracting the predicted value of Y at a given level of X from the observed value of Y at that same level. We indicate this estimation via the formula:

$$\hat{e}_i = y_i - \hat{y}_i$$

When we use least squares simple linear regression we make several assumptions. One assumption is that the fixed error is negligible (for now). Another assumption is that the random errors (ε_i's) have a mean of 0, are independent, have a common unknown variance, σ^2, and are normally distributed. This is summarized as:

$$\varepsilon_i's \sim \text{NID}\left(0, \sigma^2\right)$$

where NID means 'normally and independently distributed'. These assumptions can be easily checked via regression diagnostics (see Chapter 10). The idea of independence is that the value of one estimate of ε_i has no influence on the value of any other ε_i. We use the process known as 'least squares' for estimating β_0 and β_1. Least squares estimation minimizes the sum of the $\hat{e}_i^2$, the residual sum of squares.

In order to perform the statistical tests we need for our regression analysis, we need an estimate of the natural variability among our experimental units, i.e. we need an estimate of σ^2. This estimate is best obtained with a replicated experiment. We can also estimate it from prior experiments if available. A third estimate source is from our residual sum of squares during the modeling procedure. The problem here is that the estimate is model dependent and if we have the wrong model, we have a wrong estimate.

Least squares estimations have a few other properties besides a minimum sum of $\hat{e}_i^2$. The line determined via least squares regression will always pass through the point representing the means of X and Y. The estimates β_1 and β_0 each have their own distribution. This is important for determining if they really exist (are they different from 0?) for your data. In addition, knowing their distribution allows you to determine how close they are to the 'true' values. Let's look at a step-by-step example using SAS®.

Simple Linear Regression – No Repeated Measurements, No Replication

Suppose we have five data pairs: 1, 2; 2, 3.2; 3, 3.9; 4, 4.8; 5, 5.4. Our main questions are: 'Is there a relationship between x and y?' and, if there is, 'What is the relationship?'

The first step is to plot the data to see the shape of curve using PROC PLOT. From this point forward, supplementary files of SAS® programs with data are provided to reduce the amount of typing required to practice the examples illustrated in the text. Chapters 1–8 provide plenty of practice to get the feel for how to enter programming statements and data into the SAS® program editor. This is accomplished with the following program:

```
TITLE1 'Chapter 9 Example 1.sas';
TITLE2 'PROC PLOT';
DATA one;
INPUT x y @@;
CARDS;
1 2 2 3.2 3 3.9 4 4.8 5 5.4
RUN;
PROC PLOT DATA = ONE;
PLOT y*x;
RUN;
```

The resulting 'PLOT':

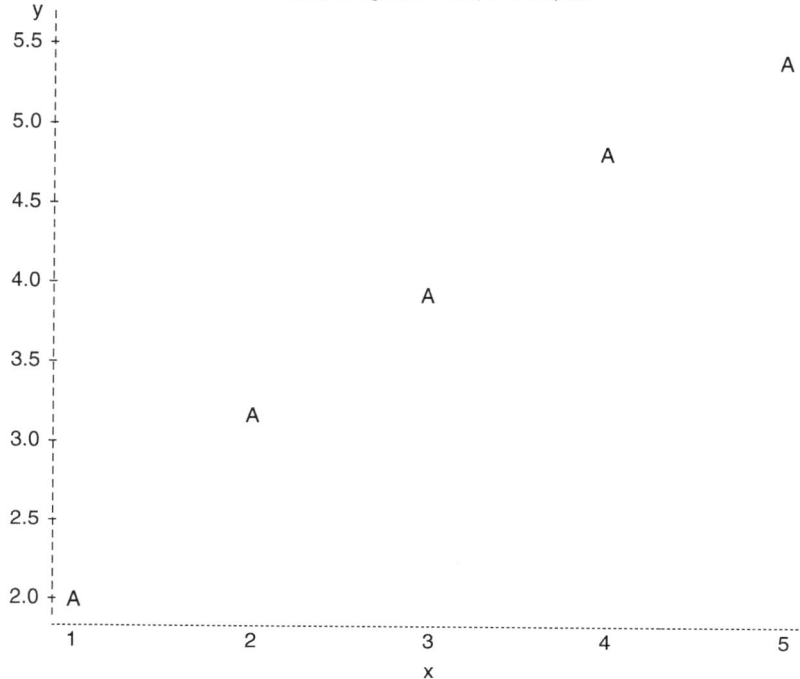

Chapter 9 Example 1.sas
PROC PLOT

Plot of y*x. Legend A – 1 obs, B – 2 obs, etc.

The PROC PLOT statement is very simple to use. We invoke the procedure with the PROC PLOT statement. This is followed by one or more PLOT requests of the form 'PLOT y*x'.

Note that there is only one observation of 'y' at each level of 'x'. This presents somewhat of a problem for regression analysis in that we cannot obtain a reliable estimate of experimental error for hypothesis testing. Even so, there are often times we wish to perform a regression analysis of this sort with observational data. We will later look at the case where we have more than one observation of 'y' at some of the 'x' values, which allows estimation of experimental error. We will also consider the best case, a regression analysis of data from a designed experiment. Thus you normally have one of three simple linear regression situations: no repeated measures, repeated measures, or a replicated experiment.

We use our plot to give us an idea as to what type of regression situation seems to be presented by our data. Is the relationship between 'x' and 'y' a straight line or does there appear to be some curvilinear response? If the relationship appears to be a straight line, then we are looking at a simple linear regression model. If the relationship appears to have a curve in it, we are looking at a polynomial regression model, which we'll cover later in this chapter.

This plot reveals that a straight line seems the best choice, thus we will look at the model:

$$Y = \beta_0 + \beta_1 X + \varepsilon_i$$

Simple Linear Regression

We make the assumption that ε_i's ~ NID$(0,\sigma^2)$, that is, the random errors are normally, identically and independently distributed with a mean of 0 and a variance of σ^2. Our chore is to estimate β_0, β_1, and σ^2 then perform some tests on β_0 and β_1 to determine whether or not each is significantly different than 0. This is easily accomplished using the regression procedure of SAS®.

We could jump right to PROC REG and the results, but illustrating the development of the regression model from one where we assume no relationship between 'x' and 'y' to one looking at a straight-line relationship is useful in understanding what regression is all about.

Let's first look at the analysis of variance assuming there is no relationship between 'y' and 'x'. This means we have no slope and the level of 'y' does not depend on the level of 'x' at which we measure 'y'. Our ANOVA table would look like this:

Source	df	SS
Total	5	81.65
μ	1	74.5
Residual	4	7.15

Now how did we get these numbers and what do they mean? First, the Total SS is the sum of the squared 'y' observations.

We begin with the total sums of squares. This value is simply the sum of the squared observations:

$$SS\,Total = \sum_{i=1}^{n} Y_i^2$$

where n is the number of observations, in this case 5. Thus:

$$SS\,Total = 2^2 + 3.2^2 + 3.9^2 + 4.8^2 + 5.4^2 = 81.65$$

The SS_{Total} is based on $n = 5$ 'x,y' pairs, thus there are 5 df. The $SS\mu$ is the correction factor seen in previous ANOVAs and is:

$$SS\mu = \frac{\left(\sum_{i=1}^{n} Y_i\right)^2}{N}$$

which is the sum of the observations, squared, divided by N, the total number of observations. The $SS\mu$ has 1 df.

$$SS\mu = \frac{(2 + 3.2 + 3.9 + 4.8 + 5.4)^2}{5} = \frac{19.3^2}{5} = 74.498$$

The Residual SS and df are those left over after modeling. The residual mean square (Residual SS/Residual df) is our best estimate at this point of random error. In this case our model is that there is no relationship between 'x' and 'y' and the only SS we can account for is that due to the mean. Even though this is a valid ANOVA, by itself, it is not very informative.

Let's build on this and take the next step and assume that there is a straight-line relationship between 'x' and 'y'. Normally we would use the PROC REG procedure in SAS® for our regression analysis; however, in order to get SS and some other information for our expanded ANOVA table, we will use the general linear models (GLM) procedure (PROC GLM).

Our SAS® program is as follows:
```
TITLE1 'Chapter 9 Example 2.sas';
TITLE2 'PROC GLM';
DATA one;
INPUT x y @@;
CARDS;
1 2 2 3.2 3 3.9 4 4.8 5 5.4
RUN;
PROC GLM DATA = ONE;
MODEL y = x / SOLUTION;
RUN;
```
The output:

Chapter 9 Example2.sas
PROC GLM

The GLM Procedure

Number of Observations Read	5
Number of Observations Used	5

Chapter 9 Example2.sas
PROC GLM

The GLM Procedure

Dependent Variable: y

Source	DF	Sum of Squares	Mean Square	F Value	Pr > F
Model	1	7.05600000	7.05600000	220.50	0.0007
Error	3	0.09600000	0.03200000		
Corrected Total	4	7.15200000			

R-Square	Coeff Var	Root MSE	y Mean
0.986577	4.634338	0.178885	3.860000

Source	DF	Type I SS	Mean Square	F Value	Pr > F
x	1	7.05600000	7.05600000	220.50	0.0007

Source	DF	Type III SS	Mean Square	F Value	Pr > F
x	1	7.05600000	7.05600000	220.50	0.0007

Parameter	Estimate	Standard Error	t Value	Pr > |t|
Intercept	1.340000000	0.18761663	7.14	0.0057
x	0.840000000	0.05656854	14.85	0.0007

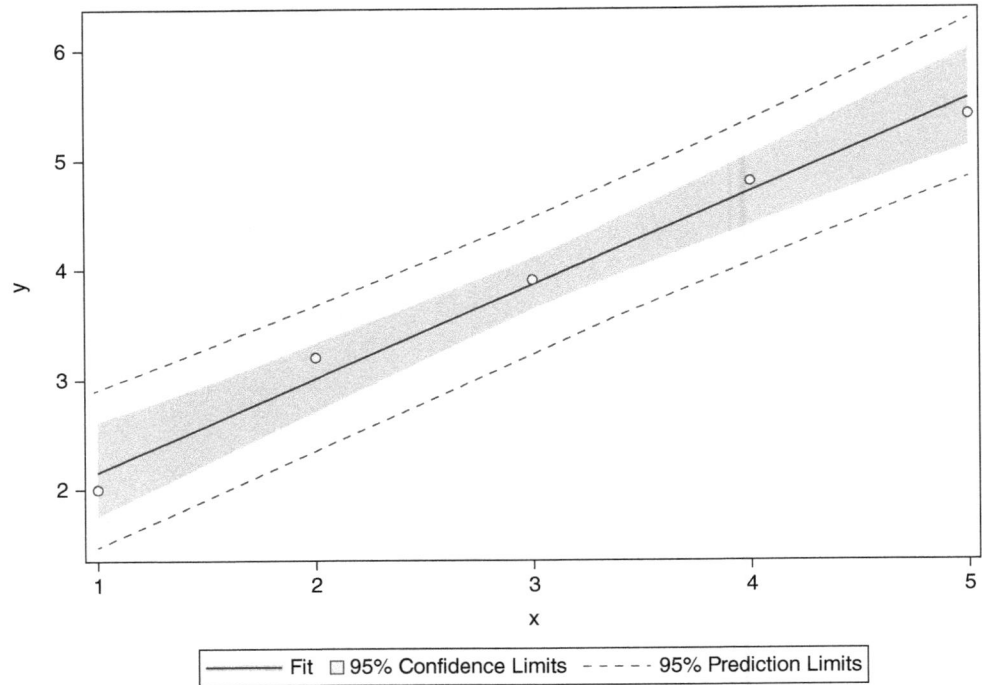

The PROC GLM procedure with the option we requested provides the information we need to fully construct our ANOVA table as well as a plot of the estimated line with a confidence band (more on this later) around the predicted line.

Let's look at this program and the output more closely. First, note that when PROC GLM is used for regression, it does not require a CLASSES statement. By not including a CLASSES statement, PROC GLM will provide estimates of our slope and intercept. The option on the model statement specifically requests that the parameter estimates are included in the output. Options on the model statement are indicated with '/' followed by the option name. Our model statement with the option is:

```
MODEL y = x / SOLUTION;
```

What about the output? Our purpose here is to construct a new ANOVA table including a component for the relationship between 'x' and 'y', which we call $SS\beta_1$, or the SS attributable to the slope. The $SS\mu$ is the SS attributable to the intercept and is also called $SS\beta_0$. We extract this information from the SAS® output.

The Total SS and $SS\mu$ ($SS\beta_0$) are obtained exactly as before via formulas previously presented. We can also retrieve them from the GLM output. Note that the GLM output presents the corrected Total SS, that is the Total SS with the correction factor ($SS\beta_0$) already subtracted off. We can easily calculate the $SS\beta_0$ from the GLM output via the following:

$$SS\beta_0 = \frac{(\overline{Y} * (Corrected\ total\ df + 1))^2}{n}$$

In this case:

$$SS\beta_0 = \frac{(3.86*(4+1))^2}{5} = \frac{19.3^2}{5} = 74.498$$

We can now calculate the Total SS via:

$$SS_{Total} = SS_{Corrected\ total} + SS_{\beta_0}$$

from this example:

$$SS_{Total} = 7.152 + 74.498 = 81.650$$

The Residual SS ($SS_{Residual}$ or SS_{Res}) is the Error SS directly from the GLM output and its corresponding df, thus the $SS_{Res} = 0.096$ with 3 df. The last SS, SS_{β_1}, is the SS_X from the GLM output, 7.056 and it has 1 df.

Our new and expanded ANOVA table looks like this:

Source	df	SS	MS	F	Pr>F
Total	5	81.65			
β_0	1	74.5	74.5	?	?
β_1	1	7.056	7.056	?	?
Residual	3	0.096	0.032		

Note that we have replaced μ with β_0 in our source column, since now we are considering a regression ANOVA. Also note how the Residual SS has decreased by exactly the SS_{β_1}. We've also added three columns: MS, F and Pr>F. We've calculated the MS for β_0, β_1 and Residual by dividing the appropriate SS by the df for each source. Now let's calculate an appropriate F-statistic to test whether or not there is a linear relationship between 'x' and 'y'.

We test this by:

$$F = \frac{MS_X}{MS_{Res}}$$

with ($df\ X$, df RES).

With an F-statistic, the df associated with the numerator and denominator of the test are often presented. They are needed in order to look up Pr>F values in probability tables (which we don't need since we're using SAS®, which provides probabilities associated with all tests) and they are useful for informing the reader of the number of observations used for estimating SS_{Res}. Residuals based on more observations are more reliable than those based on fewer observations, thus more observations instills greater confidence in the results of the test.

In this case the F-value for our test of 'x' is 220.5 with (1,3) df. The Pr>F taken directly off of the GLM output is 0.0007, indicating that there is a relationship between 'x' and 'y' at the 0.0007 level.

Look at the estimates of experimental error from our model assuming no relationship ($SS_{Res} = 7.15$ with 4 df; $MS_{Res} = 1.79$) and from the model assuming a linear relationship ($SS_{Res} = 0.096$ with 3 df; $MS_{Res} = 0.032$). Our estimate of error changed drastically from 1.79 to 0.032 as our model changed. This is a major problem. It can be addressed by having more than one observation at some of the levels of 'x', or by having more than one observation at all levels of 'x' as in a replicated experiment. We'll visit both cases shortly. But back to this example with only one observation at each level of 'x'.

Now that we know that there is a relationship between 'x' and 'y', we need to describe that relationship with the equation of a line. We need to retrieve the estimates of β_0 and β_1 from the GLM output. The estimate for β_0 is listed as the estimate for the parameter labeled 'Intercept' on the output, while β_1 is listed as the estimate for the parameter labeled 'x' on the output. In our case $\beta_0 = 1.34$ and $\beta_1 = 0.84$. The estimated equation of the line that describes the relationship between 'x' and 'y' in our data is:

$$\hat{Y} = 1.34 + 0.84 * X$$

Let's examine the parameter estimate section of the SAS® output a little more closely.

Parameter	Estimate	Standard Error	t Value	Pr > |t|
Intercept	1.340000000	0.18761663	7.14	0.0057
x	0.840000000	0.05656854	14.85	0.0007

There are five columns in this part of the output: Parameter, Estimate, Standard Error, *t*-value and Pr >\|t\|. The 'Parameter' column lists the parameters that were estimated by the SAS® program based on the model statement. Our model statement was 'MODEL y = x', thus the parameters estimated are those for a simple linear model, that is, β_0 and β_1. The 'Estimate' column provides the estimate for each parameter listed based on the data provided. The 'Standard Error' column provides estimates of the standard error for each parameter estimate. Standard errors are important for testing whether or not the estimates differ from 0 (testing the null hypothesis that there is no relationship between *x* and *y*) and also for developing confidence bands around the predicted line or confidence intervals around the parameter estimates.

The estimates of the parameters we obtain from our SAS® analysis are called point estimates. Since these values are estimated from a sample of data rather than an entire population of data, there is variability associated with our estimate. In other words, if we randomly sampled five *x,y* pairs from our population and generated a set of point estimates of our parameters and then sampled five different *x,y* pairs and generated another set of parameter estimates, the two sets of estimates would likely be different from each other. The standard errors of each estimate provided by SAS® give an indication of how variable the point estimate is, based on our data. Less variable estimates, i.e. smaller standard errors, are more desirable. This is just like the case with estimating the population mean μ with a sample mean $\bar{Y}$. The sample mean provides a point estimate of μ and we often provide confidence intervals around $\bar{Y}$ (remember Chapter 3?). It's the same for regression analysis: we estimate parameters and often provide additional information regarding variability with either confidence intervals around those estimates or confidence bands around the line generated from the equation we develop in our analysis. Much of the time, confidence bands are used since they provide an instant visual representation of the generated equation and the variability associated with the estimates.

Before we get back to our parameter estimate table discussion, let's look at the predicted line and associated confidence bands which SAS® automatically generates when using PROC GLM.

We often use regression equations to predict the values of the response variable at various levels of x. Remember: never go beyond the range of x that was used for developing the regression equation when deriving predicted values. Predicted dependent variable values can take two different values. A predicted value for a single observation is the predicted value for one individual experimental unit at a given level of x. A predicted value for a mean observation is the predicted value of the average response at a given level of x. Most times we are interested in the mean response at a given level of x. We really are not interested in the predicted response for one individual experimental unit, but rather the average response at a given level of x.

Now let's examine our plot. The output is labeled 'Fit Plot for y'. The x and y axes are similar to any other x,y plot. The plot itself consists of three important features. The first is the predicted line from the equation generated by the analysis. It is a plot of the line generated if an estimate of y is calculated for each level of x in the analysis, the x,y values plotted and the points joined with a line. The small open circles are the actual observations of y at each level of x contained in the data. The line represents the connected predicted values of y at each x. So far, this type of plot should be familiar to most. SAS® provides a nice visual representation of the variability in our data by providing two types of confidence bands around the predicted line. One (the shaded area) represents the confidence bands for predicted mean values at each level of x. The other (the dotted line) represents the confidence bands for predicted individual estimates at each level of x. Most of the time we are interested in the confidence bands for mean values. For example, we would likely be interested in an estimate of the *average* yield per plant at a specific level of x rather than an estimate of *one specific* plant at a specific level of x.

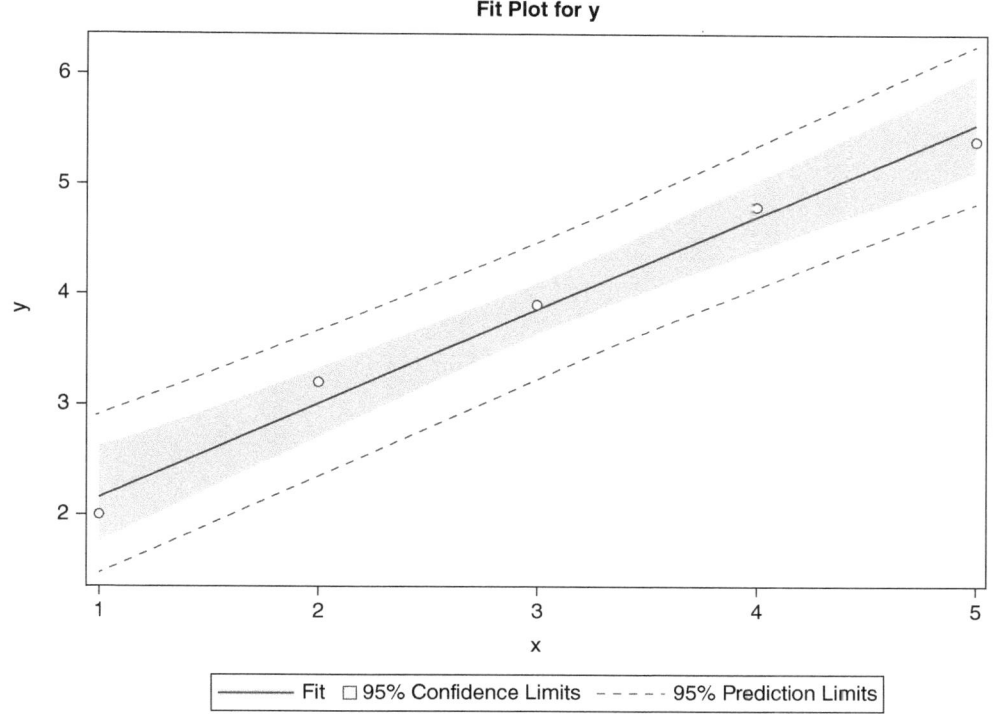

Fit Plot for y

——— Fit □ 95% Confidence Limits - - - - 95% Prediction Limits

Now, back to our parameter estimate table discussion. The fourth column of our table is a column of t-values for tests of the hypothesis that each parameter estimate is = 0. If the t-test indicates that we should reject this hypothesis, it means that there is a detectable relationship between x and y. If we fail to reject this hypothesis, then that indicates that there is no detectable relationship between x and y based on our sample of data. We look at the final column 'Pr >|t|' to determine at which level we should reject the null hypothesis. In our example the t-value for β_0 is 7.14 with an associated α of 0.0057, thus we reject the null hypothesis at the 0.0057 level. The t-value for β_1 is 14.85 with an associated α of 0.0007, thus we reject the null hypothesis at the 0.0007 level. In other words, both β_0 and β_1 are different from 0, thus indicating that there is a relationship between x and y.

Our parameter estimates are derived from one set of data. As previously mentioned, if we had another dataset, we would likely get different estimates due to the natural variability in data. So in order to get an idea of how good our estimates are, and/or what the range of each might be, we develop confidence limits around our parameter estimates.

The confidence intervals for parameter estimates are defined as:

$$\beta_0 \pm t_{(\alpha, n-2)} * se(\hat{\beta}_0)$$

and

$$\beta_1 \pm t_{(\alpha, n-2)} * se(\hat{\beta}_1)$$

where t is a tabular value at α with $n-2$ df, and $se(\hat{\beta}_0)$ and $se(\hat{\beta}_1)$ are standard errors of the estimates obtained from the PROC GLM output. Instead of calculating them by hand, let's use the following short SAS program (complete with comments!) to generate confidence intervals for our example:

Program title statements:

```
TITLE1 'Chapter 9 Example3.sas';
TITLE2 'Confidence Intervals for B0 and B1';
```

Create a dataset named 'one'. The variables labeled b0, seB0, b1 and seB1 represent the parameter estimates and standard errors for each:

```
DATA one;
INPUT b0 seB0 b1 seB1;
CARDS;
1.34 0.18762 0.84 0.05657
RUN;
```

Create a dataset named 'two':

```
DATA two;
```

from the dataset named 'one':

```
SET one;
```

Generate a t-value using the TINV function of SAS for a two-sided t-test at $\alpha = 0.05$ with 3 df:

```
t = TINV(0.975, 3);
```

Generate the upper bound for β_0 using the formula to the right of the '=' sign where b0 is the intercept estimate and seb0 is the standard error of the intercept:

```
ciub0 = b0 + (t*(seb0));
```

Generate the lower bound for β_0 using the formula to the right of the '=' sign where b0 is the intercept estimate and seb0 is the standard error of the intercept:

```
cilb0 = b0 - (t*(seb0));
```

Generate the upper bound for β_1 using the formula to the right of the '=' sign where b1 is the slope estimate and seb0 is the standard error of the slope:

```
ciub1 = b1 + (t*(seb1));
```

Generate the lower bound for β_1 using the formula to the right of the '=' sign where b1 is the slope estimate and seb0 is the standard error of the slope:

```
cilb1 = b1 - (t*(seb1));
```

Print out dataset 'two':

```
PROC PRINT DATA = two;
RUN;
```

Here's the output:

'Chapter 9 Example3.sas'
'Confidence Intervals for B0 and B1'

Obs	b0	seB0	b1	seB1	t	ciub0	cilb0	ciub1	cilb1
1	1.34	0.18762	0.84	0.05657	3.18245	1.93709	0.74291	1.02003	0.65997

The first five columns following the 'Obs' column are data we supplied, while the final four columns are the values for the confidence interval bounds that were calculated by SAS® as requested in the program. If we want to present confidence intervals for our estimates, we would present them like this:

$$CI\beta_0 : P(0.74 \leq \hat{\beta}_0 \leq 1.94) = 0.95$$

$$CI\beta_1 : P(0.66 \leq \hat{\beta}_1 \leq 1.02) = 0.95$$

Both presentations indicate that we are 95% confident that our parameter estimates lie within the indicated ranges.

Many of the procedures in SAS® produce graphics and plots as part of their output using the Output Delivery System (ODS Graphics system) which is part of SAS®. Often these plots are not really needed for an effective analysis and you can request that SAS® refrain from producing them. However, for most of the procedures used in this book, we will simply allow production of default plots rather than introduce more code into programs. If the additional superfluous plots annoy you, you can look up the appropriate code needed to turn them off by using SAS® help.

Suppose you would like to print the predicted values from your line equation and generate the line and confidence bands yourself, rather than using the default SAS® plot produced by PROC GLM. In order to produce nicer looking plots than the stick figure-like plots that PROC PLOT produces, we will introduce PROC SGPLOT, part of the SG plotting environment of SAS®. The SG plots in SAS® interface with the ODS graphics engine in SAS® and produce plots which look just like the automatically produced plots but with user modifications. They are designed with the principles of good graphic design built in, which maximizes clarity and minimizes graphic clutter. Because they look just like automatic plots, the user can mix and match different plots from different procedures as needed. The SG methodology starts with a basic graph of a given type (line, bar, pie, etc.) created from user-supplied data and allows user modification for incorporating additional information. As with other SAS® procedures, we will introduce aspects of the SG plotting environment as needed, keeping it as simple as possible.

Remember that there usually are multiple ways to achieve the same goal in SAS® and I am presenting the methods that I have found easiest and most useful.

We can obtain predicted values by using an option in the model statement, create a dataset from that option, then print it out using the following program:

Program title statements:

```
TITLE1 'Chapter 9 Example 4.sas';
TITLE2 'Predicted Values from GLM Regression';
```

Create dataset one:

```
DATA one;
INPUT x y;
CARDS;
1 2
2 3.2
3 3.9
4 4.8
5 5.4
```

Fit a simple linear model to the data. The '/ CLM' requests that SAS® calculate predicted values for a mean response (as opposed to CLI which would request the predicted value for an individual response) at each level of 'x':

```
PROC GLM NOPRINT;
MODEL y = x / CLM;
```

Create a dataset called 'graph' from the output of the MODEL statement. The dataset should include: the predicted values (P) for each observation and label it 'yhat'; the lower confidence bound for a mean response (L95M) and label it 'lm'; and the upper confidence bound for a mean response (U95M) and label it 'um':

```
OUTPUT OUT = graph P = yhat L95M = lm U95M = um;
```

Print out the dataset 'graph':

```
PROC PRINT DATA = graph;
RUN;
```

This produces:

'Chapter 9 Example 4.sas'
'Predicted Values from GLM Regression'

Obs	x	y	yhat	lm	um
1	1	2.0	2.18	1.73903	2.62097
2	2	3.2	3.02	2.70819	3.33181
3	3	3.9	3.86	3.60540	4.11460
4	4	4.8	4.70	4.38819	5.01181
5	5	5.4	5.54	5.09903	5.98097

Note that since we only want a printout of predicted values, we don't need SAS® to create output from the GLM procedure in the results window. We accomplish this by adding NOPRINT to the end of the PROC GLM request.

Simple Graphing Using PROC SGPLOT

In this section we will generate a graph of our predicted line with confidence bands using the PROC SGPLOT procedure. An excellent reference for creating graphs in SAS is *Statistical Graphics Procedures by Example: Effective Graphs Using SAS®* (Matange and Heath, 2011).

To produce our line with confidence bands from our data we submit the following program:

Program title statements:

```
TITLE1 'Chapter 9 Example 5.sas';
TITLE2 'Linear Regression Plot With Confidence Bands';
TITLE3 'PROC SGPLOT';
```

Create dataset 'one':

```
DATA one;
INPUT x y;
CARDS;
1 2
2 3.2
3 3.9
4 4.8
5 5.4
```

Invoke the SGPLOT procedure to plot data in dataset 'one':

```
PROC SGPLOT DATA = one;
```

Create a regression plot of 'x' vs 'y' where the variable names for 'x' and 'y' are given to the right of the '=' sign for each. In our dataset we coded x as 'x' and y as 'y'. In addition to the default prediction line with observations plotted, include the confidence bands for the mean response (as opposed to individual responses, CLI):

```
REG x = x y = y / CLM;
RUN;
```

which produces:

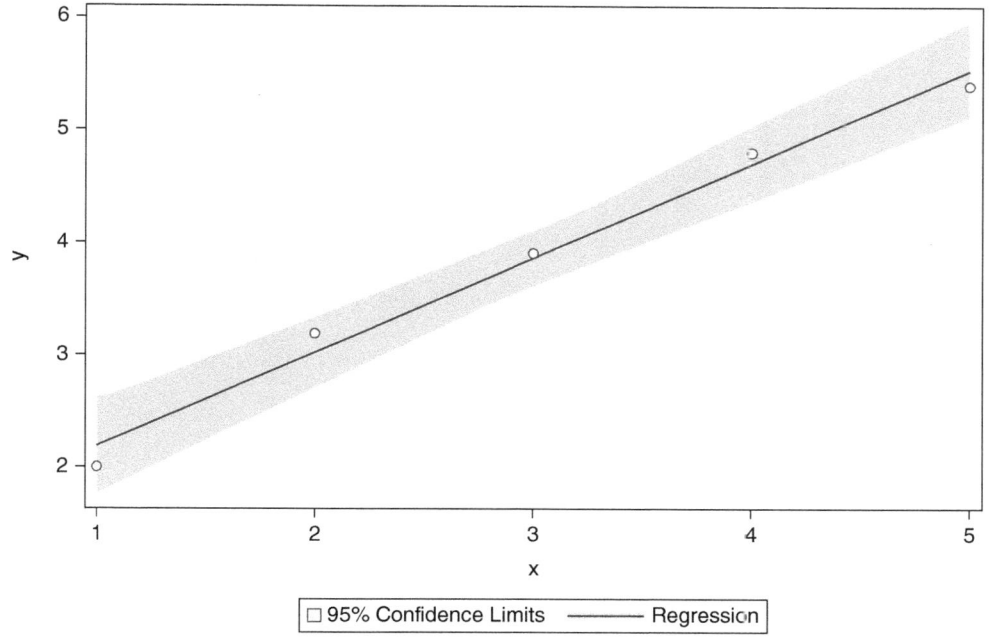

This graph is very simple and not really very attractive, so let's spruce it up a bit with some modifications to the data set and some of the options in SGPLOT.

Program title statements:

```
TITLE1 'Chapter 9 Example 6.sas';
TITLE2 'Seedling Weight With Confidence Bands';
```

Let's give the data meaning. We measured the weight of a radish seedling over the course of 5 days after radicle emergence. Let's change the data from a simple 'x y' to day and weight:

```
DATA one;
INPUT day weight;
```

Let's label our variables which will improve the look of any subsequent graphs we produce. Anywhere 'day' would be printed, we'll see 'Day', and anywhere 'weight' would be printed, we'll see 'Weight (g)':

```
LABEL day = Day;
LABEL weight = Weight (g);
CARDS;
1 2
2 3.2
3 3.9
4 4.8
5 5.4
```

Let's graph day as the 'x' variable vs weight as the 'y' variable and display both types of confidence bands: one for mean predicted values and one for individual predicted values. We don't want legends for the confidence bands displayed, so we use the NOLEGCLI NOLEGCLM options. We also do not want the legend for the regression line, so we use the NOLEGFIT option:

```
PROC SGPLOT DATA = one;
REG x = day y = weight / CLM CLI NOLEGCLI NOLEGCLM NOLEGFIT;
```

Let's add a grid to our plot. We want gridlines for both the x and y axes:

```
XAXIS grid;
YAXIS grid;
RUN;
```

which produces the plot:

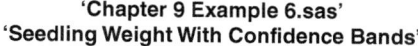

'Chapter 9 Example 6.sas'
'Seedling Weight With Confidence Bands'

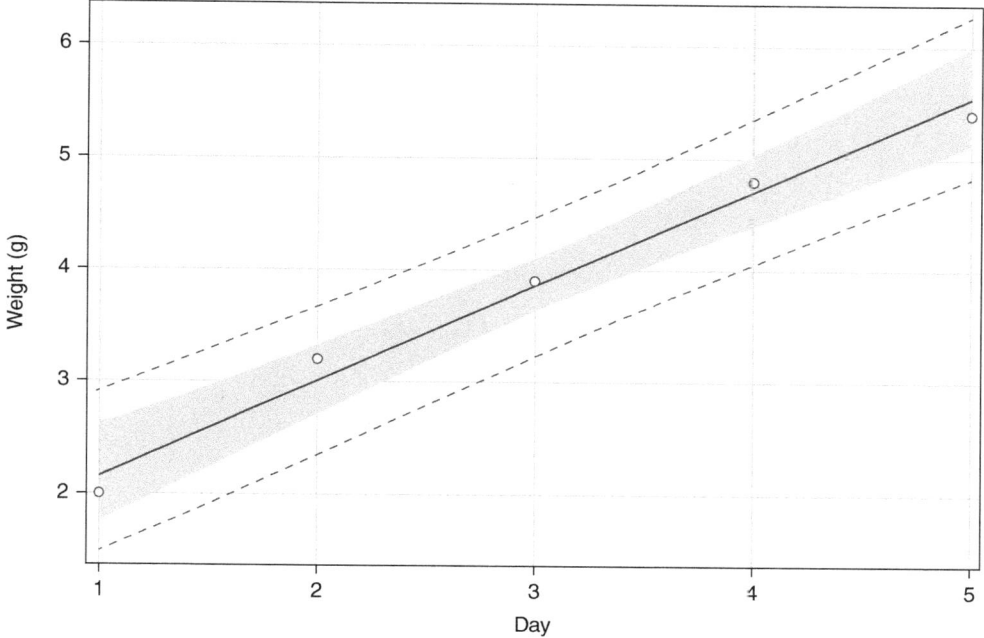

There are many more options available for this simple regression plot, but notice how just a few modifications to the code will produce a much nicer looking graph.

In this section we illustrated how to estimate the equation of a line for a set of data, generate the ANOVA table, produce estimated values and plot the regression line with confidence bands.

The problem with this example is that we don't have a good estimate of statistical error for hypothesis testing. If we alter the model we use, our 'estimate' of error changes, perhaps drastically. We need an estimate of error that is stable and does not depend on what model we use in our analysis. One option for obtaining a more stable estimate of error is to collect more than one observation at one or more levels of x. This option is presented in the next section.

Another procedure, PROC REG, is available in SAS® for regression analysis, but not as much information is presented in the output as when using PROC GLM. We will use PROC REG later in this book, particularly in Chapter 10 which deals with regression diagnostics. The general statements for using PROC REG are:

```
TITLE1 'Chapter 9 Example 7.sas';
TITLE2 'PROC REG';
DATA one;
INPUT x y;
CARDS;
1 2
2 3.2
3 3.9
4 4.8
```

```
5 5.4
PROC REG DATA = one;
MODEL y = x;
RUN;
```
The output from this procedure:

'Chapter 9 Example 7.sas'
'PROC REG'

The REG Procedure
Model: MODEL1
Dependent Variable: y

Number of Observations Read	5
Number of Observations Used	5

Analysis of Variance					
Source	DF	Sum of Squares	Mean Square	F Value	Pr > F
Model	1	7.05600	7.05600	220.50	0.0007
Error	3	0.09600	0.03200		
Corrected Total	4	7.15200			

Root MSE	0.17889	R-Square	0.9866
Dependent Mean	3.86000	Adj R-Sq	0.9821
Coeff Var	4.63434		

Parameter Estimates					
Variable	DF	Parameter Estimate	Standard Error	t Value	Pr > |t|
Intercept	1	1.34000	0.18762	7.14	0.0057
x	1	0.84000	0.05657	14.85	0.0007

Notice our estimates are the same as with PROC GLM. Also notice in the output window that there are quite a few graphs and charts produced automatically by PROC REG. We will examine them in Chapter 10 when we cover regression diagnostics.

SAS® also provides us with a measure of the strength of the relationship between x and y called R^2 or the coefficient of determination. It is always a number between 0 and 1. The closer R^2 is to 1, the stronger the relationship between x and y. The closer R^2 is to 0, the weaker the relationship between x and y. In our example, R^2 is 0.99, which would indicate a close relationship between x and y. In practice, with 'real' data R^2 values are rarely that high. While R^2 is a useful statistic, I rely on parameter hypothesis tests combined with diagnostics (Chapter 10) to determine the validity of my regression analyses.

Simple Linear Regression with Repeated Measurements at Some Levels of X

The problem with the regression in the last section was that the estimate of error was model dependent. Our error estimate was poor and we really didn't have much confidence in the results. If we change the model, our answers will change. But now suppose you had some levels of x where you had more than one observation on y. With this situation SAS® can estimate error at these levels of x and put them together (pool them) to get an estimate of error that is not model dependent. This will provide much more confidence in our results. This independent estimate of error will allow us to perform a test called the 'lack of fit' test which will guide us in determining if we have chosen the correct model to fit our data. Let's see how we do this with an example.

First, we need a simple dataset with some repeated observations for several of the levels of x. We have measured pH and subsequent biomass production. The data are:

pH	Biomass
4.5	23
4.9	41
4.9	40
4.8	38
4.8	34
4.8	37
5.1	47
5.2	52
5.2	50
5.3	58
5.4	61
5.5	68

We need to generate a SAS® dataset and add a few things to our program to perform the analysis. We need to define a new variable we'll call 'lackfit'. By including a 'lackfit' variable, we can have SAS® test for a lack of fit of the chosen model to our data. This variable is defined after the INPUT statement and before the CARDS statement.

To define the 'lackfit' variable, we simply assign it the value of x for each observation. For example, if the observation is $x = 7$, $y = 10$, 'lackfit' will take on the value of 7. Our dataset will thus have three variables: 'x', 'y' and 'lackfit'. If the test for lack of fit is significant, it will indicate that the model chosen does not fit our data, and that we need a different model to accurately describe the relationship between 'x' and 'y' in our data. If the test is not significant, it indicates that the model adequately describes the relationship between 'x' and 'y' in our data.

We need to add a few more statements to our SAS® program. First, we need to include a CLASSES statement which includes only the 'lackfit' variable:

```
CLASSES lackfit;
```

We also modify our PROC GLM statement to include a PLOTS = NONE option. This prevents SAS® from producing automatic plots which are not needed at this time and clutter up the output.

```
PROC GLM PLOTS = NONE;
```

Finally, our MODEL statement is modified to:

```
MODEL y = x lackfit / SS1;
```

The 'lackfit' must come after the 'x'. Include the SS1 in the OPTIONS portion (following the '/') to instruct SAS® to use the Type I sums of squares. Don't worry about what SS1 means right now. So here's the program and the output:

```
TITLE1 'Chapter 9 Example 8.sas';
TITLE2 'Regression lack of fit';
DATA one;
INPUT x y;
lackfit = x;
CARDS;
4.5 23
4.9 41
4.9 40
4.8 38
4.8 34
4.8 37
5.1 47
5.2 52
5.2 50
5.3 58
5.4 61
5.5 68
PROC GLM PLOTS = NONE;
CLASSES lackfit;
MODEL y = x lackfit / SS1;
RUN;
```

which produces the following output:

The GLM Procedure

Class Level Information		
Class	Levels	Values
lackfit	8	4.5 4.8 4.9 5.1 5.2 5.3 5.4 5.5

Number of Observations Read	12
Number of Observations Used	12

'Chapter 9 Example 8.sas'
'Regression lack of fit'

The GLM Procedure

Dependent Variable: y

Source	DF	Sum of Squares	Mean Square	F Value	Pr > F
Model	7	1773.083333	253.297619	90.73	0.0003
Error	4	11.166667	2.791667		
Corrected Total	11	1784.250000			

R-Square	Coeff Var	Root MSE	y Mean
0.993742	3.652083	1.670828	45.75000

Source	DF	Type I SS	Mean Square	F Value	Pr > F
x	1	1755.972414	1755.972414	629.01	<.0001
lackfit	6	17.110920	2.851820	1.02	0.5160

Verify that the number of observations used is the correct number of observations in our dataset. This also corresponds to the corrected total df + 1. This is just a check to make sure we are working with the correct data. You could also request a printout of the data just prior to the GLM request to verify this. It might seem a bit overdone to verify the correct dataset with this example, but it's a good habit to develop since in many instances you will be working with multiple datasets during any particular session with SAS®.

Next, let's look at the lack of fit test. The lack of fit test indicates that we have an appropriately fitting model for our data, since we failed to reject the lack of fit test ($\alpha = 0.51$). This significance level is located on the 'lackfit' source line in the ANOVA table output. There appears to be a relationship between x and y since we reject $H_0: \beta_1=0$ at the 0.0001 level. This significance level is located on the 'x' source line in the ANOVA table. Notice that this first analysis did not produce estimates of β_0 or β_1. We need to re-run the model without the 'lackfit' to obtain the slope and intercept estimates:

```
TITLE1 'Chapter 9 Example 9.sas';
TITLE2 'Regression lack of fit';
DATA one;
INPUT x y;
lackfit = x;
CARDS;
4.5 23
4.9 41
4.9 40
```

```
4.8 38
4.8 34
4.8 37
5.1 47
5.2 52
5.2 50
5.3 58
5.4 61
5.5 68
PROC GLM PLOTS = NONE;
MODEL y = x;
RUN;
```
which produces the output:

'Chapter 9 Example 9.sas'
'Regression lack of fit'

The GLM Procedure

Number of Observations Read	12
Number of Observations Used	12

'Chapter 9 Example 9.sas'
'Regression lack of fit'

The GLM Procedure

Dependent Variable: y

Source	DF	Sum of Squares	Mean Square	F Value	Pr > F
Model	1	1755.972414	1755.972414	620.98	<.0001
Error	10	28.277586	2.827759		
Corrected Total	11	1784.250000			

R-Square	Coeff Var	Root MSE	y Mean
0.984152	3.675615	1.681594	45.75000

Source	DF	Type I SS	Mean Square	F Value	Pr > F
x	1	1755.972414	1755.972414	620.98	<.0001

Source	DF	Type III SS	Mean Square	F Value	Pr > F
x	1	1755.972414	1755.972414	620.98	<.0001

Parameter	Estimate	Standard Error	t Value	Pr > |t|
Intercept	-168.7741379	8.62239378	-19.57	<.0001
x	42.6206897	1.71034135	24.92	<.0001

Our estimate of β_0 is (-168.77) and our estimate of β_1 is (42.62). The equation for our estimated line is:

$$\hat{Y} = -168.77 + 42.62 * X$$

Our plot of these results can be produced with the following program:

```
TITLE1 'Chapter 9 Example 10.sas';
TITLE2 'Regression lack of fit plot using PROC SGPLOT';
DATA one;
INPUT x y;
LABEL x = pH;
LABEL y = Biomass (kg);
CARDS;
4.5 23
4.9 41
4.9 40
4.8 38
4.8 34
4.8 37
5.1 47
5.2 52
5.2 50
5.3 58
5.4 61
5.5 68
RUN;
PROC SGPLOT DATA = one;
REG x = x y = y / CLM CLI NOLEGCLI NOLEGCLM NOLEGFIT;
XAXIS grid;
YAXIS grid;
RUN;
```

which produces:

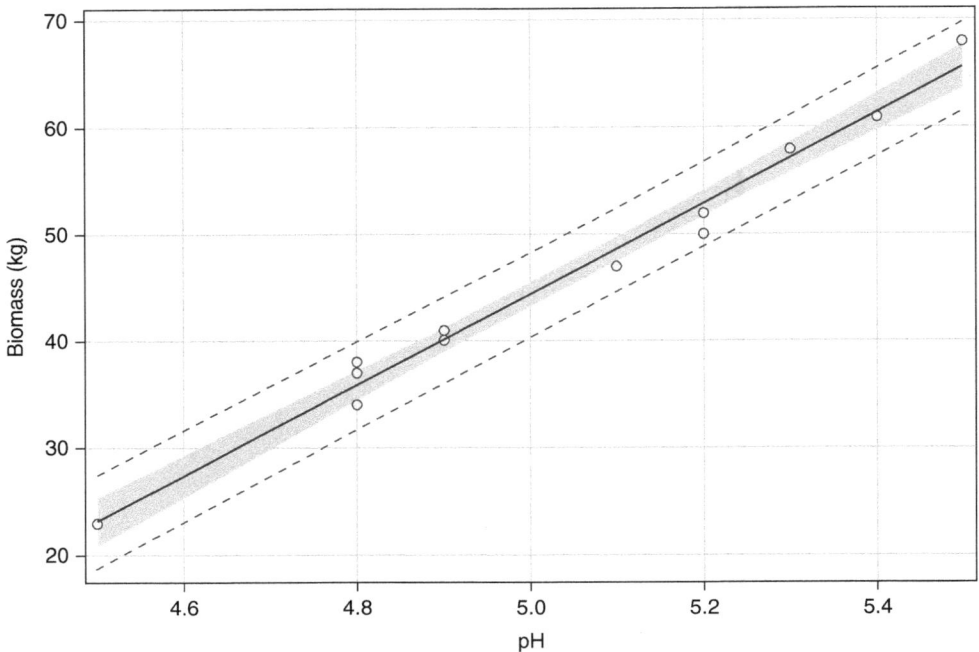

Even though we were able to test for a lack of fit, there is still a problem with our estimate of the error term. The estimate of error is based on multiple observations at only some of the levels of 'x'. What if there was a much different level of variability at the levels of 'x' where only one observation of 'y' was made? A better estimate of error is one derived from a dataset that had multiple observations of 'y' at every level of 'x', for example in a replicated experiment.

Simple Linear Regression with Replication

Suppose we had data from an experiment considering the rate of nitrogen fertilization and subsequent yield. There were four levels of nitrogen ('nit') (0, 10, 20, 30) replicated three times in a completely random design. Our data is:

X	Y	X	Y
0	13.8	20	21
0	13.5	20	22.7
0	13.2	20	22.3
10	15.5	30	18.9
10	15	30	18.3
10	15.2	30	19.6

The ANOVA, not considering the regression (yet), would be:

```
TITLE1 'Chapter 9 Example 11.sas';
TITLE2 'Regression with a replicated experiment';
DATA one;
INPUT nit yld @@;
CARDS;
0 13.8 0 13.5 0 13.2 10 15.5 10 15 10 15.2
20 21 20 22.7 20 22.3 30 18.9 30 18.3 30 19.6
RUN;
PROC ANOVA;
CLASSES nit;
MODEL yld = nit;
RUN;
```

which produces:

'Chapter 9 Example 11.sas'
'Regression with a replicated experiment'

The ANOVA Procedure

Class Level Information		
Class	Levels	Values
nit	4	0 10 20 30

Number of Observations Read	12
Number of Observations Used	12

'Chapter 9 Example 11.sas'
'Regression with a replicated experiment'

The ANOVA Procedure

Dependent Variable: yld

Source	DF	Sum of Squares	Mean Square	F Value	Pr > F
Model	3	130.2433333	43.4144444	127.07	<.0001
Error	8	2.7333333	0.3416667		
Corrected Total	11	132.9766667			

R-Square	Coeff Var	Root MSE	yld Mean
0.979445	3.356111	0.584523	17.41667

Source	DF	Anova SS	Mean Square	F Value	Pr > F
nit	3	130.2433333	43.4144444	127.07	<.0001

Our ANOVA table would look like this:

Source	df	SS	MS	F	Pr>F
Total	12	3773.06			
μ	1	3640.08			
Nitrogen	3	130.24	43.41	127.07	<0.0001
Error	8	2.73	0.34		

With a very significant *F* for nitrogen, we reject the null hypothesis of no effect of nitrogen fertilization on yield in favor of the alternate, that nitrogen influences yield. (Note that even though we generally assume that nitrogen fertilization will increase yield, we do not assume this is our alternate hypothesis. Our alternate hypothesis is that there is an effect of nitrogen, but we don't indicate whether the effect is positive or negative.)

Since nitrogen rate is a quantitative factor, the next step would be regression. With only four levels of nitrogen, we don't have a very good experiment for regression analysis, but it will do. When you anticipate performing a regression analysis, you want to have as many levels of the factor as you can, within reason. A good rule of thumb is to include at least five levels of the factor you are studying. We have four, so we're not too far off. Additionally, we only have three replicates, which is generally not enough. You normally want your df error to be at least 30. Remember, this example is kept small to make things easier to follow without cluttering things up with lots of observations.

Let's run the regression analysis with a lack of fit test. Look at the following program to see how we do it:

```
TITLE1 'Chapter 9 Example 12.sas';
TITLE2 'Regression, lack of fit, replicated data';
DATA one;
INPUT nit yld @@;
lackfit = nit;
CARDS;
0 13.8 0 13.5 0 13.2 10 15.5 10 15 10 15.2
20 21 20 22.7 20 22.3 30 18.9 30 18.3 30 19.6
RUN;
PROC GLM DATA = one PLOTS = NONE;
CLASSES lackfit;
MODEL yld = nit lackfit / ss1;
RUN;
```

The output:

'Chapter 9 Example 12.sas'
'Regression, lack of fit, replicated data'

The GLM Procedure

Class Level Information		
Class	Levels	Values
lackfit	4	0 10 20 30

Number of Observations Read	12
Number of Observations Used	12

'Chapter 9 Example 12.sas'
'Regression, lack of fit, replicated data'

The GLM Procedure

Dependent Variable: yld

Source	DF	Sum of Squares	Mean Square	F Value	Pr > F
Model	3	130.2433333	43.4144444	127.07	<.0001
Error	8	2.7333333	0.3416667		
Corrected Total	11	132.9766667			

R-Square	Coeff Var	Root MSE	yld Mean
0.979445	3.356111	0.584523	17.41667

Source	DF	Type I SS	Mean Square	F Value	Pr > F
nit	1	79.81066667	79.81066667	233.59	<.0001
lackfit	2	50.43266667	25.21633333	73.80	<.0001

Notice that we have a significant lack of fit ($\alpha = 0.0001$) indicating that something has been left out of the model and the model does not fit our data. Before we take the next step in trying to fit a model to our data, let's look at our ANOVA table so far to see where the df for our lack of fit test come from (they have to be taken from somewhere in our original ANOVA, they just don't magically appear):

Source	df	SS	MS	=	Pr>F
Total	12	3773.06			
μ	1	3640.08			
Nitrogen	3	130.24	43.41	101.29	0.0001
$\beta1$	1	79.81	79.81	233.59	0.0001
LOF	2	50.43	25.22	73.8	0.0001
Error	8	2.73	0.34		

If you consider that the 'Nitrogen' factor has 3 df associated with it, we can partition it into meaningful sources. In the above table, we have accounted for that due to a linear regression component. The 2 df remaining are associated with other sources: the lack of fit. Since this example deals with linear regression of four equally spaced levels of an independent variable (nitrogen), the remaining 2 df associated with a lack of fit are quadratic and cubic components. We haven't considered these yet since we haven't covered polynomial regression yet.

Since the LOF (lack of fit) test indicates that perhaps quadratic and/or cubic components are missing from our model, what can we do to try to find the model which fits our data? The most frequently used methodology is to include a 'bend' in the line we are modeling by adding a quadratic component (X^2) to our model. This is polynomial regression.

Polynomial Regression

Polynomial regression consists of adding components to our model by adding powers of X (X^2, X^3, etc.) to our model. We usually limit ourselves to a cubic model (a line with two bends in it) when using polynomial regression. If we can't fit our data to our model with a cubic polynomial model, we probably need to gather more data because we have excessive variability, or we should perform some type of data transformation to linearize a non-linear response. (Perhaps we should consider non-linear regression models – but that's beyond the scope of this book.) The maximum power of X that can be included in a model is (Levels of X) – 1. In this case, we have four levels of X, thus we could fit up to a cubic (X^3) model.

The quadratic and cubic polynomial models look like this:

$$Y = \beta_0 + \beta_1 + \beta_2 X^2 + \varepsilon_i$$

$$Y = \beta_0 + \beta_1 + \beta_2 X^2 + \beta_3 X^3 + \varepsilon_i$$

In our regression analysis we would need to estimate β_0, β_1 and β_2 for the quadratic model and β_0, β_1, β_2 and β_3 for the cubic model.

A very important rule in polynomial regression is that you must include a linear component in a quadratic model (i.e. you can't omit X from your model) and you must include the linear and quadratic components (X and X^2) in a cubic model.

Ok, so how do we perform a quadratic polynomial regression analysis on our data? We need to define the X^2 component much like we previously defined the lack of fit variable, and then include it in our model statement. Here it is:

```
TITLE1 'Chapter 9 Example 13.sas';
TITLE2 'Regression, quadratic model, lack of fit, replicated
    data';
DATA one;
INPUT nit yld @@;
CARDS;
0 13.8 0 13.5 0 13.2 10 15.5 10 15 10 15.2
20 21 20 22.7 20 22.3 30 18.9 30 18.3 30 19.6
RUN;
DATA two;
SET one;
lackfit = nit;
DATA three;
SET two;
nit2 = nit*nit;
PROC GLM DATA = two;
CLASSES lackfit;
MODEL yld = nit nit2 lackfit / ss1;
RUN;
```

(Note that the basic DATA statements are repeated for each program for completeness.)

Here's the output:

The GLM Procedure

Class Level Information		
Class	Levels	Values
lackfit	4	0 10 20 30

Number of Observations Read	12
Number of Observations Used	12

'Chapter 9 Example 13.sas'
'Regression, quadratic model, lack of fit, replicated data'

The GLM Procedure

Dependent Variable: yld

Source	DF	Sum of Squares	Mean Square	F Value	Pr > F
Model	3	130.2433333	43.4144444	127.07	<.0001
Error	8	2.7333333	0.3416667		
Corrected Total	11	132.9766667			

R-Square	Coeff Var	Root MSE	yld Mean
0.979445	3.356111	0.584523	17.41667

Source	DF	Type I SS	Mean Square	F Value	Pr > F
nit	1	79.81066667	79.81066667	233.59	<.0001
nit2	1	17.28000000	17.28000000	50.58	0.0001
lackfit	1	33.15266667	33.15266667	97.03	<.0001

We still have a significant lack of fit. Notice how the lack of fit now only has 1 df. This is because we have used 1 df from nit for estimating β_1 and 1 df from nit for estimating β_2. This leaves us with only 1 df left for testing lack of fit. If we went to a cubic model, we would have no df left for testing for a lack of fit.

Running a cubic model on this data is somewhat questionable since we have no test to determine a lack of fit. Additionally, a cubic model will always provide a good fit for a regression study with four levels of X. A cubic model is the best statistical fit for the data, but it may not really reflect the biology of the system. To determine if a cubic model is a good reflection of the biology, another experiment should be conducted adding a few more levels of nitrogen and perhaps increasing the replication. Altering the experimental design might be wise if sources of variation can be identified that might be contributing to experimental error.

To be complete, here is the cubic model:

```
TITLE1 'Chapter 9 Example 14.sas';
TITLE2 'Regression, cubic model, lack of fit, replicated
    data';
DATA one;
```

```
INPUT nit yld @@;
CARDS;
0 13.8 0 13.5 0 13.2 10 15.5 10 15 10 15.2
20 21 20 22.7 20 22.3 30 18.9 30 18.3 30 19.6
RUN;
DATA two;
SET one;
lackfit = nit;
DATA three;
SET two;
nit2 = nit*nit;
DATA four;
SET three;
nit3 = nit2*nit;
PROC GLM DATA = four;
CLASSES lackfit;
MODEL yld = nit nit2 nit3 lackfit / ss1;
RUN;
```
The output:

'Chapter 9 Example 14.sas'
'Regression, cubic model, lack of fit, replicated data'

The GLM Procedure

Class Level Information

Class	Levels	Values
lackfit	4	0 10 20 30

Number of Observations Read	12
Number of Observations Used	12

'Chapter 9 Example 14.sas'
'Regression, cubic model, lack of fit, replicated data'

The GLM Procedure

Dependent Variable: yld

Source	DF	Sum of Squares	Mean Square	F Value	Pr > F
Model	3	130.2433333	43.4144444	127.07	<.0001
Error	8	2.7333333	0.3416667		
Corrected Total	11	132.9766667			

R-Square	Coeff Var	Root MSE	yld Mean
0.979445	3.356111	0.584523	17.41667

Source	DF	Type I SS	Mean Square	F Value	Pr > F
nit	1	79.81066667	79.81066667	233.59	<.0001
nit2	1	17.28000000	17.28000000	50.58	0.0001
nit3	1	33.15266667	33.15266667	97.03	<.0001
lackfit	0	0.00000000	.	.	.

To produce the parameter estimates with PROC GLM, simply remove the CLASSES statement and modify the MODEL statement to:

```
MODEL yld = nit nit2 nit3;
```

which produces:

'Chapter 9 Example 14.sas'
'Regression, cubic model, lack of fit, replicated data'

The GLM Procedure

Dependent Variable: yld

Source	DF	Sum of Squares	Mean Square	F Value	Pr > F
Model	3	130.2433333	43.4144444	127.07	<.0001
Error	8	2.7333333	0.3416667		
Corrected Total	11	132.9766667			

R-Square	Coeff Var	Root MSE	yld Mean
0.979445	3.356111	0.584523	17.41667

Source	DF	Type I SS	Mean Square	F Value	Pr > F
nit	1	79.81066667	79.81066667	233.59	<.0001
nit2	1	17.28000000	17.28000000	50.58	0.0001
nit3	1	33.15266667	33.15266667	97.03	<.0001

Source	DF	Type III SS	Mean Square	F Value	Pr > F
nit	1	6.71125157	6.71125157	19.64	0.0022
nit2	1	25.82673913	25.82673913	75.59	<.0001
nit3	1	33.15266667	33.15266667	97.03	<.0001

Parameter	Estimate	Standard Error	t Value	Pr > |t|
Intercept	13.50000000	0.33747428	40.00	<.0001
nit	-0.57388889	0.12948736	-4.43	0.0022
nit2	0.09950000	0.01144431	8.69	<.0001
nit3	-0.00247778	0.00025154	-9.85	<.0001

Using PROC REG:

```
TITLE1 'Chapter 9 Example 15.sas';
TITLE2 'Regression, cubic model, using PROC REG';
DATA one;
INPUT nit yld @@;
nit2 = nit*nit;
nit3 = nit2*nit;
CARDS;
0 13.8 0 13.5 0 13.2 10 15.5 10 15 10 15.2
20 21 20 22.7 20 22.3 30 18.9 30 18.3 30 19.6
```

```
PROC REG DATA = one PLOTS = NONE;
MODEL yld = nit nit2 nit3;
   RUN;
```
The output:

'Chapter 9 Example 15.sas'
'Regression, cubic model, using PROC REG'

The REG Procedure
Model: MODEL1
Dependent Variable: yld

Number of Observations Read	12
Number of Observations Used	12

Analysis of Variance					
Source	DF	Sum of Squares	Mean Square	F Value	Pr > F
Model	3	130.24333	43.41444	127.07	<.0001
Error	8	2.73333	0.34167		
Corrected Total	11	132.97667			

Root MSE	0.58452	R-Square	0.9794
Dependent Mean	17.41667	Adj R-Sq	0.9717
Coeff Var	3.35611		

Parameter Estimates					
Variable	DF	Parameter Estimate	Standard Error	t Value	Pr > |t|
Intercept	1	13.50000	0.33747	40.00	<.0001
nit	1	-0.57389	0.12949	-4.43	0.0022
nit2	1	0.09950	0.01144	8.69	<.0001
nit3	1	-0.00248	0.00025154	-9.85	<.0001

Notice that the parameter estimates are the same for PROC GLM and PROC REG. So are the tests of significance for the linear, quadratic and cubic components to the regression equation. That is because we are evaluating the full model, also called a full rank model. This means we have a cubic model investigating four levels of X. A quadratic model would be the full model investigating three levels of X and the linear model would be the full model investigating two levels of X (even though performing a regression with two levels of X doesn't really make much sense).

When models less than full rank are examined, the tests for the parameters and their associated levels of significance will not change in PROC GLM as long as the lack of fit is being considered. Parameter estimates can be requested when performing a lack of fit with PROC GLM; however, parameter estimates requested through the 'solution' option often have the letter 'B' next to them, indicating that they are biased. In PROC REG, the tests and their associated levels of significance will change as the model changes because the error estimate changes with the model, but the parameter estimates will be correct and not biased. The solution is to use PROC GLM to test the significance of parameter estimates and the lack of fit, followed by a run of PROC REG using the chosen model to produce the correct parameter estimates.

Even though PROC REG does have a 'lackfit' option in the MODEL statement, all statistical tests are performed using the model-dependent error estimate. It is therefore important to perform lack of fit testing with PROC GLM and use the parameter estimate tests and levels of significance it generates.

Consider the previous example using a quadratic model:

$$Y = \beta_0 + \beta_1 + \beta_2 X^2 + \varepsilon_i$$

and perform the regression analysis using both PROC GLM and PROC REG for linear, quadratic and cubic models to compare the results.

	PROC GLM	PROC REG
Linear model		
Error *df*	8	10
Error SS	2.7	53.2
LOF df	2	
LOF SS	50.5	
Quadratic model		
Error *df*	8	9
Error SS	2.7	35.9
LOF df	1	
LOF SS	33.2	
Cubic model		
Error *df*	8	8
Error SS	2.7	2.7
LOF df	0	
LOF SS	0	

Notice that, in the table above, if you subtract the LOF df for the PROC GLM run from the error df from the PROC REG run for both the linear and quadratic models, you get the error df from the PROC GLM run considering the lack of fit test. Similarly, if you subtract the LOF SS for the PROC GLM run from the error SS from the PROC REG run for both the linear and quadratic models, you get the error SS from the PROC GLM run considering the lack of fit test. By utilizing the lack of fit test in the PROC GLM runs you are removing the variability associated with the choice of the wrong model from the error term. Also note how the error df and SS for PROC GLM runs do not change as the model changes while they do with PROC REG.

Your best estimate for the model is the cubic model:

$$\hat{Y} = 13.5 - 0.57 * X + 0.099 * X^2 - 0.0023 * X^3 + \hat{e}_i$$

The R^2 value is 0.98, which is not surprising given that you are fitting a cubic model to a dataset with four levels of X.

Comparing Two Regression Lines

When you have a factorial treatment structure with quantitative and qualitative factors, you are often interested in determining if the response to the quantitative factor is the same for the different qualitative factors. Let's say you have cultivar as the qualitative factor and rate of nitrogen fertilization as the quantitative factor. Do the cultivars respond similarly to the changing rates of nitrogen fertilizer? If you only have two levels of cultivar, one would simply look at the interaction between cultivar and rate of nitrogen in the ANOVA table. If the interaction is significant, this tells us that the two cultivars are not responding to nitrogen similarly. You would proceed to develop an appropriate regression equation for each cultivar, and because the interaction was significant, you can say that the two lines are different (but still, the problem is: how are the two lines different? Slopes? Intercepts? Both?).

When there are three or more levels of the qualitative factor, a significant interaction indicates that they don't all respond similarly, but it doesn't tell you which ones are different from each other. You need to determine if there are different lines for the different cultivars, i.e. are their β's different?

When comparing two or more regression lines, the four possible comparisons are:

1. The lines are coincident, i.e. the lines represent the same response and therefore their β's are estimating the same parameter, even though they might be slightly different. For example, suppose we had two lines:

$$Y = 0.45 + 3.5 * X$$

$$Y = 0.39 + 3.1 * X$$

If we test the two lines and find that they are coincident, 0.45 and 0.39 are estimating the same β_0 and 3.5 and 3.1 are estimating the same β_1.

2. The slopes (β_1) for the lines are different but the intercepts (β_0) for the lines are the same. The lines cross the $X = 0$ axis at the same level of Y, but their rate of change in response to a change in the level of X is different.

3. The slopes (β_1) for the lines are the same but the intercepts (β_0) are different. The lines would be parallel but intercept the $X = 0$ axis at different levels of Y.

4. Both the slopes (β_1) and the intercepts (β_0) of the lines differ.

Since the reason we are performing a regression analysis is to determine whether or not there is a relationship between X and Y and estimate it if there is one, it makes most sense to be interested in the rate of response of Y to X and therefore the slopes. Interest in the intercepts is usually much less common.

These hypotheses are easily tested in SAS® by using what are called dummy variables. Let's look at an example with two lines first, followed by an example investigating three lines. The methodology can be expanded to consider any number of lines.

Suppose we have two cultivars, each at six levels of nitrogen, replicated three times in a CRD. Here is the code for the dataset and an initial ANOVA:

```
TITLE1 'Chapter 9 Example 16.sas';
TITLE2 'Comparison of two regression lines';
DATA one;
INPUT rep cv nit yld;
CARDS;
1 1 10 21.2
1 1 20 41.6
```

```
1 1 30 63
1 1 40 80
1 1 50 100
1 1 60 120
2 1 10 20
2 1 20 46
2 1 30 64
2 1 40 81
2 1 50 106
2 1 60 130
3 1 10 21
3 1 20 40
3 1 30 68
3 1 40 82
3 1 50 100
3 1 60 135
1 2 10 36
1 2 20 81
1 2 30 115
1 2 40 153
1 2 50 200
1 2 60 235
2 2 10 36
2 2 20 76
2 2 30 119
2 2 40 162
2 2 50 201
2 2 60 246
3 2 10 43
3 2 20 86
3 2 30 132
3 2 40 156
3 2 50 209
3 2 60 253
RUN;
PROC ANOVA;
CLASSES cv nit;
MODEL yld = cv nit cv*nit;
RUN;
```

We first perform an ANOVA to identify cultivar and nitrogen main effects as well as to determine if there is a cultivar × nitrogen interaction.

'Chapter 9 Example 16.sas'
'Comparison of two regression lines'

The ANOVA Procedure

Class Level Information		
Class	Levels	Values
cv	2	1 2
nit	6	10 20 30 40 50 60

Number of Observations Read	36
Number of Observations Used	36

'Chapter 9 Example 16.sas'
'Comparison of two regression lines'

The ANOVA Procedure

Dependent Variable: yld

Source	DF	Sum of Squares	Mean Square	F Value	Pr > F
Model	11	152598.0656	13872.5514	494.86	<.0001
Error	24	672.8000	28.0333		
Corrected Total	35	153270.8656			

R-Square	Coeff Var	Root MSE	yld Mean
0.995610	4.940833	5.294651	107.1611

Source	DF	Anova SS	Mean Square	F Value	Pr > F
cv	1	41358.0011	41358.0011	1475.32	<.0001
nit	5	100673.5656	20134.7131	718.24	<.0001
cv*nit	5	10566.4989	2113.2998	75.39	<.0001

A significant interaction of cultivar × nitrogen indicates that the response to nitrogen varies between the two cultivars. Our job now is to develop a regression equation to describe the response to nitrogen for each cultivar. We can use PROC REG to do this:

```
PROC SORT; BY cv;
PROC REG PLOTS = none; BY cv;
MODEL yld = nit;
RUN;
```

Note that we have requested exclusion of the plots normally produced by PROC REG since we don't need them yet. Also notice that we have used PROC SORT to sort the data set by cultivar so that we can run two separate regression analysis on the data, one for each cultivar. The program produces the output:

'Chapter 9 Example 16.sas'
'Comparison of two regression lines'

The REG Procedure
Model: MODEL1
Dependent Variable: yld

cv=1

Number of Observations Read	18
Number of Observations Used	18

Analysis of Variance					
Source	DF	Sum of Squares	Mean Square	F Value	Pr > F
Model	1	22989	22989	1420.55	<.0001
Error	16	258.93029	16.18314		
Corrected Total	17	23248			

Root MSE	4.02283	R-Square	0.9889
Dependent Mean	73.26667	Adj R-Sq	0.9882
Coeff Var	5.49067		

Parameter Estimates					
Variable	DF	Parameter Estimate	Standard Error	t Value	Pr > |t|
Intercept	1	0.02667	2.16220	0.01	0.9903
nit	1	2.09257	0.05552	37.69	<.0001

The REG Procedure
Model: MODEL1
Dependent Variable: yld

cv=2

Number of Observations Read	18
Number of Observations Used	18

Analysis of Variance

Source	DF	Sum of Squares	Mean Square	F Value	Pr > F
Model	1	88089	88089	2445.34	<.0001
Error	16	576.36825	36.02302		
Corrected Total	17	88665			

Root MSE	6.00192	R-Square	0.9935
Dependent Mean	141.05556	Adj R-Sq	0.9931
Coeff Var	4.25500		

Parameter Estimates

Variable	DF	Parameter Estimate	Standard Error	t Value	Pr > |t|
Intercept	1	-2.31111	3.22593	-0.72	0.4841
nit	1	4.09619	0.08283	49.45	<.0001

These results indicate that the simple linear equation for each cultivar is thus:

$$\hat{Y} = 0.027 + 2.09 * X$$

$$\hat{Y} = -2.311 + 4.09 * X$$

The question remains as to whether these two equations represent different responses of each cultivar to nitrogen. Since we detected a significant interaction, we can automatically indicate that the cultivars do not respond similarly to nitrogen, i.e. the two equations are different. However, we don't know if the β_0's, β_1's or both differ. To figure this out, there are two null hypotheses you need to test to determine whether or not these equations are different:

(1) H_0: $\beta_0 1 = \beta_0 2$

and

(2) H_0: $\beta_1 1 = \beta_1 2$

Are the slopes for the two lines the same (2) and are the intercepts for the two lines the same (1)? We could perform t-tests on the β's but we will approach the question a little differently. We address these questions via a step-by-step procedure in SAS® creating dummy variables and performing an analysis utilizing them. The test we illustrate can be extrapolated to comparing 'm' different regression lines. In all cases you will need to create ($m-1$) dummy variables and follow the approach illustrated here.

We define our dummy variable ('A1') for this example as such:

```
A1 = 1 if cultivar = 1, otherwise A1 = 0;
```

We do not have to define a dummy variable for cultivar 2 since 'A1 = 0' is associated with any cultivar other than 1, and we only have two cultivars in this example.

In addition to the dummy variable A1, we must create the dummy variable indicating the interaction of A1 with the independent variable used in the regression analysis, in our case 'nit'.

The following SAS® program generates the dummy variables for us, using the previously created dataset 'one' above.

```
DATA two;
SET one;
A1 = . ;
IF cv = 1 then A1 = 1; ELSE A1 = 0;
A1nit = A1*nit;
RUN;
```

We initially define our dummy variable A1 as a missing value (line 3) and give A1 a value of 1 if the cultivar is 1 and 0 if the cultivar is any other value, in this case, 2. We define our second dummy variable 'A1nit' as the interaction of A1 and 'nit' by giving it the value of 'A1*nit'.

The model we are using to make the comparison between the two lines utilizing a dummy variable is:

$$Y = \beta_0 + \beta_1 X_i + \beta_2 A_i + \beta_3 X_i * A + \varepsilon_i$$

We assume that the variances from the two cultivars are equal. Since our examples will always use dummy variables of the form $A_i = 1$ if the observation is in a specific group, otherwise $A_i = 0$, the above model provides a model for observation in either of our two cultivar groups simply by letting $A_i = 0$ or 1 depending on group:

$$Cultivar\,1\,(A1 = 1): Y = \beta_0 + \beta_1 X_i + \beta_2 A_i + \beta_3 X_i * A + \varepsilon_i$$

becomes:

$$Cultivar\,1\,(A1 = 1): Y = \beta_0 + \beta_1 X_i + \beta_2 * 1 + \beta_3 X_i * 1 + \varepsilon_i$$

which is:

$$Cultivar\,1\,(A1 = 1): Y = \beta_0 + \beta_1 X_i + \beta_2 + \beta_3 X_i + \varepsilon_i$$

and moving a few components of the equation around a bit:

$$Cultivar\,1\,(A1 = 1): Y = \beta_0 + \beta_2 + \beta_1 X_i + \beta_3 X_i + \varepsilon_i$$

and finally:

$$Cultivar\,1\,(A1 = 1): Y = \beta_0 + \beta_2 + (\beta_1 + \beta_3) * X + \varepsilon_i$$

Similarly, for cultivar 2:

$$Cultivar\ 2\ (A1 = 0): Y = \beta_0 + \beta_1 X_i + \beta_2 A_i + \beta_3 X_i * A + \varepsilon_i$$

becomes:

$$Cultivar\ 2\ (A1 = 0): Y = \beta_0 + \beta_1 X_i + \beta_2 * 0 + \beta_3 X_i * 0 + \varepsilon_i$$

and simplifying and moving a few components around:

$$Cultivar\ 2\ (A1 = 0): Y = \beta_0 + \beta_1 X_i + \varepsilon_i$$

Using the model:

$$Y = \beta_0 + \beta_1 X_i + \beta_2 A_i + \beta_3 X_i * A + \varepsilon_i$$

we can substitute our notation and obtain:

$$Y = \beta_0 + \beta_1 nit + \beta_2 A + \beta_3 nit * A + \varepsilon_i$$

1. The test for coincidence is obtained with the test

$$H_0: \beta_2 = \beta_3 = 0$$

If we fail to reject H_0, then the model that corresponds to both cultivars is:

$$Y = \beta_0 + \beta_1 nit + \varepsilon_i$$

2. The test for equal slopes (parallelism) is obtained with the test

$$H_0: \beta_3 = 0$$

If we fail to reject H_0, the model for cultivar 1 (where $A = 1$) is:

$$Y = (\beta_0 + \beta_2) + \beta_1 nit + \varepsilon_i$$

and the model for cultivar 2 is:

$$Y = \beta_0 + \beta_1 nit + \varepsilon_i$$

3. The hypothesis for equal intercepts is obtained with the test

$$H_0: \beta_2 = 0$$

If we fail to reject H_0, the model for cultivar 1 is

$$Y = \beta_0 + (\beta_1 + \beta_3) * nit + \varepsilon_i$$

and the model for cultivar 2 is:

$$Y = \beta_0 + \beta_1 nit + \varepsilon_i$$

4. If we reject all three hypotheses, then both lines have different β_0's and β_1's.

We now use 'A1', 'nit' and 'A1nit' as predictors (independent variables) in a regression analysis using PROC REG to perform the above tests and get estimates of our β's.

We will include a TEST statement ('TEST A1nit=0, nit=0;') in our program to test for coincidence: the hypothesis of $H_0: \beta_2 = \beta_3 = 0$. Since we observed a significant interaction between cultivar and nitrogen in our overall ANOVA, we already know that the lines are not coincident, but we will run the test anyway just to illustrate how it's done:

```
PROC REG PLOTS = none DATA = two;
MODEL yld = A1 nit A1nit;
```

```
TEST A1nit=0, nit=0;
RUN;
```
This produces the following output:

'Chapter 9 Example 16.sas'
'Comparison of two regression lines'

The REG Procedure
Model: MODEL1
Dependent Variable: yld

Number of Observations Read	36
Number of Observations Used	36

Analysis of Variance					
Source	DF	Sum of Squares	Mean Square	F Value	Pr > F
Model	3	152436	50812	1946.58	<.0001
Error	32	835.29854	26.10308		
Corrected Total	35	153271			

Root MSE	5.10912	R-Square	0.9946
Dependent Mean	107.16111	Adj R-Sq	0.9940
Coeff Var	4.76770		

Parameter Estimates					
Variable	DF	Parameter Estimate	Standard Error	t Value	Pr > |t|
Intercept	1	-2.31111	2.74607	-0.84	0.4063
A1	1	2.33778	3.88353	0.60	0.5514
nit	1	4.09619	0.07051	58.09	<.0001
A1nit	1	-2.00362	0.09972	-20.09	<.0001

'Chapter 9 Example 16.sas'
'Comparison of two regression lines'

The REG Procedure
Model: MODEL1

Test 1 Results for Dependent Variable yld				
Source	DF	Mean Square	F Value	Pr > F
Numerator	2	55539	2127.67	<.0001
Denominator	32	26.10308		

The test for coincidence (H_0: $\beta_2 = \beta_3 = 0$) produces an F-value of 2127.67 with $\alpha < 0.0001$, which is highly significant. We would therefore reject the null hypothesis that the two lines are coincident.

From the above printout, note that our β estimates are as follows:

$\beta_0 = -2.31111$

$\beta_1 = 4.09619$

$\beta_2 = 2.33778$

$\beta_3 = -2.00362$.

The test for equal slopes (parallelism) (H_0: $\beta_3 = 0$) produces a t-value of -20.09 with $\alpha < 0.0001$. We would therefore reject the hypothesis of equal slopes. Note that we use the t-value rather than F for testing the null hypothesis. The F-value would be $10538/26.10308 = 403.7$ (which is equal to the t-value squared).

The hypothesis for equal intercepts (H_0: $\beta_2 = 0$) produces a t-value of 0.60 and $\alpha = 0.5514$, thus we fail to reject the hypothesis of equal intercepts.

The model for cultivar 1 is:

$$Y = \beta_0 + (\beta_1 + \beta_3) * nit + \varepsilon_i$$

$$Y = -2.31111 + (4.09619 - 2.00362) * nit + \varepsilon_i$$

$$Y = -2.31111 + 2.09257 * nit + \varepsilon_i$$

The model for cultivar 2 is

$$Y = \beta_0 + \beta_1 nit + \varepsilon_i$$

$$Y = -2.31111 + 4.09619 * nit + \varepsilon_i$$

While there are other methods available for performing these same tests, I think that this one is fairly simple and straightforward.

Comparing Three Regression Lines

The methodology used for comparing two regression lines is easily adapted to comparing three or more lines. Let's consider an example comparing three lines. Suppose we have three cultivars, each at six levels of nitrogen, replicated three times in a CRD. The data is as follows:

```
TITLE1 'Chapter 9 Example 17.sas';
TITLE2 'Comparison of three regression lines';
DATA one;
INPUT rep cv nit yld;
CARDS;
1   1       10      21.2
1   1       20      41.6
1   1       30      63
1   1       40      80
1   1       50      100
1   1       60      120
2   1       10      20
```

2	1	20	46
2	1	30	64
2	1	40	81
2	1	50	106
2	1	60	130
3	1	10	21
3	1	20	40
3	1	30	68
3	1	40	82
3	1	50	100
3	1	60	135
1	2	10	36
1	2	20	81
1	2	30	115
1	2	40	153
1	2	50	200
1	2	60	235
2	2	10	36
2	2	20	76
2	2	30	119
2	2	40	162
2	2	50	201
2	2	60	246
3	2	10	43
3	2	20	86
3	2	30	132
3	2	40	156
3	2	50	209
3	2	60	253
1	3	10	52
1	3	20	104
1	3	30	136
1	3	40	199
1	3	50	241
1	3	60	298
2	3	10	46
2	3	20	99
2	3	30	153
2	3	40	169
2	3	50	236
2	3	60	288
3	3	10	36
3	3	20	105
3	3	30	162
3	3	40	193
3	3	50	260
3	3	60	287

```
RUN;
PROC ANOVA;
CLASSES cv nit;
MODEL yld = cv nit cv*nit;
RUN;
```

'Chapter 9 Example 17.sas'
'Comparison of three regression lines'

The ANOVA Procedure

Class Level Information		
Class	Levels	Values
cv	3	1 2 3
nit	6	10 20 30 40 50 60

Number of Observations Read	54
Number of Observations Used	54

'Chapter 9 Example 17.sas'
'Comparison of three regression lines'

The ANOVA Procedure

Dependent Variable: yld

Source	DF	Sum of Squares	Mean Square	F Value	Pr > F
Model	17	324170.9548	19068.8797	331.40	<.0001
Error	36	2071.4667	57.5407		
Corrected Total	53	326242.4215			

R-Square	Coeff Var	Root MSE	yld Mean
0.993651	5.917829	7.585561	128.1815

Source	DF	Anova SS	Mean Square	F Value	Pr > F
cv	2	89078.4459	44539.2230	774.05	<.0001
nit	5	213596.2437	42719.2487	742.42	<.0001
cv*nit	10	21496.2652	2149.6265	37.36	<.0001

The test for interaction is significant, thus there is evidence that the three cultivars do not respond to nitrogen the same way and we need to examine the regression response for each cultivar separately and determine if we should have separate lines to describe each cultivar.

```
PROC SORT DATA = one; BY cv;
PROC REG DATA = one PLOT = none; BY cv;
MODEL yld = nit;
RUN;
```

produces the output:

'Chapter 9 Example 17.sas'
'Comparison of three regression lines'

The REG Procedure
Model: MODEL1
Dependent Variable: yld

cv=1

Number of Observations Read	18
Number of Observations Used	18

Analysis of Variance

Source	DF	Sum of Squares	Mean Square	F Value	Pr > F
Model	1	22989	22989	1420.55	<.0001
Error	16	258.93029	16.18314		
Corrected Total	17	23248			

Root MSE	4.02283	R-Square	0.9889
Dependent Mean	73.26667	Adj R-Sq	0.9882
Coeff Var	5.49067		

Parameter Estimates

Variable	DF	Parameter Estimate	Standard Error	t Value	Pr > |t|
Intercept	1	0.02667	2.16220	0.01	0.9903
nit	1	2.09257	0.05552	37.69	<.0001

'Chapter 9 Example 17.sas'
'Comparison of three regression lines'

The REG Procedure
Model: MODEL1
Dependent Variable: yld

cv=2

Number of Observations Read	18
Number of Observations Used	18

Analysis of Variance

Source	DF	Sum of Squares	Mean Square	F Value	Pr > F
Model	1	88089	88089	2445.34	<.0001
Error	16	576.36825	36.02302		
Corrected Total	17	88665			

Root MSE	6.00192	R-Square	0.9935
Dependent Mean	141.05556	Adj R-Sq	0.9931
Coeff Var	4.25500		

Parameter Estimates

Variable	DF	Parameter Estimate	Standard Error	t Value	Pr > |t|
Intercept	1	-2.31111	3.22593	-0.72	0.4841
nit	1	4.09619	0.08283	49.45	<.0001

'Chapter 9 Example 17.sas'
'Comparison of three regression lines'

The REG Procedure
Model: MODEL1
Dependent Variable: yld

cv=3

Number of Observations Read	18
Number of Observations Used	18

Analysis of Variance

Source	DF	Sum of Squares	Mean Square	F Value	Pr > F
Model	1	123469	123469	1108.44	<.0001
Error	16	1782.23492	111.38968		
Corrected Total	17	125251			

Root MSE	10.55413	R-Square	0.9858
Dependent Mean	170.22222	Adj R-Sq	0.9849
Coeff Var	6.20021		

Parameter Estimates

Variable	DF	Parameter Estimate	Standard Error	t Value	Pr > |t|
Intercept	1	0.48889	5.67267	0.09	0.9324
nit	1	4.84952	0.14566	33.29	<.0001

'Chapter 9 Example 17.sas'
'Comparison of three regression lines'

The REG Procedure
Model: MODEL1
Dependent Variable: yld

cv=3

Number of Observations Read	18
Number of Observations Used	18

Analysis of Variance

Source	DF	Sum of Squares	Mean Square	F Value	Pr > F
Model	1	123469	123469	1108.44	<.0001
Error	16	1782.23492	111.38968		
Corrected Total	17	125251			

Root MSE	10.55413	R-Square	0.9858
Dependent Mean	170.22222	Adj R-Sq	0.9849
Coeff Var	6.20021		

Parameter Estimates

Variable	DF	Parameter Estimate	Standard Error	t Value	Pr > |t|
Intercept	1	0.48889	5.67267	0.09	0.9324
nit	1	4.84952	0.14566	33.29	<.0001

These results indicate that the simple linear equation for each cultivar is:

Cultivar 1: $\hat{Y} = 0.027 + 2.09 * nit$

Cultivar 2: $\hat{Y} = -2.311 + 4.09 * nit$

Cultivar 3: $\hat{Y} = 0.489 + 4.85 * nit$

The question remains as to whether these three equations represent different responses by each cultivar to nitrogen. There are two hypotheses you need to test to determine whether or not these equations are different:

H_0: $\beta_0 1 = \beta_0 2 = \beta_0 3$

and

H_0: $\beta_1 1 = \beta_1 2 = \beta_1 3$

Are the slopes for the three lines the same and are the intercepts for the three lines the same? We address these questions again via the step-by-step procedure in SAS® creating dummy variables and performing an analysis utilizing them.

Our dummy variables are defined as such:

A1 = 1 if cultivar = 1, otherwise A1 = 0

A2 = 1 if cultivar = 2, otherwise A2 = 0

If the cultivar is '3' then A1=A2=0, thus if both A1 and A2 are = 0, we know we have cultivar '3'. Therefore, we do not have to define a dummy variable to indicate that cultivar = '3'. The logic is similar for more than three cultivars or lines.

In addition to the dummy variables A1 and A2, we must create the dummy variables indicating the interaction of each of these two (A1 and A2) with the independent variable used in the regression analysis, in our case 'nit'.

The following SAS® program generates the dummy variables using the previously created dataset 'one' above.

```
DATA two;
SET one;
A1 = .;
A2 = .;
IF cv = 1 then A1 = 1; ELSE A1 = 0;
IF cv = 2 then A2 = 1; ELSE A2 = 0;
A1nit = A1*nit;
A2nit = A2*nit;
RUN;
```

The model we are using to make the comparison among the three lines utilizing our dummy variables is:

$$Y = \beta_0 + \beta_1 * X_i + \beta_2 * A_1 + \beta_3 * A_2 + \beta_4 * X_i * A_1 + \beta_5 * X_1 * A_2 + \varepsilon_i$$

Again, we assume that the variances among the three cultivars are equal. Since our examples will always use dummy variables of the form $A_i = 1$ if the observation is in a specific group, otherwise $A_i = 0$, the above model provides a model for observation for all three cultivar groups simply by letting $A_i = 0$ or 1 depending on group:

Cultivar 1 ($A1 = 1$, $A2 = 0$):

$$Y = \beta_0 + \beta_1 * X_i + \beta_2 * A_1 + \beta_3 * A_2 + \beta_4 * X_i * A_1 + \beta_5 * X_1 * A_2 + \varepsilon_i$$

$$Y = \beta_0 + \beta_1 * X_i + \beta_2 * 1 + \beta_3 * 0 + \beta_4 * X_i * 1 + \beta_5 * X_1 * 0 + \varepsilon_i$$

$$Y = \beta_0 + \beta_1 * X_i + \beta_2 + \beta_4 * X_i + \varepsilon_i$$

$$Y = (\beta_0 + \beta_2) + (\beta_1 + \beta_4) * X_i + \varepsilon_i$$

Cultivar 2 ($A1 = 0$, $A2 = 1$):

$$Y = \beta_0 + \beta_1 * X_i + \beta_2 * A_1 + \beta_3 * A_2 + \beta_4 * X_i * A_1 + \beta_5 * X_1 * A_2 + \varepsilon_i$$

$$Y = \beta_0 + \beta_1 * X_i + \beta_2 * 0 + \beta_3 * 1 + \beta_4 * X_i * 0 + \beta_5 * X_1 * 1 + \varepsilon_i$$

$$Y = \beta_0 + \beta_1 * X_i + \beta_3 + \beta_5 * X_1 + \varepsilon_i$$

$$Y = (\beta_0 + \beta_3) + (\beta_1 + \beta_5) * X_i + \varepsilon_i$$

Cultivar 3 ($A1 = 0$, $A2 = 0$):

$$Y = \beta_0 + \beta_1 * X_i + \beta_2 * A_1 + \beta_3 * A_2 + \beta_4 * X_i * A_1 + \beta_5 * X_1 * A_2 + \varepsilon_i$$

$$Y = \beta_0 + \beta_1 * X_i + \beta_2 * 0 + \beta_3 * 0 + \beta_4 * X_i * 0 + \beta_5 * X_1 * 0 + \varepsilon_i$$

$$Y = \beta_0 + \beta_1 * X_i + \varepsilon_i$$

Using the model:

$$Y = \beta_0 + \beta_1 * X_i + \beta_2 * A_1 + \beta_3 * A_2 + \beta_4 * X_i * A_1 + \beta_5 * X_1 * A_2 + \varepsilon_i$$

we can substitute our notation and obtain

$$Y = \beta_0 + \beta_1 * nit + \beta_2 * A_1 + \beta_3 * A_2 + \beta_4 * nit * A_1 + \beta_5 * nit * A_2 + \varepsilon_i$$

1. The test for coincidence is obtained with the test

H_0: $\beta_2 = \beta_3 = \beta_4 = \beta_5 = 0$

If we fail to reject H_0, then the model that corresponds to all cultivars is:

$$Y = \beta_0 + \beta_1 * nit + \varepsilon_i$$

2. The test for equal slopes (parallelism) is obtained with the test

H_0: $\beta_4 = \beta_5 = 0$

If we fail to reject H_0, the model for cultivar 1 is:

$$Y = (\beta_0 + \beta_2) + \beta_1 * nit + \varepsilon_i$$

and the model for cultivar 2 is:

$$Y = (\beta_0 + \beta_3) + \beta_1 * nit + \varepsilon_i$$

and the model for cultivar 3 is:

$$Y = \beta_0 + \beta_1 * nit + \varepsilon_i$$

3. The hypothesis for equal intercepts is obtained with the test

$H_0: \beta_2 = \beta_3 = 0$

If we fail to reject H_0, the model for cultivar 1 is:

$$Y = \beta_0 + (\beta_1 + \beta_4) * nit + \varepsilon_i$$

and the model for cultivar 2 is:

$$Y = \beta_0 + (\beta_1 + \beta_5) * nit + \varepsilon_i$$

and the model for cultivar 3 is:

$$Y = \beta_0 + \beta_1 * nit + \varepsilon_i$$

We now use 'A1', 'A2', 'nit', 'A1nit' and 'A2nit' as predictors (independent variables) in a regression analysis using PROC REG to perform the above tests and get estimates of our β's.

We include TEST statements as well. The statement 'TEST A1 = 0, A2 = 0, A1nit = 0, A2nit=0;' tests the $H_0: \beta_2 = \beta_3 = \beta_4 = \beta_5 = 0$, the test for coincidence. The statement 'TEST A1nit = 0, A2nit=0;' tests the $H_0: \beta_4 = \beta_5 = 0$, the test for parallelism or equal slopes. The statement 'TEST A1 = 0, A2 = 0;' tests the $H_0: \beta_2 = \beta_3 = 0$, the test for equal intercepts.

```
PROC REG DATA = two PLOT = none;
MODEL yld = A1 A2 nit A1nit A2nit ;
TEST A1 = 0, A2 = 0, A1nit = 0, A2nit=0;
TEST A1nit=0, A2nit=0 ;
TEST A1=0, A2=0;
RUN;
```

It produces the following output:

'Chapter 9 Example 17.sas'
'Comparison of three regression lines'

The REG Procedure
Model: MODEL1
Dependent Variable: yld

Number of Observations Read	54
Number of Observations Used	54

Analysis of Variance					
Source	DF	Sum of Squares	Mean Square	F Value	Pr > F
Model	5	323625	64725	1186.92	<.0001
Error	48	2617.53346	54.53195		
Corrected Total	53	326242			

Root MSE	7.38457	R-Square	0.9920
Dependent Mean	128.18148	Adj R-Sq	0.9911
Coeff Var	5.76103		

Parameter Estimates					
Variable	DF	Parameter Estimate	Standard Error	t Value	Pr > |t|
Intercept	1	0.48889	3.96909	0.12	0.9025
A1	1	-0.46222	5.61314	-0.08	0.9347
A2	1	-2.80000	5.61314	-0.50	0.6202
nit	1	4.84952	0.10192	47.58	<.0001
A1nit	1	-2.75695	0.14413	-19.13	<.0001
A2nit	1	-0.75333	0.14413	-5.23	<.0001

'Chapter 9 Example 17.sas'
'Comparison of three regression lines'

The REG Procedure
Model: MODEL1

Test 1 Results for Dependent Variable yld				
Source	DF	Mean Square	F Value	Pr > F
Numerator	4	27600	506.12	<.0001
Denominator	48	54.53195		

The REG Procedure
Model: MODEL1

Test 2 Results for Dependent Variable yld				
Source	DF	Mean Square	F Value	Pr > F
Numerator	2	10660	195.48	<.0001
Denominator	48	54.53195		

'Chapter 9 Example 17.sas'
'Comparison of three regression lines'

The REG Procedure
Model: MODEL1

Test 3 Results for Dependent Variable yld				
Source	DF	Mean Square	F Value	Pr > F
Numerator	2	7.79934	0.14	0.8671
Denominator	48	54.53195		

The test for coincidence (H_0: $\beta_2 = \beta_3 = \beta_4 = \beta_5 = 0$) produced an F-value of 506.12 with $\alpha < 0.0001$, therefore we reject the hypothesis of coincidence.

The test for equal slopes (H_0: $\beta_4 = \beta_5 = 0$) produced an F-value of 195.48 with $\alpha < 0.0001$, therefore we reject the hypothesis of equal slopes.

The test for intercepts (H_0: $\beta_2 = \beta_3 = 0$) produced an F-value of 0.14 with $\alpha = 0.8671$, thus failed to reject the hypothesis of equal intercepts.

Our β estimates are as follows:

$\beta_0 = 0.48889$

$\beta_1 = 4.84952$

$\beta_2 = -0.46222$

$\beta_3 = -2.80000$

$\beta_4 = -2.75695$

$\beta_5 = -0.75333$

The model for cultivar 1 is:

$$Y = \beta_0 + (\beta_1 + \beta_4) * nit + \varepsilon_i$$

$$Y = 0.48889 + (4.84952 - 2.75695) * nit + \varepsilon_i$$

$$Y = 0.48889 + 2.09257 * nit + \varepsilon_i$$

The model for cultivar 2 is:

$$Y = \beta_0 + (\beta_1 + \beta_5) * nit + \varepsilon_i$$

$$Y = 0.48889 + (4.84952 - 0.75333) * nit + \varepsilon_i$$

$$Y = 0.48889 + 4.09619 * nit + \varepsilon_i$$

The model for cultivar 3 is:

$$Y = \beta_0 + \beta_1 * nit + \varepsilon_i$$

$$Y = 0.48889 + 4.84952 * nit + \varepsilon_i$$

Our three equations are:

Cultivar 1: $Y = 0.48889 + 2.09257 * nit + \varepsilon_i$
Cultivar 2: $Y = 0.48889 + 4.09619 * nit + \varepsilon_i$
Cultivar 3: $Y = 0.48889 + 4.84952 * nit + \varepsilon_i$

Thus the three cultivars have a common intercept but have different rates of response to nitrogen (slopes).

Let's compare the above three equations with the equations obtained from our initial regression for each cultivar run separately. The main difference between the two sets of equations is that in the set above, all cultivars have a common intercept of 0.489. Their individual slopes are the same as when analyzed separately (equations below).

Cultivar 1: $Y = 0.027 + 2.09 * Nit$
Cultivar 2: $Y = -2.311 + 4.09 * Nit$
Cultivar 3: $Y = 0.489 + 4.85 * Nit$

Interpreting these results is fairly easy. The fact that the cultivars have the same intercept indicates that, with no fertilization, all three yield pretty much the same. When nitrogen is supplied, cultivars 2 and 3 respond more favorably than cultivar 1 with respect to increased yield.

Comparing Subsets to Determine Which Lines Differ

We have determined that the cultivars respond differently to nitrogen but we have not determined which cultivars are different. To determine which slopes are different, we could compare pairs of lines (cultivar 1 with 2, cultivar 1 with 3, and cultivar 2 with 3) using the same approach as above that was outlined for comparing two lines. Since we have already determined that the intercepts do not differ, we would only have to test for different slopes. Even if comparison of two lines indicates different intercepts, stick with the original conclusion of similar intercepts since it is based on data from three lines rather than two. Greater confidence is normally associated with a greater amount of data.

To compare line 1 with line 2, for example, we would use only the data for those two cultivars. We could manipulate the original data set to include only the correct data, or we could start afresh with a newly generated dataset, which is the way I prefer and the way we'll do it here. I use the entire original dataset (all three cultivars) but subset it to include only the two cultivars being compared using 'IF' statements. This way I can use the same basic program for comparing different pairs of lines with only a few minor tweaks each time.

Cultivar 1 vs 2

```
TITLE1 'Chapter 9 Example 18.sas';
TITLE2 'Comparison of regression line subsets from example 17';
TITLE3 'Compare Cultivar 1 with 2';
DATA one;
INPUT rep cv nit yld;
CARDS;
1 1 10 21.2
1 1 20 41.6
1 1 30 63
1 1 40 80
1 1 50 100
1 1 60 120
2 1 10 20
2 1 20 46
2 1 30 64
2 1 40 81
2 1 50 106
2 1 60 130
3 1 10 21
3 1 20 40
3 1 30 68
3 1 40 82
3 1 50 100
3 1 60 135
1 2 10 36
1 2 20 81
1 2 30 115
1 2 40 153
1 2 50 200
1 2 60 235
2 2 10 36
2 2 20 76
2 2 30 119
2 2 40 162
2 2 50 201
2 2 60 246
3 2 10 43
3 2 20 86
3 2 30 132
3 2 40 156
3 2 50 209
3 2 60 253
1 3 10 52
1 3 20 104
1 3 30 136
```

```
1 3 40 199
1 3 50 241
1 3 60 298
2 3 10 46
2 3 20 99
2 3 30 153
2 3 40 169
2 3 50 236
2 3 60 288
3 3 10 36
3 3 20 105
3 3 30 162
3 3 40 193
3 3 50 260
3 3 60 287
DATA two; set one; IF cv <3;
A1 = .;
IF cv = 1 then A1 = 1; ELSE A1 = 0;
A1nit = A1*nit;
RUN;
PROC REG DATA = two PLOT = none;
MODEL yld = A1 nit A1nit;
TEST A1nit=0, nit=0;
RUN;
```
which produces the output:

The REG Procedure
Model: MODEL1
Dependent Variable: yld

Number of Observations Read	36
Number of Observations Used	36

Analysis of Variance

Source	DF	Sum of Squares	Mean Square	F Value	Pr > F
Model	3	152436	50812	1946.58	<.0001
Error	32	835.29854	26.10308		
Corrected Total	35	153271			

Root MSE	5.10912	R-Square	0.9946
Dependent Mean	107.16111	Adj R-Sq	0.9940
Coeff Var	4.76770		

Parameter Estimates

Variable	DF	Parameter Estimate	Standard Error	t Value	Pr > |t|
Intercept	1	-2.31111	2.74607	-0.84	0.4063
A1	1	2.33778	3.88353	0.60	0.5514
nit	1	4.09619	0.07051	58.09	<.0001
A1nit	1	-2.00362	0.09972	-20.09	<.0001

'Chapter 9 Example 18.sas'
'Comparison of regression line subsets from example 17'
Compare Cultivar 1 with 2

The REG Procedure
Model: MODEL1

Test 1 Results for Dependent Variable yld

Source	DF	Mean Square	F Value	Pr > F
Numerator	2	55539	2127.67	<.0001
Denominator	32	26.10308		

The test for coincidence (H$_0$: $\beta_2 = \beta_3 = 0$) produces an F-value of 2127.67 with $\alpha <$ 0.0001, which is highly significant. We would therefore reject the null hypothesis that the two lines are coincident.

From the above printout, note that our β estimates are as follows:

$\beta_0 = -2.3111$

$\beta_1 = 4.09619$

$\beta_2 = 2.33778$

$\beta_3 = -2.00362$

The test for equal slopes (parallelism) (H_0: $\beta_3 = 0$) produces a t-value of -20.09 with $\alpha < 0.0001$. We would therefore reject the hypothesis of equal slopes. Note that we use the t-value rather than F for testing the null hypothesis.

Even though the hypothesis for equal intercepts (H_0: $\beta_2 = 0$) produces a t-value of 58.09 and $\alpha < 0.001$, and we would normally reject the hypothesis of equal intercepts, we approach this situation a bit differently. We have already determined that the three cultivars have the same intercept ($\hat{\beta}_0 = 0.489$). We came to this conclusion when we compared all three lines. The process for evaluating differences among regression lines is a stepwise process, thus we stick to our original failure to reject the null hypothesis that all three lines have the same intercept. The likely reason for failing to reject equal intercepts with three lines while indicating rejection when comparing lines 1 and 2 is as follows. With three lines, the variability is greater than with only lines 1 and 2. With greater variability, you are less likely to detect significant differences when performing hypothesis tests. Another possibility is simply that these two disparate conclusions regarding whether or not to reject the null hypothesis occurred due to chance.

Since we fail to reject H_0 of equal intercepts, the model for cultivar 1 is

$$Y = \beta_0 + (\beta_1 + \beta_3) * nit + \varepsilon_i$$

and the model for cultivar 2 is:

$$Y = \beta_0 + \beta_1 nit + \varepsilon_i$$

The model for cultivar 1 is:

$$Y = \beta_0 + (\beta_1 + \beta_3) * nit + \varepsilon_i$$

$$Y = -2.31111 + (4.09619 - 2.00362) * nit + \varepsilon_i$$

$$Y = -2.31111 + 2.09257 * nit + \varepsilon_i$$

The model for cultivar 2 is

$$Y = \beta_0 + \beta_1 nit + \varepsilon_i$$

$$Y = -2.31111 + 4.09619 * nit + \varepsilon_i$$

However, we determined that the lines have the same intercept ($\hat{\beta}_0 = 0.489$), thus the equations are:

Cultivar 1: $Y = 0.489 + 2.09257 * nit + \varepsilon_i$

Cultivar 2: $Y = 0.489 + 4.09619 * nit + \varepsilon_i$

The response of cultivar 2 to nitrogen is nearly double that of the cultivar 1.

Cultivar 1 vs 3

```
TITLE1 'Chapter 9 Example 19.sas';
TITLE2 'Comparison of regression line subsets from example 17';
TITLE3 'Compare Cultivar 1 with 3';
DATA one;
INPUT rep cv nit yld;
CARDS;
1 1 10 21.2
```

```
1 1 20 41.6
1 1 30 63
1 1 40 80
1 1 50 100
1 1 60 120
2 1 10 20
2 1 20 46
2 1 30 64
2 1 40 81
2 1 50 106
2 1 60 130
3 1 10 21
3 1 20 40
3 1 30 68
3 1 40 82
3 1 50 100
3 1 60 135
1 2 10 36
1 2 20 81
1 2 30 115
1 2 40 153
1 2 50 200
1 2 60 235
2 2 10 36
2 2 20 76
2 2 30 119
2 2 40 162
2 2 50 201
2 2 60 246
3 2 10 43
3 2 20 86
3 2 30 132
3 2 40 156
3 2 50 209
3 2 60 253
1 3 10 52
1 3 20 104
1 3 30 136
1 3 40 199
1 3 50 241
1 3 60 298
2 3 10 46
2 3 20 99
2 3 30 153
2 3 40 169
2 3 50 236
2 3 60 288
```

```
3  3  10  36
3  3  20  105
3  3  30  162
3  3  40  193
3  3  50  260
3  3  60  287
DATA two; set one; IF cv <> 2;
A1 = .;
IF cv = 1 then A1 = 1; ELSE A1 = 0;
A1nit = A1*nit;
RUN;
PROC REG DATA = two PLOT = none;
MODEL yld = A1 nit A1nit;
TEST A1nit=0, nit=0;
RUN;
```
which produces the output:

‘Chapter 9 Example 19.sas’
‘Comparison of regression line subsets from example 17’
Compare Cultivar 1 with 3

The REG Procedure
Model: MODEL1
Dependent Variable: yld

Number of Observations Read	54
Number of Observations Used	54

Analysis of Variance					
Source	DF	Sum of Squares	Mean Square	F Value	Pr > F
Model	3	314479	104826	445.56	<.0001
Error	50	11764	235.27000		
Corrected Total	53	326242			

Root MSE	15.33851	R-Square	0.9639
Dependent Mean	128.18148	Adj R-Sq	0.9618
Coeff Var	11.96625		

Parameter Estimates					
Variable	DF	Parameter Estimate	Standard Error	t Value	Pr > |t|
Intercept	1	-0.91111	5.82953	-0.16	0.8764
A1	1	0.93778	10.09705	0.09	0.9264
nit	1	4.47286	0.14969	29.88	<.0001
A1nit	1	-2.38029	0.25927	-9.18	<.0001

The REG Procedure
Model: MODEL1

Test 1 Results for Dependent Variable yld				
Source	DF	Mean Square	F Value	Pr > F
Numerator	2	116528	495.30	<.0001
Denominator	50	235.27000		

The test for coincidence (H_0: $\beta_2 = \beta_3 = 0$) produces an F-value of 495.30 with $\alpha < 0.0001$, which is highly significant. We would therefore reject the null hypothesis that the two lines are coincident.

From the above printout, note that our β estimates are as follows:

$\beta_0 = -0.9111$

$\beta_1 = 4.47286$

$\beta_2 = 0.93778$

$\beta_3 = -2.38029$

The test for equal slopes (parallelism) (H_0: $\beta_3 = 0$) produces a t-value of -9.18 with $\alpha < 0.0001$. We would therefore reject the hypothesis of equal slopes. Note that we use the t-value rather than F for testing the null hypothesis.

The test for equal intercepts (H_0: $\beta_2 = 0$) produces a t-value of 0.09 and $\alpha < 0.9264$, thus we fail to reject null hypothesis of equal intercepts. Since we have already determined that the intercepts of the three lines do not differ ($\hat{\beta}_0 = 0.489$), we really do not need to look at this test when comparing subsets of two lines, but we look at the test anyway, just for practice.

Since we fail to reject H_0 of equal intercepts, the model for cultivar 1 is:

$$Y = \beta_0 + (\beta_1 + \beta_3) * nit + \varepsilon$$

and the model for cultivar 3 is:

$$Y = \beta_0 + \beta_1 nit + \varepsilon_i$$

The model for cultivar 1 is:

$$Y = \beta_0 + (\beta_1 + \beta_3) * nit + \varepsilon_i$$

$$Y = -0.9111 + (4.47286 - 2.38029) * nit + \varepsilon_i$$

$$Y = -0.9111 + 2.09257 * nit + \varepsilon_i$$

The model for cultivar 3 is

$$Y = \beta_0 + \beta_1 nit + \varepsilon_i$$

$$Y = -0.9111 + 4.47286 * nit + \varepsilon_i$$

Now correct the equations to reflect common intercepts determined comparing three lines ($\hat{\beta}_0 = 0.489$):

Cultivar 1: $Y = 0.489 + 2.09257 * nit + \varepsilon_i$

Cultivar 3: $Y = 0.489 + 4.47286 * nit + \varepsilon_i$

The response of cultivar 3 to nitrogen is nearly double that of the cultivar 1.

Cultivar 2 vs 3

```
TITLE1 'Chapter 9 Example 20.sas';
TITLE2 'Comparison of regression line subsets from example 17';
TITLE3 'Compare Cultivar 2 with 3';
DATA one;
INPUT rep cv nit yld;
CARDS;
1 1 10 21.2
1 1 20 41.6
1 1 30 63
1 1 40 80
1 1 50 100
1 1 60 120
2 1 10 20
2 1 20 46
2 1 30 64
2 1 40 81
2 1 50 106
2 1 60 130
3 1 10 21
3 1 20 40
3 1 30 68
3 1 40 82
3 1 50 100
3 1 60 135
1 2 10 36
1 2 20 81
1 2 30 115
1 2 40 153
1 2 50 200
1 2 60 235
2 2 10 36
2 2 20 76
2 2 30 119
2 2 40 162
2 2 50 201
2 2 60 246
3 2 10 43
3 2 20 86
3 2 30 132
3 2 40 156
3 2 50 209
3 2 60 253
```

```
1  3  10  52
1  3  20  104
1  3  30  136
1  3  40  199
1  3  50  241
1  3  60  298
2  3  10  46
2  3  20  99
2  3  30  153
2  3  40  169
2  3  50  236
2  3  60  288
3  3  10  36
3  3  20  105
3  3  30  162
3  3  40  193
3  3  50  260
3  3  60  287
DATA two; set one; IF cv >1;
A1 = .;
IF cv = 2 then A1 = 1; ELSE A1 = 0;
A1nit = A1*nit;
RUN;
PROC REG DATA = two PLOT = none;
MODEL yld = A1 nit A1nit ;
TEST A1nit=0, nit=0;
RUN;
```

Note the change indicating that the dummy variable $A1$ should be based on cultivar 2 rather than cultivar 1 as in the previous two examples comparing cultivars 1 and 2 and cultivars 1 and 3. In this example we're comparing cultivars 2 and 3, thus the dummy variable must be based on the value of either of these cultivars. It doesn't matter which of the two you pick. I picked cultivar 2. Change it to 3 and see what happens.

The program as listed above produces the output:

The REG Procedure
Model: MODEL1
Dependent Variable: yld

Number of Observations Read	36
Number of Observations Used	36

Analysis of Variance

Source	DF	Sum of Squares	Mean Square	F Value	Pr > F
Model	3	219214	73071	991.38	<.0001
Error	32	2358.60317	73.70635		
Corrected Total	35	221572			

Root MSE	8.58524	R-Square	0.9894
Dependent Mean	155.63889	Adj R-Sq	0.9884
Coeff Var	5.51613		

Parameter Estimates

Variable	DF	Parameter Estimate	Standard Error	t Value	Pr > |t|
Intercept	1	0.48889	4.61443	0.11	0.9163
A1	1	-2.80000	6.52579	-0.43	0.6707
nit	1	4.84952	0.11849	40.93	<.0001
A1nit	1	-0.75333	0.16757	-4.50	<.0001

'Chapter 9 Example 20.sas'
'Comparison of regression line subsets from example 17'
Compare Cultivar 2 with 3

The REG Procedure
Model: MODEL1

Test 1 Results for Dependent Variable yld

Source	DF	Mean Square	F Value	Pr > F
Numerator	2	105779	1435.14	<.0001
Denominator	32	73.70635		

The test for coincidence (H_0: $\beta_2 = \beta_3 = 0$) produces an *F*-value of 1435.14 with $\alpha <$ 0.0001, which is highly significant. We would therefore reject the null hypothesis that the two lines are coincident.

From the above printout, note that our β estimates are as follows:

$\beta_0 = 0.48889$

$\beta_1 = 4.84952$

$\beta_2 = -2.8$

$\beta_3 = -0.75333$

The test for equal slopes (parallelism) (H_0: $\beta_3 = 0$) produces a t-value of -4.50 with $\alpha < 0.0001$. We would therefore reject the hypothesis of equal slopes. Note that we use the t-value rather than F for testing the null hypothesis.

The hypothesis for equal intercepts (H_0: $\beta_2 = 0$) produces a t-value of 0.43 and $\alpha = 0.6707$, thus we fail to reject the hypothesis of equal intercepts.

Since we fail to reject H_0 of equal intercepts, the model for cultivar 2 is

$$Y = \beta_0 + (\beta_1 + \beta_3) * nit + \varepsilon_i$$

and the model for cultivar 3 is:

$$Y = \beta_0 + \beta_1 nit + \varepsilon_i$$

The model for cultivar 2 is:

$$Y = \beta_0 + (\beta_1 + \beta_3) * nit + \varepsilon_i$$

$$Y = 0.48889 + (4.84952 - 0.75333) * nit + \varepsilon_i$$

$$Y = 0.48889 + 4.09619 * nit + \varepsilon_i$$

The model for cultivar 3 is

$$Y = \beta_0 + \beta_1 nit + \varepsilon_i$$

$$Y = 0.48889 + 4.84952 * nit + \varepsilon_i$$

Now correct the equations to reflect common intercepts determined comparing three lines ($\hat{\beta}_0 = 0.489$):

Cultivar 2: $Y = 0.489 + 4.09619 * nit + \varepsilon_i$

Cultivar 3: $Y = 0.489 + 4.84952 * nit + \varepsilon_i$

The response of cultivar 3 to nitrogen is slightly greater than that of cultivar 2.

To summarize the entire analysis, we have three cultivars which respond differently to nitrogen fertilization. They all produce similarly with no addition nitrogen, hence the common intercept. Cultivar 1 responds the least to nitrogen ($\hat{\beta}_1 = 2.09257$) while cultivars 2 ($\hat{\beta}_1 = 4.09619$) and 3 ($\hat{\beta}_1 = 4.84952$) respond more favorably, with cultivar 3 being the most responsive.

Testing for Homogeneous Variances Among Lines

When we use dummy variables to compare regression lines, we make the assumption that the error variances from the data used to generate each line are homogeneous. This means that the estimate of error generated from the data from each line is actually estimating the same error. When we are considering only one factor in regression (as compared with multiple X factors as in multiple linear regression) we use Levene's test of variance homogeneity available in SAS®. You can only request this test when you run a model with one factor in it; we are interested in whether or not the variances of our three cultivars are homogeneous, so we would use the following code to run Levene's test.

```
TITLE1 'Chapter 9 Example 21.sas';
TITLE2 'Comparison of regression lines';
TITLE3 'Levene's test of variance homogeneity';
DATA one;
INPUT rep cv nit yld;
CARDS;
1 1 10 21.2
1 1 20 41.6
1 1 30 63
1 1 40 80
1 1 50 100
1 1 60 120
2 1 10 20
2 1 20 46
2 1 30 64
2 1 40 81
2 1 50 106
2 1 60 130
3 1 10 21
3 1 20 40
3 1 30 68
3 1 40 82
3 1 50 100
3 1 60 135
1 2 10 36
1 2 20 81
1 2 30 115
1 2 40 153
1 2 50 200
1 2 60 235
2 2 10 36
2 2 20 76
2 2 30 119
2 2 40 162
2 2 50 201
2 2 60 246
3 2 10 43
3 2 20 86
3 2 30 132
3 2 40 156
3 2 50 209
3 2 60 253
1 3 10 52
```

```
1 3 20 104
1 3 30 136
1 3 40 199
1 3 50 241
1 3 60 298
2 3 10 46
2 3 20 99
2 3 30 153
2 3 40 169
2 3 50 236
2 3 60 288
3 3 10 36
3 3 20 105
3 3 30 162
3 3 40 193
3 3 50 260
3 3 60 287
RUN;
PROC ANOVA PLOT = none;
CLASSES cv;
MODEL yld = cv;
MEANS cv / HOVTEST=LEVENE;
RUN;
```
which produces the output:

'Chapter 9 Example 21.sas'
'Comparison of regression lines'
Levene's test of variance homogeneity

The ANOVA Procedure

Class Level Information		
Class	Levels	Values
cv	3	1 2 3

Number of Observations Read	54
Number of Observations Used	54

'Chapter 9 Example 21.sas'
'Comparison of regression lines'
Levene's test of variance homogeneity

The ANOVA Procedure

Dependent Variable: yld

Source	DF	Sum of Squares	Mean Square	F Value	Pr > F
Model	2	89078.4459	44539.2230	9.58	0.0003
Error	51	237163.9756	4650.2740		
Corrected Total	53	326242.4215			

R-Square	Coeff Var	Root MSE	yld Mean
0.273044	53.20029	68.19292	128.1815

Source	DF	Anova SS	Mean Square	F Value	Pr > F
cv	2	89078.44593	44539.22296	9.58	0.0003

'Chapter 9 Example 21.sas'
'Comparison of regression lines'
Levene's test of variance homogeneity

The ANOVA Procedure

Levene's Test for Homogeneity of yld Variance ANOVA of Squared Deviations from Group Means					
Source	DF	Sum of Squares	Mean Square	F Value	Pr > F
cv	2	2.9671E8	1.4836E8	7.03	0.0020
Error	51	1.0761E9	21099688		

'Chapter 9 Example 21.sas'
'Comparison of regression lines'
Levene's test of variance homogeneity

The ANOVA Procedure

Level of cv	N	yld	
		Mean	Std Dev
1	18	73.266667	36.9800582
2	18	141.055556	72.2190070
3	18	170.222222	85.8353798

The *F*-value for Levene's test indicates that we should reject the hypothesis of equal variances at $\alpha = 0.0003$.

Even though there are problems with our dataset regarding non-homogeneous variances, the methodology is correct. It is our interpretation of the results generated that is suspect. We need to consult a statistician (ha, ha)!

Reference

Matange, S. and Heath, D. (2011) *Statistical Graphics Procedures by Example: Effective Graphs Using SAS®*. SAS Institute Inc., Cary, North Carolina.

10 | Regression Diagnostics

The development of a regression equation is only the first half of a regression analysis. The second, often overlooked part of a regression analysis is to make sure the assumptions underlying the analysis have been met. This is easily accomplished using the regression diagnostic procedures available in SAS®.

We saw in the previous chapter an example where an assumption was invalid, namely, that we had homogeneous variances among our three cultivars. You might wonder why I didn't use an example where our assumption was valid. The main reason was that I didn't know the assumption was invalid until I ran the test, just like in 'real' research. I decided to keep the example with 'flaws' since it reflected what researchers often encounter. After significant work analyzing data, some unforeseen problem arises. Unfortunately, the solution to handling the problem of heterogeneous variances in regression analysis is beyond the scope and intention of this text.

A number of assumptions are made when performing a regression analysis. We first assume that the data we are using are normally distributed. We can check this assumption using PROC UNIVARIATE NORMAL. We also make some assumptions about our errors (e_i's). We assume that these errors are normally distributed, they have a common unknown variance (σ^2), they have a mean of 0 and they are independent and uncorrelated. We indicate these assumptions like this:

$$e_i\text{'s} \sim NID(0, \sigma^2)$$

Since we don't know the e_i's, we consider their estimates, $\hat{e}_i$'s, and call them residuals. We use these residuals extensively in regression diagnostics by looking at 'residual plots'. Looking at these plots, we can uncover significant violations about our assumptions. Once we uncover any violations, we can 'fix' them, if possible, to make our regression analysis valid. If we don't fix assumption violations, our regression analysis is not valid. And if we don't look for these violations using regression diagnostics, how would we know they even exist? Many people don't look for these violations when they use regression.

When we 'fix' the violations we are usually performing some sort of transformation of the data. A transformation is a change in the value of a dependent or independent variable via a mathematical function. SAS® performs transformations very easily; you just need to know which transformation is appropriate to your situation. Some people are uncomfortable with transformations, thinking that they are 'messing' with the data to get it to do what they want. As long as you transform all of the data in a dataset, transformations are completely valid and totally acceptable.

While we're on the subject of transformations, we might as well see which ones are used in which situations. The situations will become clearer as we progress through this chapter.

DOI: 10.1079/9781789249927.0010

Potential Problems in Regression

Generally, we categorize our regression woes as one or more of the following, each detected by a specific piece of information available via SAS®:

1. Non-normal data
   a. Look at the Shapiro–Wilks test statistic (W) if $N < 2000$, or the Kolmogorov distance (D) test statistic otherwise.
2. Heterogeneous variances
   a. Detected by plotting $\hat{e}_i$'s vs $\hat{Y}_i$'s, any non-random pattern in the spread of the plotted points reveals likely heterogeneous variances.
3. Correlated errors
   a. Durbin–Watson (DW) test. Data must be ordered in time, say observations taken over time, or treatments applied in order. A value of the DW statistic near 2.0 reveals no correlation among the errors.
4. Outliers
   a. Detected by plotting $\hat{e}_i$'s vs $\hat{Y}_i$'s. An outlier in Y may or may not have an outlier in e.
   b. Cook's D influence statistic detects any data point which influences the model estimates heavily. A value of $D > 4/n$ where n = the number of observations in the analysis indicates a potentially influential data point.
5. Model inadequacies
   a. Detected by plotting $\hat{e}_i$'s vs $\hat{Y}_i$'s.
   b. A pattern in the plot reveals likely model inadequacies.

A Diagnostic Example

Let's take a look at an example and go through a typical regression diagnostic process. In order to adequately address diagnostic procedures, we revisit polynomial regression of Chapter 9 by including quadratic (X^2) and cubic (X^3) components to our model rather than limiting ourselves to a linear model (X only).

Suppose we had collected data to reflect the price per flat of strawberries and their availability on the open market. We would like to know: is there a relationship between price and availability, and, if so, what is that relationship?

We have the following data for one year: the average price per 12-pint flat versus average availability at the Philadelphia Regional Produce Market. Data were collected monthly and were as follows:

Month	Flats available	Price per flat (USD)
1	230	22
2	320	16
3	350	17
4	400	15
5	315	17
6	275	32

Continued

Continued.

Month	Flats available	Price per flat (USD)
7	260	23
8	255	24
9	200	21
10	168	43
11	99	50
12	75	65

This is an example of a regression problem involving no repeated measures and no replication, therefore our estimate of error is extremely model dependent. First, we will generate a SAS® dataset, then perform a simple linear regression, addressing potential problems and remedies as we go.

```
TITLE1 'Chapter 10 Example 1.sas';
TITLE2 'Regression Diagnostics';
TITLE3 'Linear model';
DATA one;
INPUT month flats price;
flats2 = flats*flats;
flats3 = flats2*flats;
CARDS;
1    230    22
2    320    16
3    350    17
4    400    15
5    315    17
6    275    32
7    260    23
8    255    24
9    200    21
10   168    43
11   99     50
12   75     65
RUN;
PROC PRINT;
RUN;
PROC REG DATA = one PLOT = none;
MODEL price = flats;
RUN;
```

The output:

'Chapter 10 Example 1.sas'
'Regression Diagnostics'
Linear model

Obs	month	flats	price	flats2	flats3
1	1	230	22	52900	12167000
2	2	320	16	102400	32768000
3	3	350	17	122500	42875000
4	4	400	15	160000	64000000
5	5	315	17	99225	31255875
6	6	275	32	75625	20796875
7	7	260	23	67600	17576000
8	8	255	24	65025	16581375
9	9	200	21	40000	8000000
10	10	168	43	28224	4741632
11	11	99	50	9801	970299
12	12	75	65	5625	421875

'Chapter 10 Example 1.sas'
'Regression Diagnostics'
Linear model

The REG Procedure
Model: MODEL1
Dependent Variable: price

Number of Observations Read	12
Number of Observations Used	12

Analysis of Variance

Source	DF	Sum of Squares	Mean Square	F Value	Pr > F
Model	1	2165.35816	2165.35816	35.92	0.0001
Error	10	602.89184	60.28918		
Corrected Total	11	2768.25000			

Root MSE	7.76461	R-Square	0.7822
Dependent Mean	28.75000	Adj R-Sq	0.7604
Coeff Var	27.00734		

Parameter Estimates

Variable	DF	Parameter Estimate	Standard Error	t Value	Pr > |t|
Intercept	1	63.98503	6.29213	10.17	<.0001
flats	1	-0.14347	0.02394	-5.99	0.0001

Our regression analysis suggests that the model is a fairly good fit (coefficient of determination $R^2 = 0.78$) and that there is a relationship between the number of flats available and price. That relationship is best estimated as:

$$\hat{Y} = 63.99 - 0.143 * flats$$

If we add a quadratic component to our model via:

```
TITLE1 'Chapter 10 Example 2.sas';
TITLE2 'Regression Diagnostics';
TITLE3 'Quadratic model';
DATA one;
INPUT month flats price;
flats2 = flats*flats;
flats3 = flats2*flats;
CARDS;
1       230     22
2       320     16
3       350     17
4       400     15
5       315     17
6       275     32
7       260     23
8       255     24
9       200     21
10      168     43
11      99      50
12      75      65
RUN;
PROC REG DATA = one PLOT = none;
MODEL price = flats flats2;
RUN;
```

we get:

'Chapter 10 Example 2.sas'
'Regression Diagnostics'
Quadratic model

The REG Procedure
Model: MODEL1
Dependent Variable: price

Number of Observations Read	12
Number of Observations Used	12

Analysis of Variance					
Source	DF	Sum of Squares	Mean Square	F Value	Pr > F
Model	2	2451.09865	1225.54932	34.78	<.0001
Error	9	317.15135	35.23904		
Corrected Total	11	2768.25000			

Root MSE	5.93625	R-Square	0.8854
Dependent Mean	28.75000	Adj R-Sq	0.8600
Coeff Var	20.64782		

Parameter Estimates					
Variable	DF	Parameter Estimate	Standard Error	t Value	Pr > |t|
Intercept	1	86.81020	9.34836	9.29	< 0001
flats	1	-0.38118	0.08546	-4.46	0 0016
flats2	1	0.00051467	0.00018074	2.85	0 0192

The quadratic component is significant ($\alpha = 0.0192$) and the R^2 value has increased to 0.89. It appears that our line has a bend in it and should now be:

$$\hat{Y} = 86.81 - 0.381 * flats + 0.000515 * flats^2$$

What about the cubic model? One of the main goals in regression analysis is to find the model that accounts for the greatest amount of variability in the data (often reflected in an R^2 value closer to 1 rather than 0, along with significant β estimates) while at the same time keeping the model as simple as possible and minimizing our estimate of error. This is especially true when generating models that create differences in estimates of error. We can run the cubic model paying close attention to these features to see what happens.

```
TITLE1 'Chapter 10 Example 3.sas';
TITLE2 'Regression Diagnostics';
TITLE3 'Cubic model';
DATA one;
INPUT month flats price;
```

```
flats2 = flats*flats;
flats3 = flats2*flats;
CARDS;
1     230     22
2     320     16
3     350     17
4     400     15
5     315     17
6     275     32
7     260     23
8     255     24
9     200     21
10    168     43
11    99      50
12    75      65
RUN;
PROC REG DATA = one PLOT = none;
MODEL price = flats flats2 flats3;
RUN;
```

The output:

'Chapter 10 Example 3.sas'
'Regression Diagnostics'
Cubic model

The REG Procedure
Model: MODEL1
Dependent Variable: price

Number of Observations Read	12
Number of Observations Used	12

Analysis of Variance					
Source	DF	Sum of Squares	Mean Square	F Value	Pr > F
Model	3	2479.49155	826.49718	22.90	0.0003
Error	8	288.75845	36.09481		
Corrected Total	11	2768.25000			

Root MSE	6.00790	R-Square	0.8957
Dependent Mean	28.75000	Adj R-Sq	0.8566
Coeff Var	20.89703		

Parameter Estimates					
Variable	DF	Parameter Estimate	Standard Error	t Value	Pr > |t|
Intercept	1	104.30984	21.88200	4.77	0.0014
flats	1	-0.68422	0.35245	-1.94	0.0882
flats2	1	0.00195	0.00163	1.20	0.2653
flats3	1	-0.00000200	0.00000226	-0.89	0.4010

Notice that our estimates of β_2 and β_3 ('flats2' and 'flats3', respectively) are not significant at $\alpha = 0.05$. This reveals that the quadratic model is the best of the three models we tried. Notice too that the R^2 value in the cubic model did not increase very much compared with the quadratic and the estimate of error was not reduced by including the cubic component in the model.

Many folks would be inclined to summarize the results as:

$$\hat{Y} = 86.81 - 0.381 * flats + 0.000515 * flats^2$$

and to call it a day. However, what about the assumptions made when using PROC REG? We don't know if we have satisfied the assumptions or not. The only way to provide evidence one way or the other is to perform regression diagnostics of PROC REG.

PROC REG produces numerous plots by default. To make regression diagnostics simpler and thereby encourage their use, we will forego discussion of these plots. The reader should explore their use if interested by consulting one of the many SAS® manuals or handbooks, available on line. The plots presented in this chapter are sufficient for performing a thorough diagnostic evaluation of the residuals for most plant science regression analyses.

Here is the SAS® program to perform the diagnostics on our analysis:

```
TITLE1 'Chapter 10 Example 4.sas';
TITLE2 'Regression Diagnostics';
TITLE3 'Quadratic model diagnostics';
DATA one;
INPUT month flats price;
flats2 = flats*flats;
flats3 = flats2*flats;
CARDS;
1   230    22
2   320    16
3   350    17
4   400    15
5   315    17
6   275    32
7   260    23
8   255    24
9   200    21
10  168    43
11  99     50
12  75     65
RUN;
PROC REG DATA = one PLOT = none;
MODEL price = flats flats2 / R DWPROB;
OUTPUT OUT = diag P = yhat RSTUDENT = rstud;
PROC UNIVARIATE NORMAL PLOT;
VAR rstud;
PROC SGPLOT;
SCATTER x=yhat y=rstud;
RUN;
```

The output:

'Chapter 10 Example 4.sas'
'Regression Diagnostics'
Quadratic model diagnostics

The REG Procedure
Model: MODEL1
Dependent Variable: price

Number of Observations Read	12
Number of Observations Used	12

Analysis of Variance

Source	DF	Sum of Squares	Mean Square	F Value	Pr > F
Model	2	2451.09865	1225.54932	34.78	<.0001
Error	9	317.15135	35.23904		
Corrected Total	11	2768.25000			

Root MSE	5.93625	R-Square	0.8854
Dependent Mean	28.75000	Adj R-Sq	0.8600
Coeff Var	20.64782		

Parameter Estimates

Variable	DF	Parameter Estimate	Standard Error	t Value	Pr > |t|
Intercept	1	86.81020	9.34836	9.29	<.0001
flats	1	-0.38118	0.08546	-4.46	0.0016
flats2	1	0.00051467	0.00018074	2.85	0.0192

'Chapter 10 Example 4.sas'
'Regression Diagnostics'
Quadratic model diagnostics

The REG Procedure
Model: MODEL1
Dependent Variable: price

Durbin-Watson D	2.545
Pr < DW	0.6598
Pr > DW	0.3402
Number of Observations	12
1st Order Autocorrelation	-0.326

Note: Pr<DW is the p-value for testing positive autocorrelation, and Pr>DW is the p-value for testing negative autocorrelation.

'Chapter 10 Example 4.sas'
'Regression Diagnostics'
Quadratic model diagnostics

The REG Procedure
Model: MODEL1
Dependent Variable: price

			Std Error Mean		Std Error	Student		
Obs	Dependent Variable	Predicted Value	Predict	Residual	Residual	Residual	-2-1 0 1 2	Cook's D
1	22	26.3642	2.3774	-4.3642	5.439	-0.802	| *| |	0.041
2	16	17.5340	2.1972	-1.5340	5.515	-0.278	| | |	0.004
3	17	16.4435	2.7334	0.5565	5.270	0.106	| | |	0.001
4	15	16.6846	4.8455	-1.6846	3.429	-0.491	| | |	0.161
5	17	17.8059	2.1611	-0.8059	5.529	-0.146	| | |	0.001
6	32	20.9069	2.2011	11.0931	5.513	2.012	| |**** |	0.215
7	23	22.4944	2.2734	0.5056	5.484	0.092	| | |	0.000
8	24	23.0750	2.2961	0.9250	5.474	0.169	| | |	0.002
9	21	31.1604	2.3950	-10.1604	5.432	-1.871	| ***| |	0.227
10	43	37.2975	2.4034	5.7025	5.428	1.051	| |** |	0.072
11	50	54.1173	3.5290	-4.1173	4.773	-0.863	| *| |	0.136
12	65	61.1164	4.5129	3.8836	3.857	1.007	| |** |	0.463

Sum of Residuals	0
Sum of Squared Residuals	317.15135
Predicted Residual SS (PRESS)	542.97547

Quadratic model diagnostics

The UNIVARIATE Procedure
Variable: rstud (Studentized Residual without Current Obs)

Moments			
N	12	Sum Weights	12
Mean	0.01740075	Sum Observations	0.20880899
Std Deviation	1.18512166	Variance	1.40451336
Skewness	0.35636515	Kurtosis	1.6795246
Uncorrected SS	15.4532803	Corrected SS	15.4496469
Coeff Variation	6810.75089	Std Error Mean	0.34211516

Basic Statistical Measures			
Location		Variability	
Mean	0.01740	Std Deviation	1.18512
Median	-0.02530	Variance	1.40451
Mode	.	Range	4.81348
		Interquartile Range	1.21098

Tests for Location: Mu0=0				
Test	Statistic		p Value	
Student's t	t	0.050862	Pr > |t|	0.9603
Sign	M	0	Pr >= |M|	1.0000
Signed Rank	S	-1	Pr >= |S|	0.9697

Tests for Normality				
Test	Statistic		p Value	
Shapiro-Wilk	W	0.949437	Pr < W	0.6288
Kolmogorov-Smirnov	D	0.20226	Pr > D	>0.1500
Cramer-von Mises	W-Sq	0.064357	Pr > W-Sq	>0.2500
Anderson-Darling	A-Sq	0.379609	Pr > A-Sq	>0.2500

Quantiles (Definition 5)	
Level	Quantile
100% Max	2.5576680
99%	2.5576680
95%	2.5576680
90%	1.0574632
75% Q3	0.5837214
50% Median	-0.0253036
25% Q1	-0.6272545
10%	-0.8490722
5%	-2.2558129
1%	-2.2558129
0% Min	-2.2558129

Extreme Observations			
Lowest		Highest	
Value	Obs	Value	Obs
-2.255813	9	0.0996347	3
-0.849072	11	0.1595615	8
-0.785039	1	1.0078812	12
-0.469470	4	1.0574632	10
-0.263398	2	2.5576680	6

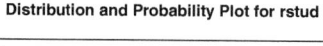

Distribution and Probability Plot for rstud

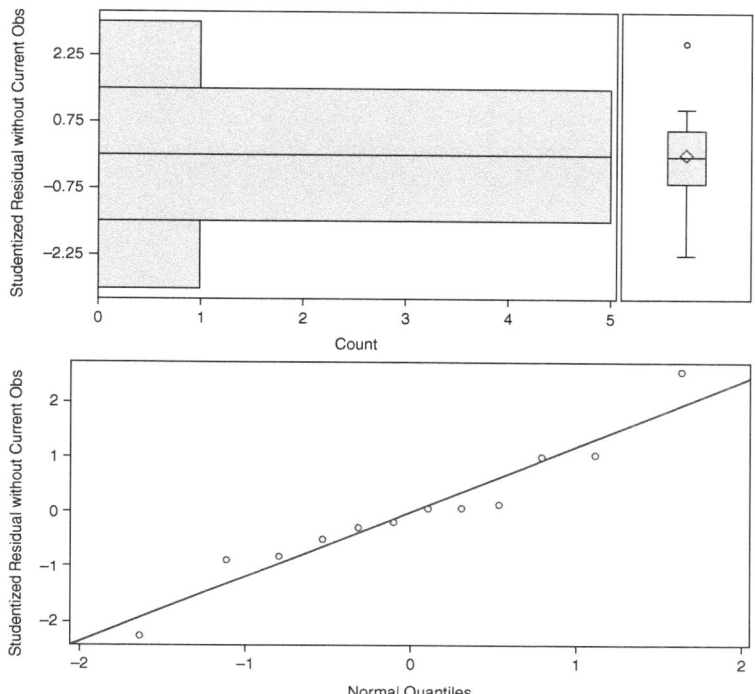

'Chapter 10 Example 4.sas'
'Regression Diagnostics'
Quadratic model diagnostics

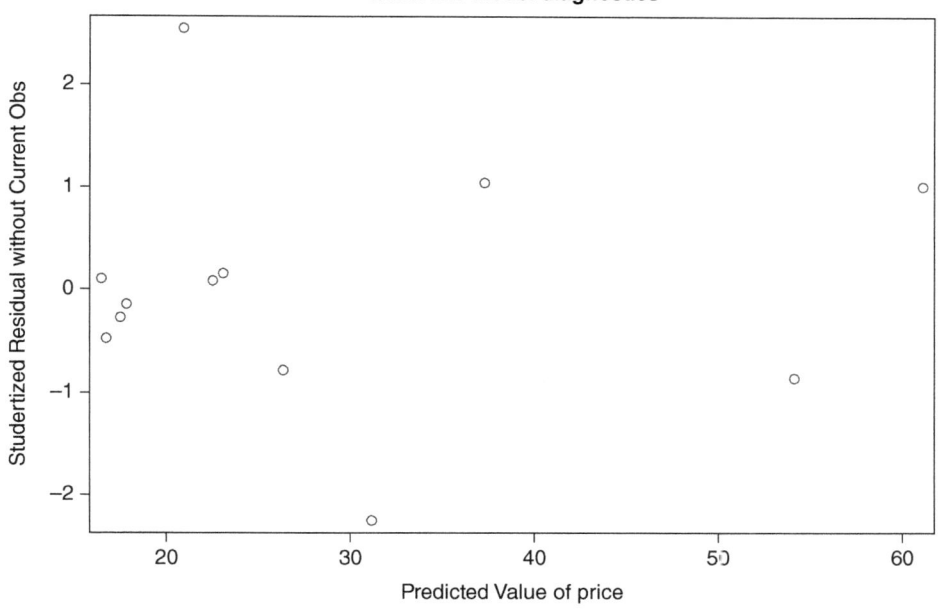

There is a large amount of information in this output. Remember we are interested in whether or not our assumption of e_i's~$NID(0, \sigma^2)$ is acceptable and we examine this assumption by examining our residuals ($\hat{e}_i = Y_i - \hat{Y}_i$). We use the studentized residual in our examination. SAS® produces three other types of residuals (common, standardized and recursive); however, studentized residuals are preferred for regression diagnostics.

We first examine the normality assumption by looking at the values of the Shapiro–Wilks test statistic (W) since $N < 2000$ (otherwise we would use the Kolmogorov D-test statistic). The W statistic = 0.949437 and Pr < W = 0.6288, indicating that we should fail to reject the hypothesis of normality; in other words, it appears our residuals are normal.

Our second assumption is that of homogeneous variance. We examine the plot produced by PROC SGPLOT of $\hat{Y}_i * \hat{e}_i$. We look for any conspicuous pattern in the spread of our plot, which we refer to as right-opening megaphone, left-opening megaphone or a combination of the two. This is a subjective decision.

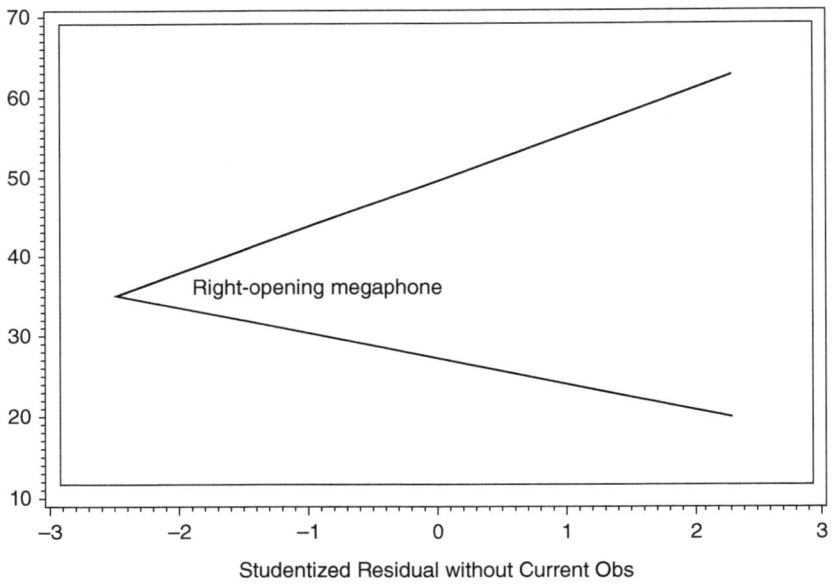

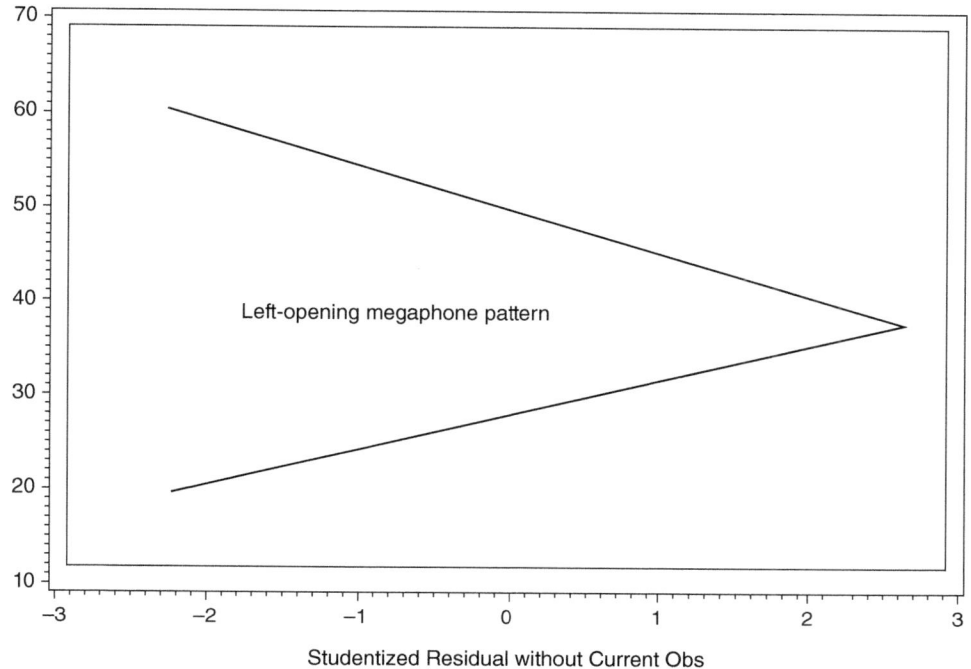

Left-opening megaphone pattern

Studentized Residual without Current Obs

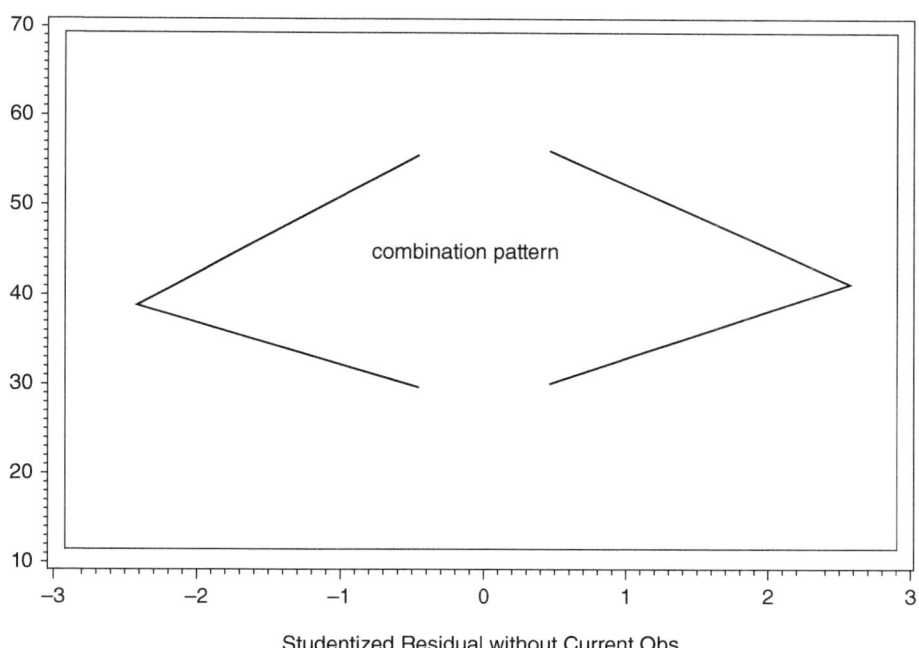

combination pattern

Studentized Residual without Current Obs

Our plot seems to have no discernible pattern to it, thus we appear to have homogeneous variances.

The third potential problem with our assumption is that our $\hat{e}_i$'s are correlated. This is a concern only if data were measured over time or treatments were applied over time. We check the Durbin–Watson test statistic (DW) to examine this. To produce the correct DW, data must be sorted in time order. Our data were input so that they exist in the dataset already in time order. SAS® provides the DW statistic along with Pr < DW and Pr > DW. Pr < DW is the P-value for testing positive autocorrelation, and Pr > DW is the P-value for testing negative autocorrelation. Positive autocorrelation is the tendency for high values of $\hat{e}_i$ to follow high values of $\hat{e}_i$ or low values to follow low values. Negative autocorrelation is the tendency for positive $\hat{e}_i$'s to follow negative $\hat{e}_i$'s and vice versa. Our data and model produced a Durbin–Watson test statistic of 2.545 with Pr < DW = 0.6598 and Pr > W = 0.3402a, indicating that our $\hat{e}_i$'s are not correlated.

The fourth potential problem with our assumption is the presence of outliers. Outliers are data points that appear to be way outside the range of the rest of the observations, an 'odd' data point. The trick with outliers is that an outlying observation (Y) may not produce an outlier $\hat{e}_i$. This is at least partially due to the fact that the particular potentially outlying data point was used in generating the model which in turn produced the $\hat{e}_i$. For this reason a plot of $\hat{Y}_i * \hat{e}_i$ may not be that useful. It is more useful to use the Cook's D influence statistic, which is a measure of how much a particular observation influences the β estimates. Cook's D measures how much β changes when a particular observation is left out of the model. Values for Cook's D > $4/n$ indicate an influential data point that should be further examined. It should not automatically be removed from the dataset. You must check the data point for possible data entry errors or values recorded that could have not been measured with the instruments used, etc. For our data, $4/n = 4/12 = 0.333$. Observation 12 produced a Cook's D of 0.463, which is > 0.333, thus it is a suspect observation. This observation is for the month of December where the price of a flat of strawberries soared to $65. This data point is correct, even though it is somewhat of an outlier. We would not get rid of this data point, but would rather remember that it is an observation that has a large influence on our model estimates. As a comparison, the model with all 12 months included is:

$$\hat{Y} = 86.81 - 0.381 * flats + 0.000515 * flats^2$$

compared with the model with December omitted:

$$\hat{Y} = 76.58 - 0.306 * flats + 0.000383 * flats^2$$

Model inadequacies can be detected by lack-of-fit tests in replicated or repeated measures experiments, while a residual plot can be used when you have non-replicated, non-repeated measures data. In a residual plot, look for a curved pattern to the residuals such as:

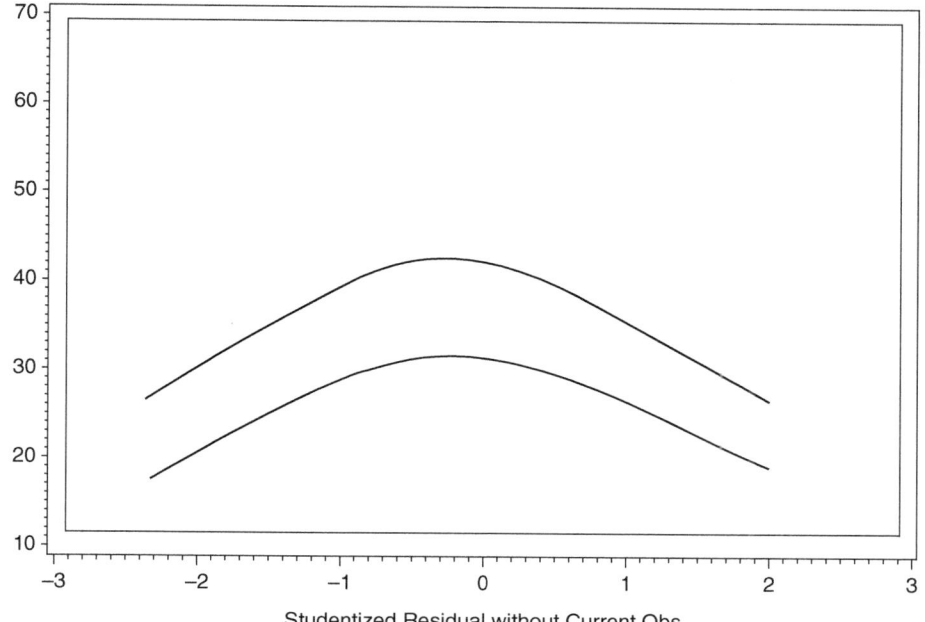

Studentized Residual without Current Obs

When problems with the assumption of $e_i's \sim NID(0, \sigma^2)$ are detected, a common approach used to fix the problem is transformation of the data and re-running the regression analysis and diagnostics to see if the problem has been taken care of.

When there are problems with the assumptions with the variance of e_i's, a variance stabilizing transformation is often used. If data consist of counts, which are often from a Poisson distribution, use the square root transformation (SQRT(Y)) or (SQRT(Y+1)) if you have some values of 0 in your dependent variables.

If your dependent variable data have a very large range, say from 1 to several thousand, use the log transformation (LOG(Y)) or (LOG(Y+1)) if the data contain any 0 values.

If data are bunched near 0 and you have fewer and fewer observations as you get farther away from 0, use the reciprocal transformation (1/Y) or (1/(Y+1)) if you have many 0 values.

If you have proportion data, often from a binomial distribution, use the arcsin of the square root transformation (SIN–1(SQRT(Y))).

When the relationship between X and Y is non-linear and non-linear regression is not an option, transform either X, Y or both via the LOG and/or reciprocal transformation.

Transforming data in SAS® is simple. Suppose we decide that we need to use the LOG transformation on our strawberry data to correct a problem with the dataset. We could use the following code to transform the price data via the LOG function to 'tprice' then run a regression analysis with diagnostics on the transformed data.

```
TITLE1 'Chapter 10 Example 5.sas';
TITLE2 'Regression Diagnostics';
```

```
TITLE3 'Data transformation';
DATA one;
INPUT month flats price;
flats2 = flats*flats;
tprice = LOG(price);
CARDS;
1       230     22
2       320     16
3       350     17
4       400     15
5       315     17
6       275     32
7       260     23
8       255     24
9       200     21
10      168     43
11      99      50
RUN;
PROC REG DATA = one PLOT = none;
MODEL tprice = flats flats2 / R DWPROB;
OUTPUT OUT = diag P = yhat RSTUDENT = rstud;
PROC UNIVARIATE NORMAL PLOT;
VAR rstud;
PROC SGPLOT;
SCATTER x=yhat y=rstud;
RUN;
```

When transformations do not correct inherent data problems, alternate solutions beyond the scope of this text are available. They include weighted least squares for heterogeneous variances, generalized least squares for correlated errors, and non-linear regression for model inadequacies. Keep in mind that correcting one problem may induce a new one, so always perform diagnostics on 'corrected' data. Always give precedence to fixing heterogeneous variances.

A fairly nice graph of the data, regression line with equation and 95% confidence bands can be produced from the following code:

```
TITLE1 'Chapter 10 Example 6.sas';
TITLE2 'Regression Plot';
TITLE3 'Quadratic model';
DATA one;
INPUT month flats price;
LABEL flats = Number of 12-pint flats available';
LABEL price = Price (USD);
flats2 = flats*flats;
CARDS;
1 230 22
2 320 16
3 350 17
```

```
4  400 15
5  315 17
6  275 32
7  260 23
8  255 24
9  200 21
10 168 43
11 99  50
12 75  65
RUN;
```
Perform the regression analysis to create an output dataset of parameter estimates called 'regdata' with the statement 'OUTEST' and request no printing of the regression results:
```
PROC REG DATA = one OUTEST = regdata NOPRINT;
MODEL price = flats flats2 / CLM;
RUN;
```
Put the regression equation in a SAS® macro variable for use later in the graph. Macro variables are a shorthand for text. You assign a text string to a variable, then anywhere you need to print the text, you reference the variable rather than typing out the text. Retrieve equation from the 'regdata' dataset created with the 'outest' statement with the 'DATA_NULL' and 'SET regdata' statements:
```
DATA _NULL_;
SET regdata;
```
Invoke MACRO variable assignment with the 'CALL SYMPUT' statement. 'eqn' is the macro variable name you assign and the text inside the double quotes is the value you assign the variable. The parameter estimates are retrieved from the 'regdata' dataset with the ||Intercept||, ||flats|| and ||flats2||. The double vertical bar operator is a concatenation indicator telling SAS® to string everything inside the double quotes together:
```
CALL  SYMPUT('eqn',"Price  =  "||Intercept||"  "||flats||"  *Flats  +
    "||flats2||"  * Flats2");
```
Note: you might have to omit either of the '+' signs in the preceding line if either β_0 or β_1 estimates are negative.
```
RUN;
PROC SGPLOT DATA = one;
REG x = flats y = price / DEGREE = 2 CLM NOLEGCLI NOLEGCLM
    NOLEGFIT;
```
Place the regression equation in the bottom left of the graph using the INSET statement, calling the 'eqn' variable you created with the '&' symbol in front of the variable name:
```
INSET "&eqn"/ POSITION = BOTTOMLEFT;
XAXIS grid;
YAXIS grid;
RUN;
```

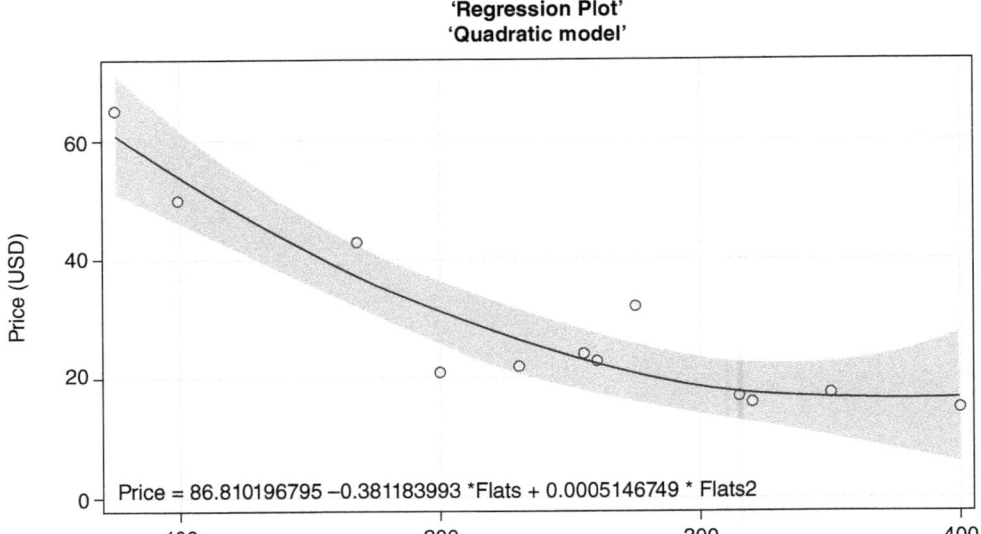

'Chapter 10 Example 6.sas'
'Regression Plot'
'Quadratic model'

Price = 86.810196795 −0.381183993 *Flats + 0.0005146749 * Flats2

Number of 12-pint flats available

You can use the preceding code as a template to generate graphs with your own data.

11 Mean Separation (Multiple Comparison) Techniques

Experiments often consist of two or more levels of one or more factors maintained in a logical and statistically acceptable design. An analysis of variance is usually performed on the data to determine whether or not there are treatment differences. Even when we determine that there are treatment differences, we don't always know which means are different from each other. When we want to determine which means in an experiment are different from each other we often utilize a mean separation technique, also called a multiple comparison procedure.

In an analysis of variance, the F-test for treatments indicates whether or not to reject the null hypothesis of equal treatment means. If we reject the null hypothesis, we know that not all means are equal, but we don't know which differ from which. If there are only two means being tested and the F-test is significant, then there is no need to use any other mean separation technique. The F-test indicates that the two levels of the factor are different. When the treatment structure is more complicated, mean separation also becomes more complicated.

Quantitative Factors

When one quantitative factor is being investigated, regression is often the most appropriate analysis. If a significant linear, non-linear, or polynomial relationship can be estimated, there is no need for any further mean separation evaluation. A significant regression relationship indicates that all values of the dependent variable are different at different levels of the independent variable. Appropriate lack-of-fit tests and the diagnostic procedures discussed in Chapter 10 should be used to assure that an appropriate model has been developed. Response surface regression is quite useful in the case of a factorial experiment with two quantitative factors. Response surface regressions produce 3-D graphs and some individuals find them hard to interpret. However, they are very useful in many situations and should be considered if appropriate.

Qualitative Factors

If three or more means of a qualitative factor are being tested in an analysis of variance, the best mean separation technique to use is a set of contrasts as discussed in Chapter 8. Recall that a contrast is a specific hypothesis test determined *before* carrying out an experiment to answer specific questions. The number of contrasts possible for any experiment is equal to the number of degrees of freedom associated with the treatment. Contrasts are most often orthogonal. Orthogonality guarantees that the sums of squares for the contrasts sum to the sums of squares for the treatments. If sensible contrasts cannot be constructed, one of many classic mean separation techniques can be used.

DOI: 10.1079/9781789249927.0011

Interpretation of Interactions

When a significant interaction between two factors or among three or more factors is detected, it is not appropriate to compare interaction means. SAS® will not perform mean separation on interaction means. If mean separation of interaction means is presented in any discussion of research results, it indicates that the researcher does not understand how a significant interaction should be interpreted. When an interaction is detected between two factors, it means that the differences among the levels of one factor depend on what level of the other factor you are considering. For example, suppose you were looking at four cultivars of maize and evaluating the effects of five different fungicides on seed rot before germination and a significant interaction was detected. The differences among the fungicides would depend upon which cultivar you were considering and vice versa. Significant interactions suggest that the effects of one factor should be evaluated within each level of the other factor. Most often we select one of the factors and compare the levels of the other factor within each level of the chosen one. Normally, one doesn't compare the levels of each factor within the levels of the other – you pick one as the factor to compare the other factor levels within. For example, you would compare fungicides within each cultivar or cultivars within each fungicide, but not both. Pick the one which makes most sense for the research objectives.

The selection of which mean separation procedure to use should consider two major things. First, it should consider whether you are comparing all possible pairs of treatment means or if you are comparing treatment means to a control. The second consideration for selection should be regarding the strength of the inference made by the technique. This brings up the subject of statistical errors.

Errors in Statistical Testing

It is important to understand the concept of errors in statistical testing in order to understand the differences among the mean separation techniques. A Type I error is the probability of rejecting the null hypothesis of no treatment effect when it is in fact true. In other words, you say that two means are different when, in fact, they are not. This probability is given as alpha (α). Since we don't want to make the mistake of saying that there are treatment differences when in fact there are none, we usually set α to a value of 0.05. A Type II error is the probability of failing to reject the null hypothesis of no treatment effect when in fact there is a treatment effect. In other words, we say two means are not different when, in fact, they are (we just didn't detect it). This probability is beta (β). We don't want to make this mistake either, so β should be set as low as possible. Ideally α and β are as small as possible. This is the difficulty in statistical testing. If we had a perfect statistical test, we would want α to be 0. The only way this could occur is if we always failed to reject the null hypothesis and accepted the alternate hypothesis no matter what the data. If the null hypothesis was in fact false, β would be 1. In the other case, if we wanted a perfect test with regard to β, we would always reject the null hypothesis. If in fact the null hypothesis was true, we would have made a Type I error with the probability of 1. In order to know whether it was β or α which was 0, we would need to know if the null hypothesis was in fact true or false. But if we knew, we wouldn't need to do any experiment! So we try to set up an experiment where both α and β can have relatively small values. This can only be done by increasing the sample size. When we define β as the probability of failing to reject the null hypothesis when it is in fact false, then $(1 - \beta)$ is the probability of rejecting the null hypothesis when it is false. $(1 - \beta)$ is also known as the power of a test. The only way we can know an actual value of β is if we know an actual outcome for the alternate hypothesis. We can set α at an exact level but not β, since we don't know the exact outcome for the alternate hypothesis. Many researchers control α but fail to give β much consideration. Making a Type II error is just as serious as making a Type I error. We can select mean separation techniques which control α and β at desirable levels. We control β in essence by selecting a powerful mean separation procedure. We control α by performing the test at a suitably low α level, usually 0.05. Remember, power is a relative term. Exact powers can only be given if exact β's are known. Exact β's are known only when exact alternate hypotheses are known.

Comparisonwise and Experimentwise Errors

There are two types of Type I errors associated with mean separation techniques. A comparisonwise Type I error rate is, simply put, the probability of declaring the two means in a single comparison as different when they in fact are not. (A more formal definition is given that says the comparisonwise Type I error rate is equal to the number of Type I errors made divided by the total number of nonsignificant comparisons. The first simpler definition suffices for all practical purposes.) A simple definition of the second type of Type I error, the experimentwise Type I error rate, is the probability of making at least one declaration of a significant difference when it in fact was not so, in an entire set of comparisons, or in other words, in an entire experiment. (Again, a more formal definition is given that says an experimentwise error rate is equal to the number of experiments with one or more Type I errors divided by total number of experiments with at least two equal means. And again, the simpler definition will suffice.) Different mean separation techniques guard against either of the two types of Type I errors. One that guards against making an experimentwise Type I error is much more stringent than one that guards against a comparisonwise Type I error. Since the experimentwise error protection is much more stringent than the comparisonwise protection, more Type II errors (failures to detect a difference) are likely to occur with experimentwise error protection. Thus β will be large and the power will not be very great.

There are many mean separation techniques available, and many of them are supported by SAS® software. The selection of an appropriate technique does not need to be confusing. The following discussion will be directed towards more of a 'which to use when' approach, rather than a description of each available procedure.

To help select the appropriate procedure, determine the following.

1. Do you want to compare treatments with each other to find 'the best' or 'the worst' among them, or do you want to compare treatments to a control or standard?
2. Do you wish to control the comparisonwise error rate or the experimentwise error rate?

The experimentwise error rate (β is high, α is low) is more conservative, has fewer rejections of equal means with lower power (since β is high) and fewer false rejections (because α is low). On the other hand, the comparisonwise error rate (β is low, α is high) is more liberal, has more rejections of equal means (greater power since β is low), but also has a greater number of false rejections (because α is high). Once you know in which direction to head, proceed.

All Pairwise Mean Comparisons

When you want to compare all treatments with each other, you are actually comparing all possible pairs of means. For t treatments, that number is:

$$\frac{(t)(t-1)}{2}$$

This number gets large rather quickly. If you consider an experiment with four levels of factor A and five levels of factor B, there are:

$$\frac{(20)(19)}{2} = 190$$

pairs of comparisons. Now imagine a much larger experiment.

All pairwise comparison procedures are based on the standardized difference between the two means being compared. 'Standardized' means that the difference between the two means is divided by the standard error of the means. We don't need to delve into this, but here's the point: all of the available procedures are based on the same foundation.

The simplest approach to mean separation is to perform all possible pairwise mean comparisons with a t-test. This approach is the classic LSD approach (least significant difference). A value is calculated such that in order to declare two means significantly different, they must differ by at least the calculated value. It's that simple. The formula for the LSD is:

$$LSD = t\left(\alpha;df\right)\sqrt{\frac{2s^2}{n}}$$

where t is a tabular value for the t-statistic with df degrees of freedom (degrees of freedom error from the ANOVA table) at the α level of significance, s^2 is the mean square for error from the ANOVA table and the two means you are trying to separate are calculated on 'n' observations each (we are assuming equal replication). If the two means have different numbers of observations, then the formula is:

$$LSD = t\left(\alpha;df\right)\sqrt{\frac{2s^2}{n_i + n_j}}$$

where one mean is based on 'i' observations and the other is based on 'ij' observations. This procedure is called Fisher's LSD.

Fisher's LSD

Fisher's LSD procedure controls the comparisonwise error rate, meaning that for each individual comparison of two means the probability of a Type I error is α. There are two types of Fisher's LSD. Fisher's Protected LSD requires significance of the ANOVA F-test for the factor whose means are being separate. If the F-test for the factor of interest is not significant, yet mean separation is still desired, the LSD is called Fisher's Unprotected LSD.

The LSD is a fixed range test, meaning that the same value for the LSD is used for all pairwise comparisons of the treatment means. The LSD allows for uneven replication and can also provide a confidence interval of the differences between treatments.

Let's look at a simple example using SAS® to examine the differences in yield of four cultivars of cabbage. SAS® will perform an ANOVA, calculate our LSD value and then produce a table with treatment means in it. Treatment means will be followed by a letter, such that means followed by the same letter are not significantly different from each other. Here are the statements for our example:

```
TITLE1 'Chapter 11 Example 1.sas';
TITLE2 'Fishers LSD';
DATA one;
INPUT cultivar $ yld @@;
CARDS;
A 13.8 A 13.5 A 13.2 B 15.5 B 15 B 15.2
C 21 C 22.7 C 22.3 D 18.9 D 18.3 D 19.6
PROC ANOVA PLOTS = none;
CLASSES cultivar;
MODEL yld = cultivar;
MEANS cultivar / LSD;
RUN;
```

which produces the following output:

'Chapter 11 Example 1.sas'
'Fishers LSD'

The ANOVA Procedure

Class Level Information		
Class	Levels	Values
cultivar	4	A B C D

Number of Observations Read	12
Number of Observations Used	12

'Chapter 11 Example 1.sas'
'Fishers LSD'

The ANOVA Procedure

Dependent Variable: yld

Source	DF	Sum of Squares	Mean Square	F Value	Pr > F
Model	3	130.2433333	43.4144444	127.07	<.0001
Error	8	2.7333333	0.3416667		
Corrected Total	11	132.9766667			

R-Square	Coeff Var	Root MSE	yld Mean
0.979445	3.356111	0.584523	17.41667

Source	DF	Anova SS	Mean Square	F Value	Pr > F
cultivar	3	130.2433333	43.4144444	127.07	<.0001

The ANOVA Procedure

t Tests (LSD) for yld

Note: This test controls the Type I comparisonwise error rate, not the experimentwise error rate.

Alpha	0.05
Error Degrees of Freedom	8
Error Mean Square	0.341667
Critical Value of t	2.30600
Least Significant Difference	1.1006

yld t Grouping for Means of cultivar (Alpha = 0.05)

Means covered by the same bar are not significantly different.

cultivar	Estimate	
C	22.0000	
D	18.9333	
B	15.2333	
A	13.5000	

You should be familiar with the ANOVA output and be able to interpret the results. The F-value of 127.07 for cultivar is significant at $\alpha < 0.0001$. Thus, from the ANOVA table we can surmise that the mean yields of our four cabbage cultivars differ, but we don't know which differ from which.

When considering our cultivar means and our hypotheses tests involving them, our null hypothesis (H_0) is simply:

H_0: $\mu_A = \mu_B = \mu_C = \mu_D$

If we reject the null hypothesis (which, in our example, we do) we are saying that we have collected enough evidence that one or more of the '=' signs above should be '≠'. We just don't know which ones. With our example, we need to make 6 ((4*3)/2) pairs of mean comparisons (six '=' sign evaluations) and they are:

H_0: $\mu_A = \mu_B$
H_0: $\mu_A = \mu_C$
H_0: $\mu_A = \mu_D$
H_0: $\mu_B = \mu_C$
H_0: $\mu_B = \mu_D$
H_0: $\mu_C = \mu_D$

This is where our mean separation procedure comes in. It helps us declare which of the three '=' signs above should be '≠'. For each comparison, however, there is a chance that we make an incorrect decision regarding our '=' sign and the probability that we have made an incorrect decision is α, the probability of a Type I error. Since we are using Fishers LSD, it is a comparisonwise Type I error.

The statement 'MEANS cultivar / LSD;' in our program requests means for all levels of 'cultivar' along with the 'LSD' 'mean separation. The top portion of the output provides the values for α, error degrees of freedom, error mean square, the t-value and the LSD.

Alpha	0.05
Error Degrees of Freedom	8
Error Mean Square	0.341687
Critical Value of t	2.30600
Least Significant Difference	1.1006

To determine whether we have '=' or '≠' for each pair of means, we simply determine whether the absolute value of the difference between the two means is ≥ the calculated LSD value. SAS® has already performed these comparisons and presents the results graphically in the table:

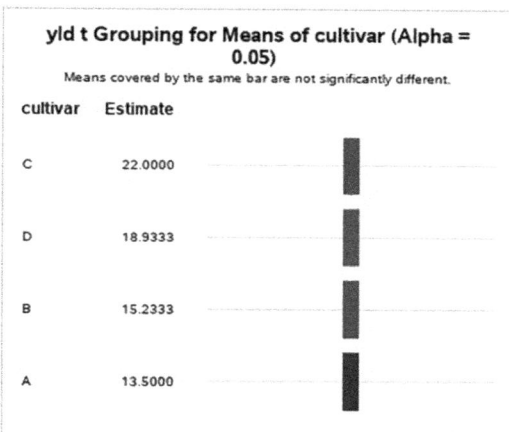

yld t Grouping for Means of cultivar (Alpha = 0.05)

Means covered by the same bar are not significantly different.

cultivar	Estimate
C	22.0000
D	18.9333
B	15.2333
A	13.5000

The interpretation is straightforward: means joined by the same-colored bar are not significantly different from each other. In our example, for each comparison of means listed:

$H_0: \mu_A = \mu_B$
$H_0: \mu_A = \mu_C$
$H_0: \mu_A = \mu_D$
$H_0: \mu_B = \mu_C$
$H_0: \mu_B = \mu_D$
$H_0: \mu_C = \mu_D$

we reject the null hypothesis of an '=' sign in favor of the alternate of '≠' with the realization that the chance we are incorrect for each comparison is 0.05. Make sure that you're clear that *each comparison* has this chance of being incorrect, since we are using a procedure (Fisher's LSD) which protects against comparisonwise Type I error.

If a graphic representation of mean differences is not appropriate for you due to journal or thesis format requirements, etc., letters can be assigned to the means. Means with the same bars in the above table would be followed by the same letter to indicate that they are not significantly different. Start with the letter 'a' (usually always lower case) for the largest mean, moving through the alphabet as needed to accommodate all mean differences. In the above example, we would use the 'a', 'b', 'c' and 'd' since all four means are different.

For the following table:

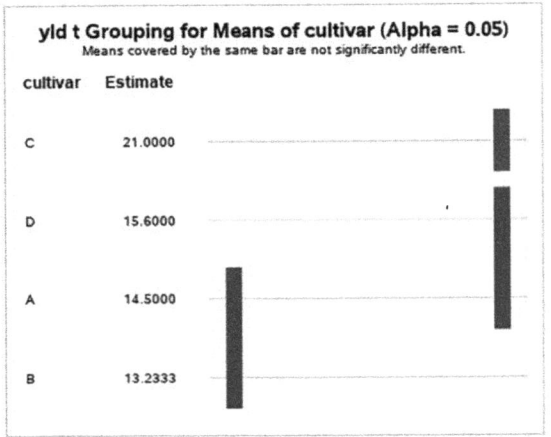

the conversion to letters would be thus:

Cultivar	Mean
A	14.5 bc
B	13.2 c
C	21.0 a
D	15.6 b

I like to align similar letters vertically following the means, to combine a 'graphical' and text-based interpretation of the mean separation.

Macros are available online to perform this conversion from a table with bars to a table with letters if you wish to pursue their use. The macros are rather complicated, therefore I prefer to perform the conversion manually.

Problems with the LSD

The major weakness of either LSD method is that with a large number of means to compare, the probability of at least one Type I error (the experimentwise error rate) approaches 1 even though our comparisonwise Type I error rate is 0.05. Suppose we set $\alpha = 0.05$ for comparing 15 treatment means in an experiment. With 15 means, there are $(n*(n-1))/2$ or $(15*14)/2 = 105$ possible pairs of means to compare. The probability that you would incorrectly make at least one claim that two of the means differed when they really didn't is equal to $1-(1-\alpha)^m$ or 0.9954 (m = the number of possible comparisons). The comparisonwise Type I error rate was set at 0.05 but the experimentwise Type I error rate ended up at 0.9954! Again, each researcher must decide which type of error rate is acceptable. Surely one would not want an almost certain chance of declaring two means different when they are in fact not different if the difference between the treatments has a large economic impact. Thus in such a case the researcher may wish to protect against such a high experimentwise error rate by using a more stringent mean separation technique. Under the same circumstances the researcher would not want to fail to detect a real difference (make a Type II error).

One way of using the LSD to ensure an experimentwise error rate of α^* is to adjust the α used in the calculation of the actual LSD value used to detect treatment differences. Recall that one must set α (the level of significance) to some level in order to look up a t-value to calculate the LSD. To ensure an experimentwise error rate of α^* for m comparisons, select a t-value with an α equal to α^*/m. For example, to ensure an experimentwise error rate of 0.05 for comparing 10 means, calculate the LSD using a t-value for $\alpha = (0.05/45) = 0.0011$. In SAS® this is accomplished by requesting the BON (Bonferroni) option rather than the LSD option. SAS® automatically adjusts the comparisonwise error rate (α) so that the experimentwise error rate is by default 0.05. Here is our previous example using BON instead of LSD:

```
TITLE1 'Chapter 11 Example 2.sas';
TITLE2 'Bonferroni Mean Separation';
DATA one;
INPUT cultivar $ yld @@;
CARDS;
A 13.8 A 13.5 A 13.2 B 15.5 B 15 B 15.2
C 21 C 22.7 C 22.3 D 18.9 D 18.3 D 19.6
PROC ANOVA PLOTS = none;
CLASSES cultivar;
MODEL yld = cultivar;
MEANS cultivar / BON;
RUN;
```

Here's the output:

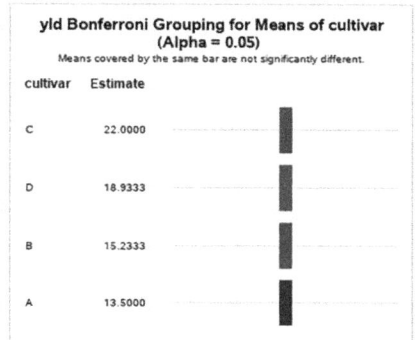

'Chapter 11 Example 2.sas'
'Bonferroni Mean Separation'

The ANOVA Procedure

Bonferroni (Dunn) t Tests for yld

Note: This test controls the Type I experimentwise error rate, but it generally has a higher Type II error rate than REGWQ.

Alpha	0.05
Error Degrees of Freedom	8
Error Mean Square	0.341667
Critical Value of t	3.47888
Minimum Significant Difference	1.6603

yld Bonferroni Grouping for Means of cultivar (Alpha = 0.05)

Means covered by the same bar are not significantly different.

cultivar	Estimate
C	22.0000
D	18.9333
B	15.2333
A	13.5000

Notice that SAS® provides a note at the very top of the output to remind you what type of Type I error the procedure protects against, as well as any other pertinent information. For the BON option, SAS® includes the note 'This test controls the Type I experimentwise error rate, but it generally has a higher Type II error rate than REGWQ'. REGWQ is another mean separation option. The higher Type II error rate of BON compared with REGWQ reminds you that BON is more conservative than REGWQ and that BON might fail to detect differences when they exist compared with differences REGWQ might detect. If you check your LSD output, you'll find the SAS® note 'This test controls the Type I comparisonwise error rate, not the experimentwise error rate.'

So now we have two mean separation techniques: the liberal LSD and the conservative BON. What if you want a mean separation that is somewhere in between these two?

The 'TUKEY' option provides just such a procedure. This option was developed by Tukey and later modified by Kramer (both statisticians) specifically for pairwise comparisons and controls the experimentwise Type I error rate with equal or unequal sample sizes without being overly conservative like the BON option.

Here's the output for our example using TUKEY (in which case it would be Example 3) in place of LSD (in Example 1) as our option in the MEANS statement:

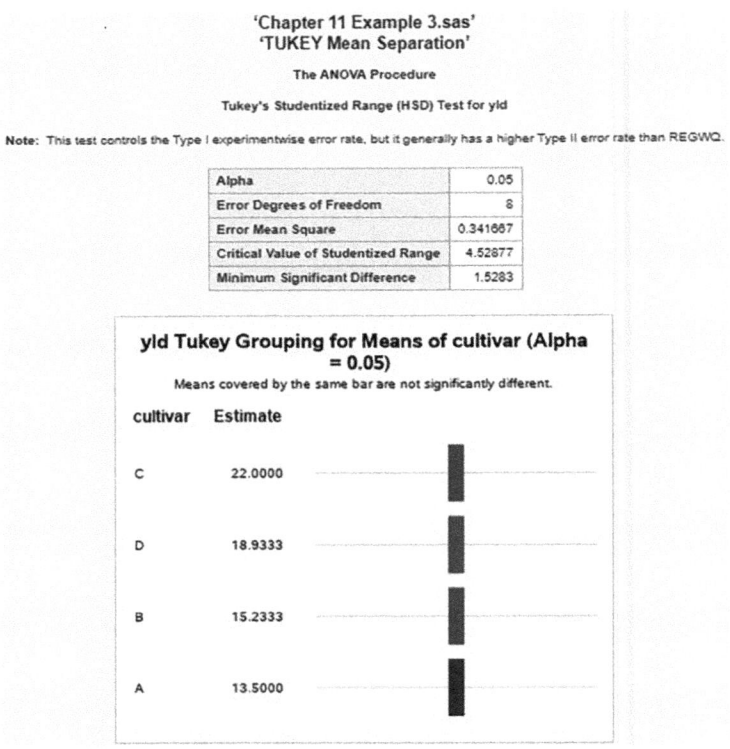

'Chapter 11 Example 3.sas'
'TUKEY Mean Separation'

The ANOVA Procedure

Tukey's Studentized Range (HSD) Test for yld

Note: This test controls the Type I experimentwise error rate, but it generally has a higher Type II error rate than REGWQ.

Alpha	0.05
Error Degrees of Freedom	8
Error Mean Square	0.341667
Critical Value of Studentized Range	4.52877
Minimum Significant Difference	1.5283

yld Tukey Grouping for Means of cultivar (Alpha = 0.05)

Means covered by the same bar are not significantly different.

cultivar	Estimate
C	22.0000
D	18.9333
B	15.2333
A	13.5000

You may have noticed that the REGWQ option has been mentioned in several of the notes SAS® provides with mean separation output. The REGWQ option was developed by a group of statisticians based on the mean separation approach which considers a changing value for the needed mean difference to indicate significance called a 'step-down' procedure. Step-down procedures compare groups of means rather than pairs. The way they work is that treatment means are arranged in descending order. The largest mean is compared with the smallest mean, and if they are different, based on a complicated function related to the LSD, the procedures begin to test subsets of means. Means which are not significantly different from each other are usually indicated with similar letters of the alphabet following equal means. Most step-down procedures should only be used with equal sample sizes. Step-down procedures are generally more powerful than those procedures based on a single mean difference value, such as the LSD, BON and TUKEY options. Of the many step-down procedures available in SAS® REGWQ is the most powerful one. The drawback of step-down procedures is that they do not provide simultaneous confidence intervals for the means. If you don't need confidence intervals, then you should use the REGWQ option rather than LSD, BON or TUKEY since it is more powerful than any of these.

Thus it's relatively easy to decide which mean separation technique you should choose if you are comparing all means in a set with each other. If you are not concerned with the increasing chance of committing an experimentwise Type I error with many individual mean comparisons and/or need confidence intervals, use the LSD option. If you want to control the experimentwise Type I error rate, and need confidence intervals, then you should use TUKEY. If you don't need confidence intervals, then use REGWQ – it's most powerful.

Treatments vs a Control

If you have a series of treatments you want to compare with a standard or a control, you should use the DUNNETT option on the MEANS line in your SAS® program. It is slightly different than the other options in that you need to give it the level of the treatment you are considering as the control. For example, suppose your treatment was cultivars A, B, C and D. Also suppose that cultivar B is the standard you wish to compare the others with. You would indicate it thus:

```
TITLE1 'Chapter 11 Example 4.sas';
TITLE2 'Dunnett Mean Separation';
DATA one;
INPUT cultivar $ yld @@;
CARDS;
A 13.8 A 13.5 A 13.2 B 15.5 B 15 B 15.2
C 21 C 22.7 C 22.3 D 18.9 D 18.3 D 19.6
PROC ANOVA PLOTS = none;
CLASSES cultivar;
MODEL yld = cultivar;
MEANS cultivar / DUNNETT(B);
RUN;
```
which produces:

The ANOVA Procedure

Dunnett's t Tests for yld

Note: This test controls the Type I experimentwise error for comparisons of all treatments against a control.

Alpha	0.05
Error Degrees of Freedom	8
Error Mean Square	0.341667
Critical Value of Dunnett's t	2.87967
Minimum Significant Difference	1.3744

Comparisons significant at the 0.05 level are indicated by ***.

cultivar Comparison	Difference Between Means	Simultaneous 95% Confidence Limits		
C - B	6.7667	5.3923	8.1410	***
D - B	3.7000	2.3256	5.0744	***
A - B	-1.7333	-3.1077	-0.3590	***

DUNNETT produces this table listing all treatment comparisons with the control providing confidence limits for the mean differences as well as significance at the 0.05 level, indicated in the rightmost column.

A few other options in the MEANS statement which one might need include the following.

1. 'CLDIFF', which presents the results of the BON, LSD or TUKEY option as confidence intervals of the differences between pairs of means.
2. 'CLM', which presents the results of BON and LSD as confidence intervals around the means.
3. 'E', which tells SAS which mean square from our ANOVA table to use as the error term in performing the mean separation. This will be need in some of the more complicated designs such as split-plot and split-block.

Summary

I once heard someone say that they run all of the mean separation procedures in SAS® and then pick out the results that they like the best! This is absurd, so please don't ever try this. Deciding which step to take after the ANOVA is really rather simple. Quantitative factors should be further analyzed with regression and diagnostics. Qualitative factors should be further analyzed using contrasts. If *a priori* questions were not posed or if contrasts are not feasible, then one of the mean separation techniques described above should be used.

12 The Randomized Complete Block Design

In the completely random design of Chapter 6, experimental units were assumed to be homogeneous, and that any variation among them prior to treatment application was due to random error. In the ANOVA, we calculated the variation due to treatments and that due to error, performed an *F*-test and determined whether or not our treatment means were different.

Suppose our experimental units vary due to some classifiable characteristic and we could group them accordingly. Characteristics might be age, height, size, position in the field, etc. We could then reduce the 'random variation' (error) by accounting for that variation due to the grouping factor.

The experimental design which accounts for such identifiable variation is called the randomized complete block design (RCBD). The RCBD can be simple, holding several levels of a single treatment, or complex, holding a complicated factorial. Remember, you have a treatment structure that needs to fit into an appropriate design. Even though a factorial treatment structure will fit well into an RCBD, this type of treatment structure is often more suited to split-plot (two factors) or split-split-plot (three factors) designs. These will be covered later.

In grouping the experimental units in an RCBD, we want to minimize the variability of experimental units within the groupings (blocks) and maximize the variability among the blocks. In addition, treatment applications or data acquisitions should be performed by block. Each block should be completed as a unit if all of the experimental units in the experiment cannot be treated or measured the same day. If there is any variation due to days, that variation will go into the block variation rather than into experimental error. The RCBD is called complete because each block contains a complete set of treatments in the experiment.

Field experiments may be blocked due to an observed or potential gradient in the field where the experiment will be performed. For example, any measured gradient such as soil pH, fertility, type or moisture can be used to set up blocking in the field. In perennial crops, blocking can be set up due to observed soil-based or plant-based gradients such as those already mentioned for the soil and vigor or other plant characteristics for plant-based gradients. Similarly, greenhouse experiments may be blocked by location in the greenhouse. Blocking in the greenhouse helps account for gradients in heat, light and moisture, since all may vary significantly in the greenhouse environment. Blocking can also be utilized in lab experiments. Time of day is an especially important factor which should be considered in lab experiments, particularly when measurements or other tedious practices are performed on experimental units. As the day wears on, researchers may become tired and not as diligent in their work. By blocking over time, you can make sure each treatment is performed or responses to treatment measured at different times of the day.

With one strong gradient you should use long narrow blocks that are perpendicular to the gradient. That way, all of the plots in one block will be at the same level of the gradient. As you move across the field, each block will be at another level of the gradient.

DOI: 10.1079/9781789249927.0012

With two strong gradients, you should probably use a Latin square design or square-shaped blocks if the Latin square is not feasible. (Latin square designs have limitations which we will discuss soon.) If there are no real predictable patterns in the field, use square blocks.

The RCBD usually provides more accurate results compared with the completely random design (CRD), especially when significant variability exists among experimental units and blocking using experimental unit characteristics can be employed. Treatments are randomized separately within each block. As in the CRD, any number of treatments can be included in an RCBD experiment.

Using PROC PLAN to Assign Treatments to an RCBD

Suppose you had an RCBD with six treatments and six blocks. You can use PROC PLAN to assign your six treatments to the six blocks:

```
TITLE1 'Chapter 12 Example 1.sas';
TITLE2 'PROC PLAN Treatment Assignment for RCBD';
TITLE3 '6 Treatments in 6 blocks';
PROC PLAN;
FACTORS block = 6 ORDERED treatment = 6;
RUN;
```

The output:

Chapter 12 Example 1.sas
Treatment Assignment for RCBD using PROC PLAN

The PLAN Procedure

Factor	Select	Levels	Order
block	6	6	Ordered
treatment	6	6	Random

block	treatment					
1	2	4	5	1	3	6
2	1	5	3	2	4	6
3	5	4	1	6	2	3
4	1	6	4	2	3	5
5	3	2	4	6	1	5
6	4	3	6	2	1	5

The output of PROC PLAN provides a table of random treatment assignments numbered from 1 to 6 for each of six blocks that are listed in order from 1 to 6 by using the 'ORDERED' option following the 'block = 6' portion of the FACTORS statement. The list of factors needing randomization are provided in the FACTORS statement, in our example blocks and treatment. The number of levels of each factor requiring processing is indicated after the '=' sign which follows the name of the factor. IF the option 'ORDERED' is not indicated, a random assignment request is assumed. Thus we are requesting a random assignment of six treatments to each of six blocks listed in order.

You might be wondering why we requested an ordered list for blocks. Since we are defining our blocks, experimental units are assigned to each block based on the chosen blocking characteristic; hence, the block value for each experimental unit is assigned, but not randomly. Treatment assignments are randomly made within each block. Remember we want to ensure that each treatment occurs once in each block. You might think of it this way as well: we are simply repeating our basic experiment with six treatments, six different times, once with each set of experimental units possessing the characteristic of each of the six separate blocks.

As an example, let's suppose our experiment was a lab experiment. Our experimental unit is a single petri dish with a colony of bacteria growing on it. We have 36 petri dishes, each with the same type of bacteria; however, the colony coverage in each dish varies considerably. We have six different bactericides we want to test on these colonies. In order to make sure that all of the dishes with weaker colonies are not assigned the same bactericide, we separate our dishes into six groups of six by visually assessing the size of the colony in each dish. Now we can randomly assign the six treatments within each block of dishes corresponding to each of the six general colony size categories. The major goal of blocking is to minimize experimental unit variability within each block while maximizing the variability among blocks.

In field or greenhouse examples, the site of the experiment is normally assessed for gradients and field or greenhouse maps are developed and the area is blocked out according to gradient. Treatments are then randomly assigned to the experimental units within each block and the randomization is done separately for each block.

In our SAS® example above, there are six blocks of six experimental units. In block 1, the first unit would get treatment 2, the second 4, the third 5, etc. Similarly, treatments would be assigned to units in blocks 2 through 6. The main difference in assigning treatments to experimental units in an RCBD vs a CRD is that in the RCBD you must assign the treatments randomly within each block, a different randomization for each block. For the CRD you randomly assigned the treatments to the experimental units as one large group.

Each block in a randomized complete block experiment is a complete replicate of the experiment – that's why it's called complete. I recommend never calling the replicates in the RCBD 'replicates' (even though it is not wrong to do so), but rather, always call them blocks. All blocks are replicates, but not all replicates are blocks. You only account for the variation due to replicates when those replicates are blocks.

Remember that the function of an ANOVA is to account for different sources of variation in an experiment and ultimately determine whether differences we observe among treatments are real or due to chance. In the CRD ANOVA, we accounted for total variation, that due to μ and that due to treatments. The remaining variability was assumed to be random and we called its source 'error'. In the RCBD we account for an additional source of variation, namely, that attributable to blocking.

The ANOVA table including only sources and df for an RCBD looks like this:

Source	df
Total	(N)
μ	1
Treatment	(t–1)
Block	(b–1)
Error	(t–1)(b–1)

You would have SS, MS, F and Prob > F columns in the final ANOVA table. The only appropriate F-test in an RCBD experiment is the test(s) for treatment(s). You do not test for a statistical effect of blocking. Even though many researchers test for a block effect to see if their blocking was effective, it is not a statistically valid test. In statistical theory involving ANOVAs, equations called 'expected mean squares' (EMS) are mathematically derived to determine appropriate and valid statistical tests for various experimental circumstances. Even though they are more theoretical in nature, we cover EMS construction in Chapter 15 since they are invaluable for experiments repeated in different years and at different locations. We don't need to construct EMS for most designed experiments, since appropriate and valid tests are 'built into' common experimental designs. If you constructed the EMS for the RCBD design, it would be apparent that you can't test for a block effect. We'll look at this when covering expected mean squares.

Relative Efficiency of RCBD to CRD

The appropriate way to determine whether or not blocking was effective is to examine a statistical measure called the relative efficiency (RE). Just as the name implies, relative efficiency is a relative measure of the efficiency of one design compared with another. One usually compares the relative efficiency of a more complicated experimental design to a simpler design. In this case, we would look at the relative efficiency of the RCBD compared with the CRD. If blocking was effective, then our estimate of error for the RCBD (σ^2_{RCBD}) should be less than the estimate of error for the CRD (σ^2_{CRD}), given the same experimental data. In other words, you compare the estimates of error for the RCBD with the error you would have gotten if you had not blocked the experiment. One can calculate this CRD error estimate readily from the RCBD ANOVA table. We define the RE of the RCBD to the CRD as:

$$\frac{(b-1)*MS_{Blocks}RCBD + b(t-1)*MS_{Error}RCBD}{(bt-1)*MS_{Error}RCBD}$$

where b is the number of blocks and t is the number of treatments in the RCBD.

The numerator is an estimate of the error you would have gotten if you had used a CRD instead of a RCBD, and the denominator is the error estimate from the RCBD. The multipliers in both the numerator and denominator are needed to account for the differences in the number of df available for estimating error in the two different designs.

Suppose we had calculated an RE of 2. This would be interpreted as meaning that for a similar experiment using a CRD, you would need twice as many replicates as in the RCBD to have the same precision in detecting differences as in the original RCBD. Similarly, if the RE was 0.5, then you would only need half the number of replicates in a CRD to have the same precision as in the RCBD. This would occur when there is really no blocking factor effect. In this case you lose precision in the RCBD due to fewer df in the RCBD error estimate compared with the CRD error estimate (since $(b-1)$ df were used for the blocking effect).

The RCBD: an Example

We will look at an example from start to finish. Suppose we examined the yield of four lettuce cultivars. We also examined the seed source. We were able to obtain the same four cultivars (two romaine type, 'Jericho' and 'Parris Island Red', and two leaf types, 'Red Oak' and 'New Red Sails') from five seed companies: two foreign (Africana Seeds, Germania Seed Company) and three domestic (Johnny's Selected Seeds, Territorial Seeds and W. Atlee Burpee Seed Company).

We have a set of specific *a priori* questions to ask as follows:

1. Is romaine different in productivity compared with leaf lettuce?
2. Within romaine, do the cultivars differ?
3. Within leaf, do the cultivars differ?
4. Is there a difference between foreign and domestic seed companies?
5. Within foreign companies, do the two companies differ?
6. Within domestic companies, do East coast companies (Johnny's and Burpee) differ from the West coast company (Territorial)?
7. Does Johnny's differ from Burpee's?

Notice that even though we have 19 df for treatment, and therefore could ask 19 specific questions, we don't have to ask all 19. The 12 df left over will provide evidence whether or not we are omitting questions from our set. If the MS leftover is significantly larger than error, it shows we have left something out of our set of questions. We can't, however, come up with additional questions after the fact. We would just know we left something(s) out.

So we have a 4 × 5 factorial we want to put in an RCBD. The field we must use is located relatively close to a wooded area, and we are concerned that the closer the plots are to the woods, the more likely they are to be munched on by critters. Thus we should use the RCBD. We are able to accommodate five replicates, thus we have an RCBD with five blocks and 20 treatments. Before we assign plot treatments, let's develop the ANOVA table and the set of contrast coefficients:

Source	df
Total	100
μ	1
Block	4
Treatment	19
Error (Block × Treatment)	76

The ANOVA table could be expanded to reveal the factorial nature of the treatment structure as such:

Source	df
Total	100
μ	1
Block	4
Treatment	19
Cultivar	3
Seed source	4
Cultivar * seed source	12
Error (Block × Treatment)	76

As a matter of fact, we can expand the ANOVA table even more to include the contrasts. We will position the contrasts under appropriate main factors to show where the SS for the contrasts are coming from.

Source	df
Total	100
μ	1
Block	4
Treatment	19
Cultivar	3
Contrast 1: Is romaine any different in productivity compared with leaf lettuce?	1
Contrast 2: Within romaine, do the cultivars differ?	1
Contrast 3: Within leaf, do the cultivars differ?	1
Seed source	4
Contrast 4: Is there a difference between foreign and domestic seed companies?	1
Contrast 5: Within foreign companies, do the two companies differ?	1
Contrast 6: Within domestic companies, are East coast companies (Johnny's and Burpee) different from the West coast (Territorial) company?	1
Contrast 7: Is Johnny's different than Burpee's?	1
Cultivar * seed source	12
Error (Block × Treatment)	76

If a significant interaction between cultivar and seed source is detected, further evaluation is needed. There would be a number of ways to further evaluate the data, depending on what information is most important to the research. I would approach a significant interaction this way. Some of the growers I work with need to know where they should purchase their specific lettuce cultivars from, thus I would perform a mean separation of seed companies within cultivar. However, I also work with some growers who are very loyal to their seed companies, thus they would prefer to know which lettuce cultivars they should purchase from their specific seed company. For these growers, I would separate cultivars within seed companies. With this approach I would be able to answer all my grower co-operator questions. A significant interaction would also indicate that I should use the information generated by the contrasts with a bit of caution. The contrasts are powerful statistically, but a significant interaction would indicate that cultivar performance varies with seed source.

Here's the table of treatments, their assignment codes and the contrast coefficients considering 20 treatments:

	Treatment code																			
	1	2	3	4	5	6	7	8	9	10	11	12	13	14	15	16	17	18	19	20
Cultivar	J					P					R					N				
Seed source	A	G	J	T	W	A	G	J	T	W	A	G	J	T	W	A	G	J	T	W
Contrast 1	1	1	1	1	1	1	1	1	1	1	−1	−1	−1	−1	−1	−1	−1	−1	−1	−1
Contrast 2	1	1	1	1	1	−1	−1	−1	−1	−1	0	0	0	0	0	0	0	0	0	0
Contrast 3	0	0	0	0	0	0	0	0	0	0	1	1	1	1	1	−1	−1	−1	−1	−1
Contrast 4	3	3	−2	−2	−2	3	3	−2	−2	−2	3	3	−2	−2	−2	3	3	−2	−2	−2
Contrast 5	1	−1	0	0	0	1	−1	0	0	0	1	−1	0	0	0	1	−1	0	0	0
Contrast 6	0	0	1	−2	1	0	0	1	−2	1	0	0	1	−2	1	0	0	1	−2	1
Contrast 7	0	0	1	0	−1	0	0	1	0	−1	0	0	1	0	−1	0	0	1	0	−1

You should check the orthogonality of these contrasts.

An alternative ANOVA table could be set up which considers only the factorial nature of the treatments rather than the set of 20 treatments. Contrasts could also be performed using this approach; however, you would need the '20 treatment approach' table if you had more complex contrasts to evaluate. More complex contrasts would likely result in a set of non-orthogonal contrasts. Non-orthogonal contrasts, simply explained, means that the contrast SS do not add up to the treatment SS. Another way to think about orthogonality is that the results of one orthogonal contrast will in no way give any idea of what the results of another orthogonal contrast will be, but the results of a non-orthogonal contrast may hint at what the results of another non-orthogonal contrast might be. The real problem associated with non-orthogonal contrasts is that the experimentwise Type I error is not controlled. If non-orthogonal contrasts are used, the α level should be adjusted (Bonferroni's adjustment) such that the 'new' α for testing each individual contrast is α/n where n is the total number of contrasts under consideration. If some important questions can be answered with non-orthogonal contrasts, then you should use them. Remember, though, they should be *a priori* questions, and you should use Bonferroni's adjustment.

The ANOVA table considering the factorial approach rather than the '20-treatment' approach is:

Source	df
Total	100
μ	1
Block	4
Cultivar	3
Seed source	4
Cultivar × seed source	12
Error	76

Here's the contrast table for this approach:

	Cultivar			
Contrast	'Jericho'	'New Red Sails'	'Parris Island'	'Red Oak'
Romaine vs leaf	1	−1	1	−1
'Jericho' vs 'Parris Island'	1	0	−1	0
'Red Oak' vs 'New Red Sails'	0	−1	0	1

	Seed source				
	A	G	J	T	W
Foreign vs domestic	3	3	−2	−2	−2
AS vs GS	1	−1	0	0	0
East vs west	0	0	1	−2	1
JS vs WAB	0	0	1	0	−1

When you look at the contrasts this way, you can see that you are using the 3 df associated with cultivar and the 4 df associated with seed company to answer your questions. The 12 df left over are from the interaction between cultivar and seed company. If there was a significant interaction between cultivar and seed company, it would mean that how well each cultivar performed depended on where you bought your seeds.

Now that we have our *a priori* questions set we can proceed with the experimental layout. To randomly assign the 20 treatments to our field, we'll use the following SAS program:

```
TITLE1 'Chapter 12 Example 2.sas';
TITLE2 'Treatment Assignment for RCBD';
TITLE3 'Five blocks, 20 treatments';
PROC PLAN;
FACTORS block = 5 ordered treatment = 20;
RUN;
```

The output:

'Chapter 12 Example 2.sas'
'Treatment Assignment for RCBD'
'Five blocks, 20 treatments'

The PLAN Procedure

Factor	Select	Levels	Order
block	5	5	Ordered
treatment	20	20	Random

block	treatment																			
1	3	19	1	12	14	16	4	6	20	7	5	18	11	13	8	10	17	2	15	9
2	2	19	20	8	7	17	5	6	4	10	13	14	11	16	9	12	18	15	3	1
3	3	12	15	20	6	4	11	7	18	8	19	9	14	16	17	5	10	1	13	2
4	8	16	2	15	1	12	11	9	6	3	7	10	19	14	4	17	5	13	18	20
5	12	4	9	17	3	15	19	13	20	16	2	5	11	10	6	7	8	14	18	1

Our field map would look like this:

Block 1	Block 2	Block 3	Block 4	Block 5
		Treatment		
3	2	3	8	12
19	19	12	16	4
1	20	15	2	9
12	8	20	15	17
14	7	6	1	3
16	17	4	12	15
4	5	11	11	19
6	6	7	9	13
20	4	18	6	20
7	10	8	3	16
5	13	19	7	2
18	14	9	10	5
11	11	14	19	11
13	16	16	14	10
8	9	17	4	6
10	12	5	17	7
17	18	10	5	8
2	15	1	13	14
15	3	13	18	18
9	1	2	20	1

Notice how you randomly assigned the 20 treatments to the 20 plots in each block, repeating the assignment for all five blocks with a different randomization, provided by the PROC PLAN output.

The data collected was the number of marketable cases produced per plot. We will assume that the data are normally distributed, but if you were concerned that the data may not be, you could use PROC UNIVARIATE NORMAL in SAS®.

Here is our basic program to generate the dataset we will use plus the code to test for normality. Note that we are including designations for treatment, cultivar and seed source in order for two different approaches to our analysis.

```
TITLE1 'Chapter 12 Example 3.sas';
TITLE2 'RCBD, 5 blocks 20 treatments';
TITLE3 'Lettuce cultivar and seed source evaluation';
DATA one;
INPUT block treatment cultivar $ ssource $ cases;
CARDS;
1 1 j a 740
2 1 j a 655
3 1 j a 1162
4 1 j a 621
5 1 j a 601
1 2 j g 869
2 2 j g 759
3 2 j g 969
4 2 j g 630
5 2 j g 714
1 3 j j 972
2 3 j j 755
3 3 j j 545
4 3 j j 724
5 3 j j 807
1 4 j t 1075
2 4 j t 662
3 4 j t 547
4 4 j t 838
5 4 j t 911
1 5 j w 1179
2 5 j w 644
3 5 j w 652
4 5 j w 931
5 5 j w 489
1 16 n a 712
2 16 n a 570
3 16 n a 474
4 16 n a 1028
5 16 n a 615
1 17 n g 810
2 17 n g 756
```

```
3  17  n  g  674
4  17  n  g  1118
5  17  n  g  611
1  18  n  j  817
2  18  n  j  805
3  18  n  j  783
4  18  n  j  918
5  18  n  j  623
1  19  n  t  354
2  19  n  t  959
3  19  n  t  773
4  19  n  t  599
5  19  n  t  713
1  20  n  w  455
2  20  n  w  1071
3  20  n  w  685
4  20  n  w  506
5  20  n  w  826
1  6  p  a  980
2  6  p  a  655
3  6  p  a  753
4  6  p  a  1031
5  6  p  a  523
1  7  p  g  521
2  7  p  g  748
3  7  p  g  851
4  7  p  g  202
5  7  p  g  692
1  8  p  j  527
2  8  p  j  860
3  8  p  j  951
4  8  p  j  510
5  8  p  j  709
1  9  p  t  629
2  9  p  t  951
3  9  p  t  455
4  9  p  t  502
5  9  p  t  895
1  10  p  w  734
2  10  p  w  1066
3  10  p  w  563
4  10  p  w  502
5  10  p  w  806
1  11  r  a  829
2  11  r  a  257
3  11  r  a  656
4  11  r  a  505
5  11  r  a  399
```

```
1  12  r  g  934
2  12  r  g  566
3  12  r  g  766
4  12  r  g  515
5  12  r  g  403
1  13  r  j  402
2  13  r  j  557
3  13  r  j  863
4  13  r  j  711
5  13  r  j  600
1  14  r  t  509
2  14  r  t  568
3  14  r  t  865
4  14  r  t  816
5  14  r  t  704
1  15  r  w  608
2  15  r  w  560
3  15  r  w  368
4  15  r  w  919
5  15  r  w  703
RUN;
PROC PRINT DATA = one;
RUN;
PROC UNIVARIATE NORMAL DATA = one;
VAR cases;
RUN;
```

The results of the normality test are:

Tests for Normality				
Test		Statistic	p Value	
Shapiro-Wilk	W	0.992503	Pr < W	0.8564
Kolmogorov-Smirnov	D	0.045644	Pr > D	>0.1500
Cramer-von Mises	W-Sq	0.035859	Pr > W-Sq	>0.2500
Anderson-Darling	A-Sq	0.237461	Pr > A-Sq	>0.2500

Since $N < 2000$, we use the Shapiro–Wilk test, $W = 0.992503$ which has an $\alpha = 0.8564$, indicating that we fail to reject the null hypothesis of normality. We can therefore proceed with the ANOVA. Since we will be performing contrasts, we need to use PROC GLM and not PROC ANOVA.

The ANOVA considering the '20-treatment' approach:

```
PROC GLM DATA = one PLOTS = none;
CLASSES block treatment;
MODEL cases = block treatment;
CONTRAST '(1) Romaine vs Leaf'
treatment
 1 1 1 1 1 1 1 1 1 1 -1 -1 -1 -1 -1 -1 -1 -1 -1 -1;
CONTRAST '(2) Within Romaine do CVs differ'
treatment
```

```
   1  1  1  1  1  -1  -1  -1  -1  -1  0  0  0  0  0  0  0  0  0  0;
CONTRAST '(3) Within Leaf do CVs differ'
treatment
   0  0  0  0  0  0  0  0  0  0  1  1  1  1  1  -1  -1  -1  -1  -1;
CONTRAST '(4) Foreign vs Domestic'
treatment
   3  3  -2  -2  -2  3  3  -2  -2  -2  3  3  -2  -2  -2  3  3  -2  -2  -2;
CONTRAST '(5) Within Foreign do COs differ'
treatment
   1  -1  0  0  0  1  -1  0  0  0  1  -1  0  0  0  1  -1  0  0  0;
CONTRAST '(6) Within Domestic East vs West'
treatment
   0  0  1  -2  1  0  0  1  -2  1  0  0  1  -2  1  0  0  1  -2  1;
CONTRAST '(7) Johnny vs Burpee'
treatment
   0  0  1  0  -1  0  0  1  0  -1  0  0  1  0  -1  0  0  1  0  -1;
RUN;
```
The output:

'Chapter 12 Example 3.sas'
'RCBD, 5 blocks 20 treatments'
'Lettuce cultivar and seed source evaluation'

The GLM Procedure

Class Level Information		
Class	Levels	Values
block	5	1 2 3 4 5
treatment	20	1 2 3 4 5 6 7 8 9 10 11 12 13 14 15 16 17 18 19 20

Number of Observations Read	100
Number of Observations Used	100

'Chapter 12 Example 3.sas'
'RCBD, 5 blocks 20 treatments'
'Lettuce cultivar and seed source evaluation'

The GLM Procedure

Dependent Variable: cases

Source	DF	Sum of Squares	Mean Square	F Value	Pr > F
Model	23	602040.750	26175.685	0.59	0.9248
Error	76	3391108.000	44619.842		
Corrected Total	99	3993148.750			

R-Square	Coeff Var	Root MSE	cases Mean
0.150768	29.79114	211.2341	709.0500

Source	DF	Type I SS	Mean Square	F Value	Pr > F
block	4	50927.2000	12731.8000	0.29	0.8866
treatment	19	551113.5500	29005.9763	0.65	0.8544

Source	DF	Type III SS	Mean Square	F Value	Pr > F
block	4	50927.2000	12731.8000	0.29	0.8866
treatment	19	551113.5500	29005.9763	0.65	0.8544

Contrast	DF	Contrast SS	Mean Square	F Value	Pr > F
(1) Romaine vs Leaf	1	104264.4100	104264.4100	2.34	0.1305
(2) Within Romaine do CVs differ	1	67344.5000	67344.5000	1.51	0.2230
(3) Within Leaf do CVs differ	1	142791.6800	142791.6800	3.20	0.0776
(4) Foreign vs Domestic	1	9922.6667	9922.6667	0.22	0.6386
(5) Within Foreign do COs differ	1	2924.1000	2924.1000	0.07	0.7986
(6) Within Domestic East vs West	1	26.1333	26.1333	0.00	0.9808
(7) Johnny vs Burpee	1	739.6000	739.6000	0.02	0.8979

So what does this output mean? First of all, when we examine our F-value for treatment we see that we have a value of 0.65 with $\alpha = 0.8544$. Thus we fail to reject the null hypothesis of equal treatment means. In other words, there was no treatment effect. However, we posed some *a priori* questions; thus we should look at the Prob > F values of our contrasts to answer the specific questions we posed. We should look at these probabilities even when our overall treatment F is non-significant.

1. Is romaine any different in productivity compared with leaf lettuce? No, $F = 2.34$, $\alpha = 0.1305$.
2. Within romaine, do the cultivars differ? No, $F = 1.51$, $\alpha = 0.2230$.
3. Within leaf, do the cultivars differ? No, $F = 3.20$, $\alpha = 0.0776$.
4. Is there a difference between foreign and domestic seed companies? No, $F = 0.22$, $\alpha = 0.6386$.
5. Within foreign companies, do the two companies differ? No, $F = 0.07$, $\alpha = 0.7986$.

6. Within domestic companies, are East coast companies (Johnny's and Burpee) different from the West coast (Territorial) company? No, $F = 0.00$, $\alpha = 0.9808$.

7. Is Johnny's different than Burpee's? No, $F = 0.02$, $\alpha = 0.8979$.

The answer to all our questions is 'No'. The α level for each contrast is > 0.05, thus the likelihood that the answer to any of our questions is 'Yes' is very low. All lettuces produced an average of 709 cases per plot.

We could insert a 'MEANS treatment' statement in our program if we wanted the mean value for each of our 20 treatments, but we don't need them for this example.

When researchers run an experiment and they detect no significant main effects, interactions or contrasts, they often think that they ran a poor experiment. This is not a good thought process for approaching any research problem, since it is just as valuable to detect that there are no differences among treatments as it is that there are.

Let's look at the experiment as a factorial structure rather than 20 treatments and see what we get.

Then we run the ANOVA with contrasts using PROC GLM again, with different MODEL and CONTRAST statements to reflect our new approach:

```
PROC GLM DATA = one PLOTS = none;
CLASSES block cultivar ssource;
MODEL cases = block cultivar ssource cultivar*ssource;
CONTRAST '(1) Romaine vs Leaf' cultivar 1 -1 1 -1;
CONTRAST '(2) Within Romaine do CVs differ' cultivar 1 0 -1 0;
CONTRAST '(3) Within Leaf do CVs differ' cultivar 0 -1 0 1;
CONTRAST '(4) Foreign vs Domestic' ssource 3 3 -2 -2 -2;
CONTRAST '(5) Within Foreign do COs differ' ssource 1 -1 0 0 0;
CONTRAST '(6) Within Domestic East vs West' ssource 0 0 1 -2 1;
CONTRAST '(7) Johnny vs Burpee' ssource 0 0 1 0 -1;
RUN;
```

The results:

The GLM Procedure

Dependent Variable: cases

Source	DF	Sum of Squares	Mean Square	F Value	Pr > F
Model	23	602040.750	26175.685	0.59	0.9248
Error	76	3391108.000	44619.842		
Corrected Total	99	3993148.750			

R-Square	Coeff Var	Root MSE	cases Mean
0.150768	29.79114	211.2341	709.0500

Source	DF	Type I SS	Mean Square	F Value	Pr > F
block	4	50927.2000	12731.8000	0.29	0.8866
cultivar	3	314400.5900	104800.1967	2.35	0.0792
ssource	4	13612.5000	3403.1250	0.08	0.9893
cultivar*ssource	12	223100.4600	18591.7050	0.42	0.9526

Source	DF	Type III SS	Mean Square	F Value	Pr > F
block	4	50927.2000	12731.8000	0.29	0.8866
cultivar	3	314400.5900	104800.1967	2.35	0.0792
ssource	4	13612.5000	3403.1250	0.08	0.9893
cultivar*ssource	12	223100.4600	18591.7050	0.42	0.9526

Contrast	DF	Contrast SS	Mean Square	F Value	Pr > F
(1) Romaine vs Leaf	1	104264.4100	104264.4100	2.34	0.1305
(2) Within Romaine do CVs differ	1	67344.5000	67344.5000	1.51	0.2230
(3) Within Leaf do CVs differ	1	142791.6800	142791.6800	3.20	0.0776
(4) Foreign vs Domestic	1	9922.6667	9922.6667	0.22	0.6386
(5) Within Foreign do COs differ	1	2924.1000	2924.1000	0.07	0.7986
(6) Within Domestic East vs West	1	26.1333	26.1333	0.00	0.9808
(7) Johnny vs Burpee	1	739.6000	739.6000	0.02	0.8979

Hey! The contrast results are exactly like the first analysis when we looked at the experiment as 20 treatments. Both ways are acceptable, but the second approach provides more information. The second analysis explicitly shows that there was no interaction between cultivar and seed company. In the first analysis, we could only infer it by looking at the 'leftover' SS if we subtracted all the contrast SS from the treatment SS and then tested the leftovers with error. But we won't even do that, since the second method is the method of choice, and it saves us from hand calculations.

Let's plug the numbers into our formula for determining whether or not blocking was effective. If blocking was effective, great. If not, a similar experiment may need blocking by some other factor than the one we used in this experiment or it may not need blocking at all. Here are our calculations:

$$\frac{(b-1)*MS_{Blocks}RCBD + b(t-1)*MS_{Error}RCBD}{(bt-1)*MS_{Error}RCBD}$$

$b = 5$
$t = 20$
$MS_{Blocks} = 12731.8$
$MS_{Error} = 44619.842$

thus:

$$RE = \frac{(4)*12731.8 + 5(19)*44619.842}{(99)*44619.842} = \frac{(4289812.19)}{(4417364.358)} = 0.97$$

Since our RE was nearly 1, this indicates that the design was as efficient as a CRD design in controlling variability. In other words, blocking wasn't really that effective and we really didn't need to do it.

In this chapter we learned how to set up a randomized complete block experiment and compare its efficiency with the CRD. We also learned how to run contrasts within a factorial treatment structure. In the next chapter we will examine the Latin square design.

13 The Latin Square Design

In the RCBD, we accounted for a known source of variation among experimental units to remove it from our estimate of experimental error. If two known sources of variation exist, the Latin square design (LS) is often a good choice with limitations on the number of treatments you can have. In essence, you are 'double blocking' with this design. We refer to the 'blocking' factors as rows and columns.

For example, suppose our experimental units were potted peach trees. Some are in 1-gallon pots others are in 2-gallon pots. We could block according to pot size. Also suppose that some of the trees are 2 years old while some are 3 years old. We could block according to tree age as well. Thus we could account for variation due to pot size and to tree age at the same time in the same experiment. The blocking may also be based on spatial associations or on the order of treatment application.

In the LS, each treatment occurs the same number of times (usually once) in each column and the same number of times (usually once) in each row. This puts a restriction on the use of this design. The number of treatments is limited by the number of rows and the number of columns. In other words, # treatments = # columns = # rows. When there are large numbers of treatments, the experiment gets too large. When there are a small number of treatments, there aren't enough df to adequately estimate experimental error. Thus the LS is optimally suited to experiments where there are between four and eight treatments.

PROC PLAN for an LS

PROC PLAN can be used to generate random treatment assignments for an LS. The program is a bit more complicated than for most other designs, but it beats having to generate random assignments by hand. This example is for an LS with four treatments:

```
01 TITLE1 'Chapter 13 Example 1.sas';
02 TITLE2 'Latin Square Design';
03 TITLE3 'PROC PLAN Treatment Assignment';
04 TITLE4 'Four treatments';
05 PROC PLAN;
06 FACTORS Row=4 ORDERED Col=4 ORDERED / NOPRINT;
07 TREATMENTS Tmt=4 CYCLIC;
08 OUTPUT OUT=LatinSquare
09 Row CVALS=('Row 1' 'Row 2' 'Row 3' 'Row 4') RANDOM
10 Col CVALS=('Col 1' 'Col 2' 'Col 3' 'Col 4') RANDOM
11 Tmt NVALS=( 1 2 3 4 ) RANDOM;
12 QUIT;
13 PROC SORT DATA=LatinSquare OUT=LatinSquare;
```

DOI: 10.1079/9781789249927.0013

```
14 BY Row Col;
15 PROC TRANSPOSE DATA = LatinSquare(RENAME=(Col=_NAME_))
16 OUT =tLatinSquare(DROP=_NAME_);
17 BY Row;
18 VAR Tmt;
19 PROC PRINT DATA=tLatinSquare NOOBS;
20 RUN;
```

The output:

'Chapter 13 Example 1.sas'
Latin Square Design
'PROC PLAN Treatment Assignment'

Row	Col_1	Col_2	Col_3	Col_4
Row 1	1	3	2	4
Row 2	3	4	1	2
Row 3	2	1	4	3
Row 4	4	2	3	1

To modify the program for different numbers of treatments, follow these guidelines:

1. Adjust line 04 to indicate the correct number of treatments.
2. Adjust line 06 to indicate the correct number of rows and columns.
3. Adjust line 07 to indicate the correct number of treatments.
4. Adjust lines 09, 10 and 11 to reflect the correct numbers of rows, columns and treatments inside the parentheses following the CVALS, CVALS and NVALS keywords.
5. Changes needed are boldface in the following program.

```
01 TITLE1 'Chapter 13 Example 1.sas';
02 TITLE2 'Latin Square Design';
03 TITLE3 'PROC PLAN Treatment Assignment';
04 TITLE4 'Four treatments';
05 PROC PLAN;
06 FACTORS Row=4 ORDERED Col=4 ORDERED / NOPRINT;
07 TREATMENTS Tmt=4 CYCLIC;
08 OUTPUT OUT=LatinSquare
09 Row CVALS=('Row 1' 'Row 2' 'Row 3' 'Row 4') RANDOM
10 Col CVALS=('Col 1' 'Col 2' 'Col 3' 'Col 4') RANDOM
11 Tmt NVALS=( 1 2 3 4 ) RANDOM;
12 QUIT;
13 PROC SORT DATA=LatinSquare OUT=LatinSquare;
14 BY Row Col;
15 PROC TRANSPOSE DATA = LatinSquare(RENAME=(Col=_NAME_))
16 OUT =tLatinSquare(DROP=_NAME_);
17 BY Row;
18 VAR Tmt;
19 PROC PRINT DATA=tLatinSquare NOOBS;
20 RUN;
```

An adjusted program for an LS with seven treatments is shown below.

```
01 TITLE1 'Chapter 13 Example 2.sas';
02 TITLE2 'Latin Square Design';
```

```
03 TITLE3 'PROC PLAN Treatment Assignment';
04 TITLE4 'Seven treatments';
05 PROC PLAN;
06 FACTORS Row=7 ORDERED Col=7 ORDERED / NOPRINT;
07 TREATMENTS Tmt=7 CYCLIC;
08 OUTPUT OUT=LatinSquare
09 Row CVALS=('Row 1' 'Row 2' 'Row 3' 'Row 4' 'Row 5' 'Row 6'
   'Row 7' ) RANDOM
10 Col CVALS=('Col 1' 'Col 2' 'Col 3' 'Col 4' 'Col 5' 'Col 6'
   'Col 7' ) RANDOM
11 Tmt NVALS=( 1 2 3 4 5 6 7 ) RANDOM;
12 QUIT;
13 PROC SORT DATA=LatinSquare OUT=LatinSquare;
14 BY Row Col;
15 PROC TRANSPOSE DATA = LatinSquare(RENAME=(Col=_NAME_))
16 OUT =tLatinSquare(DROP=_NAME_);
17 BY Row;
18 VAR Tmt;
19 PROC PRINT DATA=tLatinSquare NOOBS;
20 RUN;
```

The output:

'Chapter 13 Example 2.sas'
Latin Square Design
'PROC PLAN Treatment Assignment'
'Seven treatments'

Row	Col_1	Col_2	Col_3	Col_4	Col_5	Col_6	Col_7
Row 1	2	5	7	3	4	1	6
Row 2	3	6	1	7	5	4	2
Row 3	7	2	4	1	6	5	3
Row 4	1	3	5	4	2	6	7
Row 5	6	4	3	2	1	7	5
Row 6	5	1	2	6	7	3	4
Row 7	4	7	6	5	3	2	1

If you want to make the output a bit more informative, you can change the appropriate row, column and treatment values as illustrated below in the example for a three-treatment LS where rows are pot size, columns are plant age and the three treatments are rates of the growth regulator ethephon, indicated by Eth 0 ppm, Eth 250 ppm and Eth 500 ppm. Note the NVALS has been changed to CVALS in line 12 to indicate that treatment value assignments are character based rather than numeric. This is a poor LS due to the number of df available for estimating error (we'll get to that in a moment), but I use it to illustrate how to embellish output.

```
01 TITLE1 'Chapter 13 Example 3.sas';
02 TITLE2 'Latin Square Design';
03 TITLE3 'PROC PLAN Treatment Assignment';
04 TITLE4 'Three ethephon treatments';
05 TITLE5 'Row = pot size and Col = plant age';
```

```
06 PROC PLAN;
07 FACTORS Row=3 ORDERED Col=3 ORDERED / NOPRINT;
08 TREATMENTS Tmt=3 CYCLIC;
09 OUTPUT OUT=LatinSquare
10 Row CVALS=('1 litre' '2 litre' '3 litre' ) RANDOM
11 Col CVALS=('1 year' '2 year' '3 year' ) RANDOM
12 Tmt CVALS=( '0 ppm' '250 ppm' '500 ppm' ) RANDOM;
13 QUIT;
14 PROC SORT DATA=LatinSquare OUT=LatinSquare;
15 BY Row Col;
16 PROC TRANSPOSE DATA = LatinSquare(RENAME=(Col=_NAME_))
17 OUT =tLatinSquare(DROP=_NAME_);
18 BY Row;
19 VAR Tmt;
20 PROC PRINT DATA=tLatinSquare NOOBS;
21 RUN;
```

The output:

'Chapter 13 Example 3.sas'
Latin Square Design
'PROC PLAN Treatment Assignment'
'Three ethephon treatments'
'Row = pot size and Col = plant age'

Row	_1_year	_2_year	_3_year
1 litre	500 ppm	250 ppm	0 ppm
2 litre	0 ppm	500 ppm	250 ppm
3 litre	250 ppm	0 ppm	500 ppm

The LS ANOVA

In the LS ANOVA, we must account for one more source of variation than we did in the RCBD, namely, that variation due to rows (if we assume columns are like the blocks in the RCBD). The source and df columns for the ANOVA table would look like:

Source	df
Total	(N)
μ	(1)
Treatment	$(t-1)$
Row	$(r-1)$
Column	$(c-1)$
Error	$(t-1)(t-2)$

where t is the number of treatments, r is the number of rows and c is the number of columns. Remember, there are equal numbers of treatments, rows and columns, thus $t = r = c$. Of course, you would also have SS, MS, F and Prob > F columns in the final ANOVA table.

Relative Efficiency of the Latin Square

RE of LS to CRD

We define the RE of the LS to the CRD as:

$$\frac{\left(MSRow_{LS} + MSColumn_{LS} + (t-1)MSErr_{LS} \right)}{(t+1)MSError_{LS}}$$

The numerator is an estimate of the error you would have gotten in a CRD, and the denominator is the error estimate from the LS. The multipliers in both the numerator and denominator are needed to account for the differences in the number of df available for estimating error in the two different designs.

Again, an example helps to illustrate the interpretation of the RE. Suppose we calculated an RE of 3.25. This would mean that for a similar experiment using a CRD, you would need about three times as many replicates as in the LS to have the same precision in detecting differences. Similarly, if the RE was 0.5, then you would only need half the number of replicates in a CRD to have the same precision as in the LS. This would occur when there were no column or row factor effects. If there really are no row or column effects, you lose much precision in the LS due to many fewer df in the LS error estimate compared with the CRD error estimate (since $(r-1)$ df were used for the row effect and $(c-1)$ df were used for the column effect in the LS).

RE of LS to RCBD

We can also determine the relative efficiency of the LS to an RCBD. There are two formulas for estimating this RE. If the rows of the LS would be the block in the RCBD, we define the RE of the LS to the RCBD as:

$$\frac{(MSRow_{LS} + (t-1)MSErr_{LS})}{(t)MSError_{LS}}$$

If the columns of the LS would be the block in the RCBD, we define the RE of the LS to an RCBD as:

$$\frac{(MSColumn_{LS} + (t-1)MSErr_{LS})}{(t)MSError_{LS}}$$

In each formula, the numerator is an estimate of the error you would have gotten in an RCBD, and the denominator is the error estimate from the LS. The only difference in the two formulas is whether you consider rows or columns as the blocking factor in an RCBD. The multipliers in both the numerator and denominator are needed as explained before.

The interpretation of the RE should be clear from the previous examples. If there is either no row or no column effect, you lose precision in the LS due to fewer df in the LS error estimate compared with the RCBD error estimate (since either $(r-1)$ or $(c-1)$ df were used for the row or column effect).

If the df errors in the LS ANOVA are less than 20, then the RE calculations above must be multiplied by the factor K. K is defined as:

$$\frac{\left[(t-1)(t-2)+1\right]\left[(t-1)^2+3\right]}{\left[(t-1)(t-2)+3\right]\left[(t-1)^2+1\right]}$$

The CRD, RCBD and LS ANOVA Tables for the Same Experiment

Suppose we had an experiment with two cultivars, four rates of N, and eight reps. Let's look at the ANOVAs for a CRD, RCBD and an LS design. Only one of these designs would likely be appropriate for the experimental conditions under consideration, but we present all three for illustration.

CRD ANOVA

Source	df
Total	64
μ	1
Treatment	7
Cultivar	1
Rate of N	3
Cultivar * Rate of N	3
Error	56

RCBD ANOVA

Source	df
Total	64
μ	1
Block	7
Treatment	7
Cultivar	1
Rate of N	3
Cultivar * Rate of N	3
Error	49

LS ANOVA

Source	df
Total	64
μ	1
Row	7
Column	7
Treatment	7
Cultivar	1
Rate of N	3
Cultivar * Rate of N	3
Error	42

Notice the differences among the designs with respect to the df in the error term. With the more complex design, you have 'paid the price' for greater complexity by having fewer df in the error.

A Latin Square Example

Now let's consider an example Latin square experiment. We are going to consider a lab-based example since, even though Latin squares are useful in field and greenhouse situations, they are also useful for lab experiments. Suppose we are investigating the vegetative to floral transition of the apical meristem in main crowns of strawberry plants. We have so many apices to examine that we will need five days to complete our work. We also know that as the day wears on, we grow tired and may not be as accurate later in the day as we are in the morning, thus we want to account for the time of day. We have five treatments we want to consider: (1) a control; (2) long days (16 hours) at 25°C; (3) long days at 10°C; (4) short days (8 hours) at 25°C; and (5) short days at 10°C. Our samples will come from plants that have been growing in different greenhouses. We will randomly sample 100 plants each day, 20 from each treatment, which will be evaluated during their allotted time.

We will determine floral or vegetative status of the main crown terminal meristem, thus we have a proportion that are floral for each observation. Since this is a 5 × 5 Latin square we will have 25 observations. While proportions are often evaluated with statistical methods other than an ANOVA, we are approaching this experiment assuming the proportions are normally distributed. Of course, we will test this hypothesis of normality and see what happens.

We can set up our experiment as a Latin square if we divide each day's work schedule into five segments, thus we'd have five rows (days), five columns (time of day) and five treatments. Once we establish our experimental protocols, we need a sampling plan for our design. We create it with the following SAS® program.

```
01  TITLE1 'Chapter 13 Example 4.sas';
02  TITLE2 'Latin Square Design';
03  TITLE3 'PROC PLAN Sampling Protocol';
04  TITLE4 'Strawberry Experiment';
05  TITLE5 '5 Treatments';
06  TITLE6 'Row = day of the week and Col = time of day';
07  PROC PLAN;
08  FACTORS Row=5 ORDERED Col=5 ORDERED / NOPRINT;
09  TREATMENTS Tmt=5 CYCLIC;
10  OUTPUT OUT=LatinSquare
11  Row CVALS=('Monday' 'Tuesday' 'Wednesday' 'Thursday' 'Friday' )
    RANDOM
12  Col CVALS=('8 to 9:30am' '9:30 to 11am' '11am to 12:30pm' '1 to
    2:30pm' '2:30 to 4pm' ) RANDOM
13  Tmt CVALS=( 'Control' 'LD25' 'LD10' 'SD25' 'SD10' ) RANDOM;
14  QUIT;
15  PROC SORT DATA=LatinSquare OUT=LatinSquare;
16  BY Row Col;
17  PROC TRANSPOSE DATA = LatinSquare(RENAME=(Col=_NAME_))
18  OUT =tLatinSquare(DROP=_NAME_);
```

```
19  BY Row;
20  VAR Tmt;
21  PROC PRINT DATA=tLatinSquare NOOBS;
22  RUN;
```

Our sampling plan is:

'Chapter 13 Example 4.sas'
Latin Square Design
'PROC PLAN Sampling Protocol'
'Strawberry Experiment'
'5 Treatments'
'Row = day of the week and Col = time of day'

Row	_1_to_2_30pm	_11am_to_12_30pm	_2_30_to_4pm	_8_to_9_30am	_9_30_to_11am
Friday	Control	SD10	LD10	LD25	SD25
Monday	LD25	SD25	Control	SD10	LD10
Thursday	SD10	LD10	LD25	SD25	Control
Tuesday	LD10	LD25	SD25	Control	SD10
Wednesday	SD25	Control	SD10	LD10	LD25

We evaluate our apices and collect the following data:

Day of the week	Time of day	Treatment	% floral
Monday	8 to 9:30am	SD10	50
	9:30 to 11am	LD10	78
	11am to 12:30pm	SD25	12
	1 to 2:30pm	LD25	91
	2:30 to 4pm	Control	25
Tuesday	8 to 9:30am	Control	34
	9:30 to 11am	SD10	44
	11am to 12:30pm	LD25	87
	1 to 2:30pm	LD10	75
	2:30 to 4pm	SD25	22
Wednesday	8 to 9:30am	LD10	69
	9:30 to 11am	LD25	95
	11am to 12:30pm	Control	38
	1 to 2:30pm	SD25	31
	2:30 to 4pm	SD10	51
Thursday	8 to 9:30am	SD25	17
	9:30 to 11am	Control	26
	11am to 12:30pm	LD10	80
	1 to 2:30pm	SD10	42
	2:30 to 4pm	LD25	89
Friday	8 to 9:30am	LD25	76
	9:30 to 11am	SD25	24
	11am to 12:30pm	SD10	46
	1 to 2:30pm	Control	30
	2:30 to 4pm	LD10	79

The SAS® program for our analysis:

```
01  TITLE1 'Chapter 13, Example 5.sas';
02  TITLE2 'Latin Square';
```

```
03  DATA one;
04  INPUT dow $ tod $ trt $ pfloral;
05  CARDS;
06  Mon 8to930 SD10 55
07  Mon 930to11 LD10 73
08  Mon 11to1230 SD25 12
09  Mon 1to230 LD25 89
10  Mon 230to4 Control 25
11  Tues 8to930 Control 34
12  Tues 930to11 SD10 44
13  Tues 11to1230 LD25 87
14  Tues 1to230 LD10 75
15  Tues 230to4 SD25 22
16  Wed 8to930 LD10 69
17  Wed 930to11 LD25 95
18  Wed 11to1230 Control 38
19  Wed 1to230 SD25 31
20  Wed 230to4 SD10 51
21  Thurs 8to930 SD25 17
22  Thurs 930to11 Control 26
23  Thurs 11to1230 LD10 80
24  Thurs 1to230 SD10 42
25  Thurs 230to4 LD25 89
26  Fri 8to930 LD25 76
27  Fri 930to11 SD25 24
28  Fri 11to1230 SD10 46
29  Fri 1to230 Control 30
30  Fri 230to4 LD10 79
31  RUN;
32  PROC UNIVARIATE NORMAL;
33  VAR pfloral;
34  PROC ANOVA DATA = one PLOTS = none;
35  CLASSES trt dow tod;
36  MODEL pfloral = trt dow tod;
37  MEANS trt;
38  RUN;
```

Our normality test:

Tests for Normality				
Test		Statistic	p Value	
Shapiro-Wilk	W	0.921708	Pr < W	0.0560
Kolmogorov-Smirnov	D	0.142848	Pr > D	>0.1500
Cramer-von Mises	W-Sq	0.120998	Pr > W-Sq	0.0559
Anderson-Darling	A-Sq	0.724928	Pr > A-Sq	0.0511

The Shapiro–Wilk statistic of 0.92 fails to reject the null hypothesis of normality at the 0.06 level, thus we can proceed with our analysis.

‘Chapter 13 Example 5.sas’
‘Latin Square’

The ANOVA Procedure

Dependent Variable: pfloral

Source	DF	Sum of Squares	Mean Square	F Value	Pr > F
Model	12	16177.68000	1348.14000	29.52	<.0001
Error	12	548.08000	45.67333		
Corrected Total	24	16725.76000			

R-Square	Coeff Var	Root MSE	pfloral Mean
0.967231	12.90719	6.758205	52.36000

Source	DF	Anova SS	Mean Square	F Value	Pr > F
trt	4	16012.96000	4003.24000	87.65	<.0001
dow	4	132.16000	33.04000	0.72	0.5925
tod	4	32.56000	8.14000	0.18	0.9453

Our treatment *F*-value of 87.65 rejects the null hypothesis of equal treatment means at the 0.001 level. Since we did not pose any *a priori* questions, and we don't need confidence intervals, I would choose to separate the means using the REGWQ option.

‘Chapter 13 Example 5.sas’
‘Latin Square’

The ANOVA Procedure

Ryan-Einot-Gabriel-Welsch Multiple Range Test for pfloral

Note: This test controls the Type I experimentwise error rate.

Alpha	0.05
Error Degrees of Freedom	12
Error Mean Square	45.67333

Number of Means	2	3	4	5
Critical Range	11.423702	12.609539	12.689416	13.623788

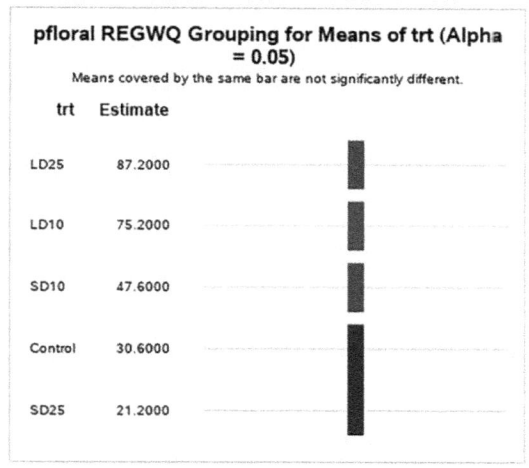

pfloral REGWQ Grouping for Means of trt (Alpha = 0.05)

Means covered by the same bar are not significantly different.

trt	Estimate
LD25	87.2000
LD10	75.2000
SD10	47.6000
Control	30.6000
SD25	21.2000

Interpretation of the mean separation is straightforward. The controls and SD25-treated apices had the lowest percentage of apices which were floral and they were not different from each other. Apices from plants grown under the SD10 treatment had the next highest floral values, followed by LD10 and LD25. All three treatments were significantly different from each other and from both of the other treatments.

We might want to calculate relative efficiencies to determine whether or not our column and row blocking was effective in case we planned on repeating this experiment.

RE of LS to CRD

We define the RE of the LS to the CRD as:

$$\frac{\left(MSRow_{LS} + MSColumn_{LS} + (t-1)MSErr_{LS}\right)}{(t+1)MSError_{LS}}$$

thus:

$$RE = \frac{\left(33.04 + 8.14 + (4*45.67)\right)}{6*45.67} = \frac{223.86}{274.02} = 0.82$$

We must adjust this value with the K factor since our df error () is < 20.

$$K = \frac{[(t-1)(t-2)+1][(t-1)^2 + 3]}{[(t-1)(t-2)+3][(t-1)^2 + 1]}$$

$$\frac{[(4*3)+1]*[4^2 + 3]}{[(4*3)+3]*[4^2 + 1]} = \frac{247}{255} = 0.97$$

Our RE of (0.82*0.97) = 0.79 indicates that you would only need 0.79, the number of replicates in a CRD, to have the same precision as in the LS. In other words, if you repeated this experiment using a CRD you would only need (0.79*5) = 4 replicates. Even so, I would still recommend that you use as many replicates as you can afford and that fit your experimental approach and design. The RE value of 0.79 is useful in the sense that it informs you that your blocking schemes with row and column were not effective. For our example, it indicates that there was not much variability associated with the day of the week or the time of day for dissection.

RE of LS to RCBD

We could also determine the relative efficiency of the LS to an RCBD. There are two formulas for estimating this RE. If the rows (day of the week) of the LS would be the block in the RCBD, we define the RE of the LS to the RCBD as:

$$\frac{\left(MSRow_{LS} + (t-1)MSErr_{LS}\right)}{(t)MSError_{LS}}$$

$$RE = \frac{\left(33.04 + (4*45.67)\right)}{(5*45.67)} = \frac{215.72}{228.35} = 0.94$$

And since df error < 20, multiply by K, thus RE = (0.94*0.97) = 0.91

If the columns (time of day) of the LS would be the block in the RCBD, we define the RE of the LS to a RCBD as:

$$\frac{\left(MSColumns_{LS} + (t-1)MSErr_{LS}\right)}{(t)MSError_{LS}}$$

$$RE = \frac{(8.14 + (4*45.67))}{(5*45.67)} = \frac{190.82}{228.35} = 0.84$$

And since df error < 20, multiply by K, thus RE = (0.84*0.97) = 0.81

The interpretation of the RE should be clear from the previous examples. If there is either no row or no column effect, you lose precision in the LS due to fewer df in the LS error estimate compared with the RCBD error estimate (since either $(r-1)$ or $(c-1)$ df were used for the row or column effect).

So far we have examined three very useful experimental designs for holding our treatments. In the next chapter, we will cover one of the most useful designs in agricultural research: the split-plot with variations. These split designs are particularly effective for factorial experiments.

14 The Split-plot and Split-split-plot Designs

The Split-plot

One of the most flexible and useful experimental designs for agricultural research is the split-plot and its variations. This group of designs is extremely convenient for holding factorial treatment structures. Previously discussed designs such as the CRD and the RCBD are useful for single-factor experiments or experiments where factorial combinations are easy to apply to the experimental units. Many agricultural factors are difficult to apply to experimental units of the size normally encountered in research situations. These factors might include irrigation, fertilization, pest control, mulch application or harvesting. They are often more easily applied to larger units. While CRD, RCBD or LS designs can be used for these experiments, the split-plot or one of its variations is usually much more effective and efficient in accounting for variation and for easing treatment application.

The best way to learn the split-plot design is with an example. Let's consider an experiment with strawberry production and see how useful this design is. Suppose we were interested in growing six different cultivars of strawberries on either raised or flat beds. This is a simple 2 × 6 factorial treatment structure with 12 experimental treatments. We have enough space to include five replications, thus we have a total of 60 experimental units, or in this case, 60 ten-foot plots. Since we always want to use the simplest design possible, let's consider putting this factorial in a CRD. This would mean that we would have to randomly assign 60 bed height/cultivar treatment combinations to the 60 experimental units. We are assuming that there is no reason to block. The ANOVA table for this experiment would be:

Source	df
Total	60
μ	1
Bed height (BH)	1
Cultivar (CV)	5
BH × CV	5
Error	48

But what if we needed to account for a gradient in our field? We would then need to consider the RCBD. We would randomly assign 12 treatments in each of five blocks for a total of 60 plots. Remember that the randomization of the 12 treatments is performed separately for each block. Our ANOVA table would now reflect this blocking:

Source	df
Total	60
μ	1

Continued

DOI: 10.1079/9781789249927.0014

Continued.

Source	df
Block (Blk)	4
Bed height (BH)	1
Cultivar (CV)	5
BH × CV	5
Error	44

This seems reasonable enough. But now consider the process of making the beds of two different heights, raised and flat. Due to the equipment involved, it would be very difficult to make ten-foot raised and flat beds. Let's consider a simple experiment with five blocks and only the bed height factor. We would randomly assign raised or flat beds in each of the five blocks. Now we can much more easily make a raised bed and a flat bed in each block, resulting in a total of ten rather large plots. Even though it seems quite limited, we do have a valid RCBD experiment with two treatments and five blocks. Our ANOVA table would be:

Source	df
Total	10
μ	1
Block (Blk)	4
Bed height (BH)	1
Error	4

This experiment is not very powerful, due to the limited df available for estimating error. Even so, we would be able to test whether or not bed height had an effect on any dependent variable and we have solved the problem of making the beds. However, we have not included cultivar in our experiment, and that was one of the factors we were originally interested in.

What if our bed height plots are large enough to fit six cultivar plots within each? Within each block, you have a raised and a flat bed. Each would have six cultivars randomly assigned to them. If you think about it, you now have a 'mini' RCBD inside each block. The 'blocks' within each 'mini' are the bed heights. Thus you would have a total of 12 plots in the original block 1, 12 plots in block 2, etc. You have split the bed height plots into six sub-plots. The ANOVA table for each 'mini' experiment would be:

Source	df
Total	12
μ	1
Block (really bed ht)	1
Cultivar (CV)	5
Error	5

This isn't a very good experiment for testing cultivars because it's not big enough. But consider that you have five of these 'mini' RCBD's . What if we can combine the information in all five of them into one coherent analysis that considers both factors, bed height and cultivar?

Before we do this, we must consider something else. Our cultivar plots are 10 ft long and our bed height plots are 60 ft long. Statistically, larger plots tend to be more variable than smaller plots. Thus there is likely to be more variability associated with the bed height plots and less variability associated with the cultivar plots, simply due to their different sizes. We want to be able to use the correct measure of variability for testing our different sized plots.

This is the split-plot design. A set of main plots (in this case the bed heights) are arranged in an acceptable design, usually the RCBD. A second set of treatments (in this case, cultivar) are randomly assigned to each of the main plots with different randomizations for each block; in our example, we would have ten unique randomizations for cultivar within main plot (two main plot randomizations in each of five blocks). Since we have two different plot sizes, we have two different estimates of error for testing effects. One error (Error a) is associated with the larger main plots and the other is associated with the smaller sub-plots (Error b).

In a split-plot analysis we are able to test for a bed height effect, a cultivar effect and any interaction between these two factors. Two-factor experiments are easily accommodated in a split-plot design.

Before we see how this is accomplished in an ANOVA, let's take a look at some important features of the split-plot design. The larger plot is called the main plot and main plot treatments are randomly assigned within each block, each block randomized separately. The smaller plot is called the sub-plot and sub-plot treatments are randomly assigned within each main plot. For sub-plot treatment assignments, you always have (number of replicates or blocks × number of main plot treatments) random assignments. Thus in our example you would have (5 blocks × 2 bed heights) or 10 separate cultivar randomizations. More statistical information is contained in the sub-plot analysis and less in the main plot analysis. Main plots can be arranged in any design (for us that includes CRD, RCBD and LS), but they are usually arranged in an RCBD.

Many researchers often consider time (i.e. multiple harvests over time, or observations taken over time) as a sub-plot in a split-plot experiment. This assumes that there is no correlation over time but this assumption is not often valid. Correlation over time means that the observation of an experimental unit at one time may influence an observation on that same experimental unit at another time. For example, a heavy harvest often follows a light harvest and vice versa. You could almost predict whether you will have a heavy or light harvest based on the previous harvest. Harvest dates or observations taken over time are more appropriately handled with a split-block design. A reference for considering a split block is provided at the end of this chapter.

Treatment assignments are obtained using PROC PLAN as such:

```
TITLE1 'Chapter 14 Example 1.sas';
TITLE2 'Split-plot Design';
TITLE3 'PROC PLAN Treatment Assignment';
TITLE4 '2 Main plots, 6 sub-plots and 5 blocks';
PROC PLAN;
FACTORS Block=5 ORDERED BedHeight=2 Cultivar=6;
RUN;
```

The output:

The PLAN Procedure

Factor	Select	Levels	Order
Block	5	5	Ordered
BedHeight	2	2	Random
Cultivar	6	6	Random

Block	BedHeight	Cultivar						
1	2	2	1	4	5	3	6
	1	3	2	1	5	4	6
2	2	3	1	2	4	5	6
	1	2	3	4	5	1	6
3	2	4	3	6	1	5	2
	1	1	6	5	2	3	4
4	2	6	1	3	2	5	4
	1	4	1	6	3	2	5
5	2	5	6	3	4	2	1
	1	5	6	3	1	4	2

Now let's examine the ANOVA table for our split-plot example. In the ANOVA table you will notice two errors, 'Error a' and 'Error b'. Remember that they are associated with the two different-sized plots we have in a split-plot experiment. You will always have two error terms in a split-plot experiment.

Source	df	
Total	60	
μ	1	
Block	4	Main plot analysis
Bed height (BH)	1	
Error a	4	
(Block × Bed height)		
Cultivar (CV)	5	Sub-plot analysis
BH × CV	5	
Error b	44	
(CV × Block +		
BH × CV × Block)		

'Error a' is used to test for a bed height main effect while 'Error b' is used to test for a cultivar main effect as well as testing for a bed height × cultivar interaction. We indicate within parentheses what the two errors include in terms of sources of variation in our experiment, because we are going to use them in our SAS® program.

Now let's see how to go from the ANOVA table of a split-plot to the complete analysis, all performed by SAS®.

First we need some data to work with. Following are total yield (g/plant) for the 60 plots in this study. Lines 22 through 28 perform the ANOVA using the appropriate error term for hypotheses tests on 'bedheight', 'cultivar' and the bedheight*cultivar interaction.

```
01  TITLE1 'Chapter 14 Example 2.sas';
02  TITLE2 'Split-plot Design';
03  TITLE3 'ANOVA';
04  DATA one;
05  INPUT block bedheight $ cultivar $ yield @@;
06  CARDS;
07  1 r a 756 1 r b 645 1 r c 555 1 r d 668 1 r e 621
08  1 r f 502 1 f a 668 1 f b 621 1 f c 502 1 f d 754
09  1 f e 644 1 f f 557 2 r a 823 2 r b 676 2 r c 601
10  2 r d 645 2 r e 601 2 r f 497 2 f a 711 2 f b 599
11  2 f c 489 2 f d 799 2 f e 711 2 f f 578 3 r a 799
12  3 r b 711 3 r c 578 3 r d 711 3 r e 599 3 r f 489
13  3 f a 645 3 f b 601 3 f c 497 3 f d 711 3 f e 599
14  3 f f 489 4 r a 812 4 r b 684 4 r c 600 4 r d 902
15  4 r e 663 4 r f 588 4 f a 722 4 f b 576 4 f c 501
16  4 f d 801 4 f e 555 4 f f 499 5 r a 902 5 r b 663
17  5 r c 588 5 r d 722 5 r e 576 5 r f 501 5 f a 801
18  5 f b 555 5 f c 499 5 f d 722 5 f e 576 5 f f 501
19  RUN;
20  PROC PRINT;
21  RUN;
22  PROC ANOVA DATA = one;
23  CLASSES block bedheight cultivar;
24  MODEL yield = block bedheight block*bedheight
25              cultivar cultivar*bedheight
    cultivar*bedheight(block);
26  TEST H = bedheight E = block*bedheight;
27  TEST H = cultivar cultivar*bedheight E =
    cultivar*bedheight(block);
28  RUN;
29  MEANS bedheight / LSD E = block*bedheight PLOTS = none;
30  MEANS cultivar cultivar*bedheight / REGWQ E =
    cultivar*bedheight(block) PLOTS = none;
31  RUN;
```

Let's look carefully at each step in the program.

```
22  PROC ANOVA DATA = one;
23  CLASSES block bedheight cultivar;
```

These two lines are your standard 'PROC ANOVA' and 'CLASSES' statements we've seen before. Notice that in the CLASSES statement we include the main effect factors that are in our ANOVA table, including 'block', since our split plot is 'housed' in an RCBD.

```
24  MODEL yield = block bedheight block*bedheight
25  cultivar cultivar*bedheight    cultivar*bedheight(block);
```

The 'MODEL' statement is a little different than what we are used to seeing as it does not follow our original rules for translating our 'Source' column in our ANOVA table into our 'MODEL' statement. For a split-plot design, we include everything except 'Total' and 'µ' from our 'Source' column in our model statement. This means including the error terms.

We need to know how to enter the error terms correctly in the SAS® MODEL statement. 'Error a' is easily included as 'block × main plot', in this case, 'block*bedheight'. 'Error b', which consists of two pieces, 'block × sub-plot' plus 'block × sub-plot × main plot', is coded in SAS® as 'cultivar*bedheight(block)'. This tells SAS® to compute (cultivar*block + cultivar*bedheight*block). Note that by including the error terms in the model, SAS® has no 'ERROR' with which to perform any testing. This will lead to many '.' in our output. Notice in the output above how the columns for 'F' and 'Pr>F' both have periods (the symbol for missing values) in every location normally occupied by a numerical value.

This is a good thing for it will automatically remind us that we need to tell SAS® the correct denominator for the tests we want to perform in the ANOVA. Remember, because we have two different-sized plots in our experiment, we have two different estimates of variability, one for each plot size. The correct testing is accomplished with the next two lines of code:

```
26  TEST H = bedheight E = block*bedheight;
27  TEST H = cultivar cultivar*bedheight E =
    cultivar*bedheight(block);
```

Additionally, if we are performing any main effect mean separations, we need to indicate the correct error term as well. We could add the following code to the program above to give us our mean separations:

```
28  MEANS bedheight / LSD E = block*bedheight;
29  MEANS cultivar cultivar*bedheight / REGWQ E =
    cultivar*bedheight(block);
30  RUN;
```

Note that even though we code for an LSD mean separation of CV*BH means, SAS® will not perform the mean separation because it is inappropriate. SAS® will, however, give us the means.

And here are the results from SAS®:

'Chapter 14 Example 2.sas'
Split-plot Design
ANOVA

The ANOVA Procedure

Class Level Information		
Class	Levels	Values
block	5	1 2 3 4 5
bedheight	2	f r
cultivar	6	a b c d e f

Number of Observations Read	60
Number of Observations Used	60

Split-plot Design
ANOVA

The ANOVA Procedure

Dependent Variable: yield

Source	DF	Sum of Squares	Mean Square	F Value	Pr > F
Model	59	690580.9833	11704.7624	.	.
Error	0	0.0000	.		
Corrected Total	59	690580.9833			

R-Square	Coeff Var	Root MSE	yield Mean
1.000000	.	.	636.0167

Source	DF	Anova SS	Mean Square	F Value	Pr > F
block	4	12020.9000	3005.2250	.	.
bedheight	1	23800.4167	23800.4167	.	.
block*bedheight	4	23182.1667	5795.5417	.	.
cultivar	5	508431.2833	101686.2567	.	.
bedheight*cultivar	5	45186.0833	9037.2167	.	.
bedhei*cultiv(block)	40	77960.1333	1949.0033	.	.

Tests of Hypotheses Using the Anova MS for block*bedheight as an Error Term

Source	DF	Anova SS	Mean Square	F Value	Pr > F
bedheight	1	23800.41667	23800.41667	4.11	0.1127

Tests of Hypotheses Using the Anova MS for bedhei*cultiv(block) as an Error Term

Source	DF	Anova SS	Mean Square	F Value	Pr > F
cultivar	5	508431.2833	101686.2567	52.17	<.0001
bedheight*cultivar	5	45186.0833	9037.2167	4.84	0.0020

'Chapter 14 Example 2.sas'
Split-plot Design
ANOVA

The ANOVA Procedure

t Tests (LSD) for yield

Note: This test controls the Type I comparisonwise error rate, not the experimentwise error rate.

Alpha	0.05
Error Degrees of Freedom	4
Error Mean Square	5795.542
Critical Value of t	2.77645
Least Significant Difference	54.575

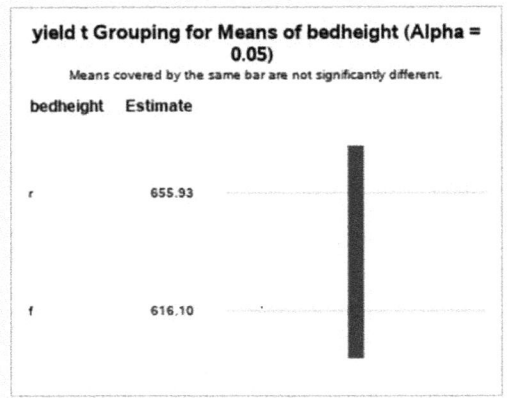

yield t Grouping for Means of bedheight (Alpha = 0.05)
Means covered by the same bar are not significantly different.

bedheight	Estimate
r	655.93
f	616.10

Split-plot Design
ANOVA

The ANOVA Procedure

Ryan-Einot-Gabriel-Welsch Multiple Range Test for yield

Note: This test controls the Type I experimentwise error rate.

Alpha	0.05
Error Degrees of Freedom	40
Error Mean Square	1949.003

Number of Means	2	3	4	5	6
Critical Range	49.194061	53.752379	56.214208	56.38887	59.076425

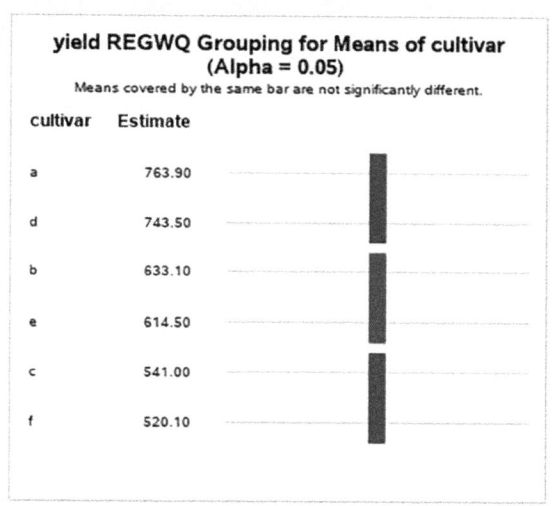

yield REGWQ Grouping for Means of cultivar (Alpha = 0.05)

Means covered by the same bar are not significantly different.

cultivar	Estimate
a	763.90
d	743.50
b	633.10
e	614.50
c	541.00
f	520.10

'Chapter 14 Example 2.sas'
Split-plot Design
ANOVA

The ANOVA Procedure

Level of bedheight	Level of cultivar	N	yield Mean	yield Std Dev
f	a	5	709.400000	60.027494
f	b	5	590.400000	25.412595
f	c	5	497.600000	5.176872
f	d	5	757.400000	41.979757
f	e	5	617.000000	62.076566
f	f	5	524.800000	39.939955
r	a	5	818.400000	53.200564
r	b	5	675.800000	24.590649
r	c	5	584.400000	18.955210
r	d	5	729.600000	101.337555
r	e	5	612.000000	32.664966
r	f	5	515.400000	40.905990

Remember to check the 'CLASS LEVELS INFORMATION' to verify that SAS has read your data correctly.

Notice all the missing value indicators '.' in the ANOVA table? That's because we have included 'everything' in the MODEL statement, therefor SAS® doesn't have anything left over with which to perform any F-tests. All the correct tests (with the correct denominators) are given under the 'Test of Hypotheses' sections. If you did not include the error terms in the model statement, both errors would have been pooled into one, and the tests SAS® would have performed automatically would have been incorrect.

To interpret our ANOVA, we first look at the interaction term. There is a significant cultivar × bed height interaction (0.0020 level), therefore we don't look at main effects. The interaction is indicating to us that cultivar performance depends on whether they are grown on raised or flat beds. A logical approach would be to look at cultivar differences within bed height. We need some sort of mean separation among cultivars within bed height.

Mean separation of interaction means is a little more complicated with a split-plot. Assuming you want to perform an LSD mean separation of the main plot by sub-plot interaction, you must do this by hand. The formulas for appropriate LSD values are presented in the following table.

Comparison	LSD formula
Compare main plot means	$t_a \dfrac{\sqrt{2(Ea)}}{rb}$
Compare sub-plot means	$t_b \dfrac{\sqrt{2(Eb)}}{ra}$
Compare sub-plot means in the same main plot	$t_b \dfrac{\sqrt{2(Eb)}}{r}$
Compare sub-plot means in different main plots	$t_{ab} \dfrac{\sqrt{2[(b-1)Eb+Ea]}}{rb}$

where r = number of replications, a = number of levels of main plot factor, b = number of levels of sub-plot factor, Ea is MS Error for main plots, Eb is MS Error for sub-plot, t_a is t-value for df Ea, t_b is t-value for df Eb and t_{ab} is a weighted t-value computed from the following formula:

$$t_{ab} = \frac{(b-1)Eb(t_b) + Ea(t_a)}{(b-1)Eb + Ea}.$$

Here are our interaction means in table form.

Cultivar	Flat bed	Raised bed
A	709.4	818.4
B	590.4	675.8
C	497.6	584.4
D	757.4	729.6
E	617.0	612.0
F	524.8	515.4

We must decide whether we want to separate cultivar means within bed heights or bed height means within cultivar. Actually we will look at both separations. You could have the situation where you have a grower who likes a particular cultivar and needs to know if there is a difference in performance on flat beds versus raised beds. And there might be another grower who specifically uses raised beds and needs to know how the cultivars compare with each other. Note that we are performing both separations because we have clients who need different questions answered. We are not performing both separations just to see which gives us more treatment differences!

First we will look at cultivars within bed height (sub-plots within main). We need to calculate our LSD value:

$$t_b \sqrt{\frac{2*1949.0033}{5}}$$

and t_b is the *t*-value for 40 df. We can get that with this little SAS® program. As a matter of fact, we can calculate our LSD value while we're at it.

```
01  TITLE1 'Chapter 14 Example 3.sas';
02  TITLE2 'Split-plot Design';
03  TITLE3 'ANOVA';
04  TITLE4 'LSD calculations';
05  DATA one;
08  t = tinv(0.975,40);
09  lsd = (t*sqrt((2*1949.0033)/5));
10  PROC PRINT DATA = one;
11  RUN;
```

which produces the output:

'Chapter 14 Example 3.sas'
Split-plot Design
ANOVA
LSD calculations

Obs	t	lsd
1	2.02108	56.4311

Next we will look at bed height differences for each cultivar (sub-plot means for different main plots). Again, calculate our LSD value:

$$t_{ab} \sqrt{\frac{2[(5)1949.0033 + 5795.5417]}{5*6}}$$

and t_{ab} is the *t*-value calculated from the formula:

$$t_{ab} = \frac{(b-1)Eb(t_b) + Ea(t_a)}{(b-1)Eb + Ea}$$

$$t_{ab} = \frac{(5)1949.0033(t_b) + 5795.5417(t_a)}{(5)1949.0033 + 5795.5417}$$

We can get all of our information from this program:

```
01  TITLE1 'Chapter 14 Example 4.sas';
02  TITLE2 'Split-plot Design';
03  TITLE3 'ANOVA';
```

```
04 TITLE4 'LSD calculations';
05 DATA one;
12 ta = tinv(0.975,4);
13 tb = tinv(0.975,40);
14 tab=((5*1949.0033*tb)+(5795.5417*ta))/(5*1949.0033+5795.5417);
15 lsd = tab*(sqrt((2*((5*1949.0033)+5795.5417))/(5*6)));
16 PROC PRINT DATA = one;
17 RUN;
```

which produces the output:

'Chapter 14 Example 4.sas'
Split-plot Design
ANOVA
LSD calculations

Obs	ta	tb	tab	lsd
1	2.77645	2.02108	2.30278	74.1207

Now we can include these LSD values in our table as footnotes.

Cultivar	Flat bed	Raised bed
A	709.4	818.4
B	590.4	675.8
C	497.6	584.4
D	757.4	729.6
E	617.0	612.0
F	524.8	515.4

$LSD_{(0.05)}$ for mean separation within column is 56.4
$LSD_{(0.05)}$ for mean separation within row is 74.1

If you don't want to perform all the hand calculations, you could obtain a relatively close estimate of the separation using SAS®. It involves performing two smaller ANOVAs, one for each bed height, and performing the mean separation of cultivars within each bed height. The drawback of doing it this way is that in the larger, complete ANOVA, the bedheight*cultivar interaction is tested with 'Error b' that has 40 df. In each of the smaller ANOVAs, each only has 20 df, thus making them less powerful. However, we already know that cultivar response is different on raised beds compared with flat beds, since we had a significant interaction in the larger ANOVA. We run the risk of committing a Type II error (we might miss a difference that really exists) with the smaller ANOVAs. But we save a huge effort in hand calculations!

We get SAS® to perform the two smaller ANOVAs with the following program:

```
01 TITLE1 'Chapter 14 Example 5.sas';
02 TITLE2 'Split-plot Design';
03 TITLE3 'ANOVA';
04 TITLE4 'Mean separation of cultivars within bedheight';
05 DATA one;
06 INPUT block bedheight $ cultivar $ yield @@;
07 CARDS;
08 1 r a 756 1 r b 645 1 r c 555 1 r d 668 1 r e 621
```

```
09 1 r f 502 1 f a 668 1 f b 621 1 f c 502 1 f d 754
10 1 f e 644 1 f f 557 2 r a 823 2 r b 676 2 r c 601
11 2 r d 645 2 r e 601 2 r f 497 2 f a 711 2 f b 599
12 2 f c 489 2 f d 799 2 f e 711 2 f f 578 3 r a 799
13 3 r b 711 3 r c 578 3 r d 711 3 r e 599 3 r f 489
14 3 f a 645 3 f b 601 3 f c 497 3 f d 711 3 f e 599
15 3 f f 489 4 r a 812 4 r b 684 4 r c 600 4 r d 902
16 4 r e 663 4 r f 588 4 f a 722 4 f b 576 4 f c 501
17 4 f d 801 4 f e 555 4 f f 499 5 r a 902 5 r b 663
18 5 r c 588 5 r d 722 5 r e 576 5 r f 501 5 f a 801
19 5 f b 555 5 f c 499 5 f d 722 5 f e 576 5 f f 501
20 RUN;
21 PROC PRINT;
22 RUN;
23 PROC SORT DATA = one;
24 BY bedheight;
25 PROC ANOVA DATA = one;
26 BY bedheight;
27 CLASSES block cultivar;
28 MODEL yield = block cultivar;
29 MEANS cultivar / REGWQ PLOTS = none;
30 RUN;
```

Here's the OUTPUT from the run. SAS® provides separate output for each level of the factor specified in the 'BY' statement. Thus we have two separate ANOVAs, one for the raised beds and one for the flat beds.

Flat beds:

'Chapter 14 Example 5.sas'
Split-plot Design
ANOVA
Mean separation of cultivars within bedheight

The ANOVA Procedure

bedheight=f

Class Level Information		
Class	Levels	Values
block	5	1 2 3 4 5
cultivar	6	a b c d e f

Number of Observations Read	30
Number of Observations Used	30

Split-plot Design
ANOVA
Mean separation of cultivars within bedheight

The ANOVA Procedure

Dependent Variable: yield

bedheight=f

Source	DF	Sum of Squares	Mean Square	F Value	Pr > F
Model	9	269586.3000	29954.0333	17.16	<.0001
Error	20	34910.4000	1745.5200		
Corrected Total	29	304496.7000			

R-Square	Coeff Var	Root MSE	yield Mean
0.885350	6.781273	41.77942	616.1000

Source	DF	Anova SS	Mean Square	F Value	Pr > F
block	4	11037.2000	2759.3000	1.58	0.2181
cultivar	5	258549.1000	51709.8200	29.62	<.0001

'Chapter 14 Example 5.sas'
Split-plot Design
ANOVA
Mean separation of cultivars within bedheight

The ANOVA Procedure

Ryan-Einot-Gabriel-Welsch Multiple Range Test for yield

bedheight=f

Note: This test controls the Type I experimentwise error rate.

Alpha	0.05
Error Degrees of Freedom	20
Error Mean Square	1745.52

Number of Means	2	3	4	5	6
Critical Range	68.823795	75.45997	79.012265	79.069389	83.056251

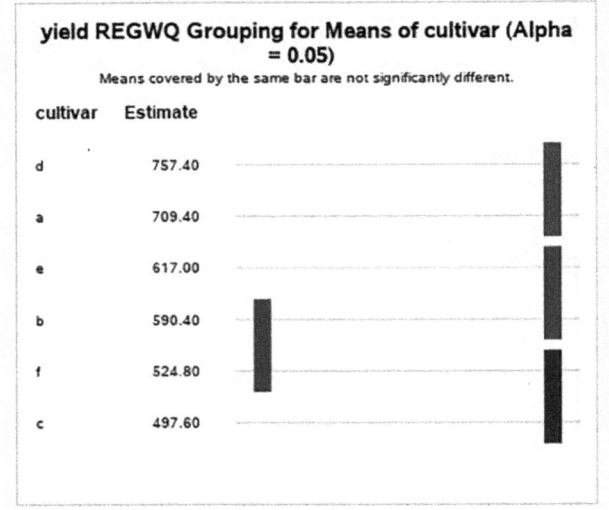

yield REGWQ Grouping for Means of cultivar (Alpha = 0.05)
Means covered by the same bar are not significantly different.

cultivar	Estimate
d	757.40
a	709.40
e	617.00
b	590.40
f	524.80
c	497.60

Raised beds:

'Chapter 14 Example 5.sas'
Split-plot Design
ANOVA
Mean separation of cultivars within bedheight

, The ANOVA Procedure

bedheight=r

Class Level Information		
Class	Levels	Values
block	5	1 2 3 4 5
cultivar	6	a b c d e f

Number of Observations Read	30
Number of Observations Used	30

'Chapter 14 Example 5.sas'
Split-plot Design
ANOVA
Mean separation of cultivars within bedheight

The ANOVA Procedure

Dependent Variable: yield

bedheight=r

Source	DF	Sum of Squares	Mean Square	F Value	Pr > F
Model	9	319234.1333	35470.4593	16.48	<.0001
Error	20	43049.7333	2152.4867		
Corrected Total	29	362283.8667			

R-Square	Coeff Var	Root MSE	yield Mean
0.881171	7.073112	46.39490	655.9333

Source	DF	Anova SS	Mean Square	F Value	Pr > F
block	4	24165.8667	6041.4667	2.81	0.0534
cultivar	5	295068.2667	59013.6533	27.42	<.0001

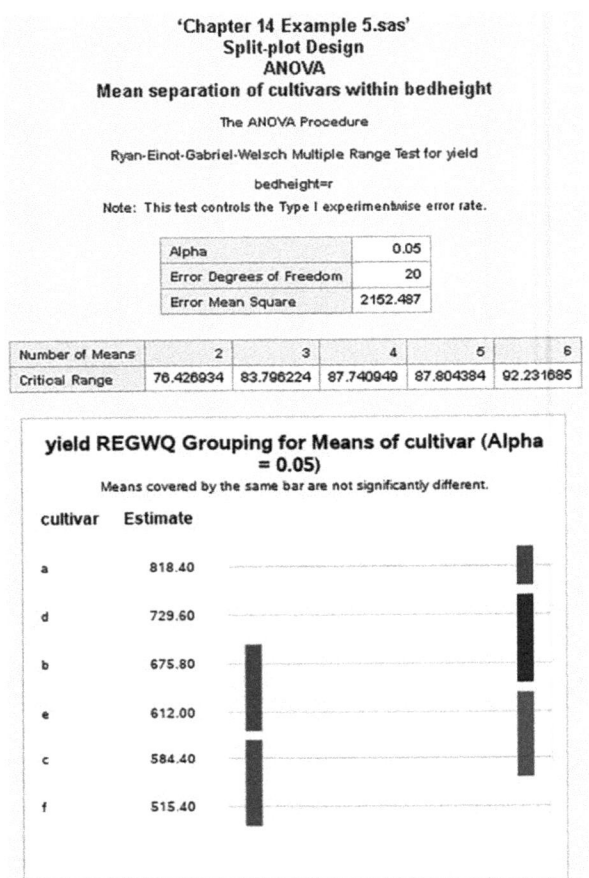

'Chapter 14 Example 5.sas'
Split-plot Design
ANOVA
Mean separation of cultivars within bedheight

The ANOVA Procedure

Ryan-Einot-Gabriel-Welsch Multiple Range Test for yield

bedheight=r

Note: This test controls the Type I experimentwise error rate.

Alpha	0.05
Error Degrees of Freedom	20
Error Mean Square	2152.487

Number of Means	2	3	4	5	6
Critical Range	76.426934	83.796224	87.740949	87.804384	92.231685

yield REGWQ Grouping for Means of cultivar (Alpha = 0.05)

Means covered by the same bar are not significantly different.

cultivar	Estimate
a	818.40
d	729.60
b	675.80
e	612.00
c	584.40
f	515.40

Now let's see what the table of means looks like using letters to separate means which are significantly different from each other. The first table is showing letters using the hand calculations. The second shows the mean separation using the SAS® approximation. Only separations within bed heights are presented. If you wanted the SAS® approximation separating bed height within cultivar, you should try to write the SAS® program to do it.

But how do we assign letters to means when we are using our hand calculations? First, create a column of the means you wish to separate, for example a column of means from the flat beds, and list the means in order from largest to smallest like this:

757.4
709.4
617.0
590.4
524.8
497.6

Assign the largest mean (757.4) the letter 'a'. Now subtract the second largest mean (709.4) from the largest mean (757.4). Is the difference (757.4 – 709.4 = 48) as large or larger than the previously calculated LSD (56.4)? No. Since it isn't, assign the second largest mean the letter 'a' since it is not significantly different than the largest mean. Now subtract the third largest mean (617.0) from the largest mean. The difference (140.4) is larger than the LSD, thus we are finished with the comparisons of all smaller means to the largest. Since the third largest mean is different from the largest, all means smaller than the third largest mean are also different.

Now we repeat this process comparing the second largest mean and compare all smaller means with it. Is the third largest mean (617.0) different than the second largest mean? Yes, since (709.4 – 617 = 92.4) is larger than the LSD. Since 617.0 is different than 709.4, we assign the letter 'a' to the third largest mean, then move on to compare the third largest mean (617.0) with all smaller means.

Is 617.0 different from 590.4? No, since (617.0 – 590.4 = 26.6) is less than the LSD. Thus 617.0 and 590.4 would both be followed by the letter 'b'. Are 617.0 and 524.8 different? Yes (you do the math). Ok, then move on to the comparisons of 590.4 to all smaller means. Is 590.4 different from 524.8? Yes. Assign 524.8 the letter 'c'. Is 524.8 different from 497.6? No. Assign 489.6 the letter 'c' as well. Now we have our means from largest to smallest, each followed by a letter to indicate which means are different:

    757.4 a
    709.4 a
    617.0 b
    590.4 b
    524.8 c
    497.6 c

While in our example each mean is followed by a single letter, other examples might have means followed by two or more letters. It depends on the magnitude of mean differences and the variability in each particular example. Separate the means within the raised bed plots in a similar manner. One final step is to place the correct letter after each mean in the table below.

Hand calculations:

Cultivar	Flat bed	Raised bed
A	709.4 a	818.4 a
B	590.4 b	675.8 b
C	497.6 c	584.4 c
D	757.4 a	729.6 b
E	617.0 b	612.0 c
F	524.8 c	515.4 d

Mean separation within column by Fisher's Protected LSD, $\alpha = 0.05$.

SAS® calculations:

Cultivar	Flat bed	Raised bed
A	709.4 a	818.4 a
B	590.4 b	675.8 b
C	497.6 c	584.4 c
D	757.4 a	729.6 b
E	617.0 b	612.0 c
F	524.8 c	515.4 d

Mean separation within column by REGWQ, $\alpha = 0.05$.

There is no difference between the two methods, hand calculation vs SAS® approximation. In my experience, this is often the case, thus I usually use the SAS® approximation when a large number of hand calculations are required. I am willing to assume the slightly greater risk of committing a Type II error (we might miss a difference that really exists) with the smaller ANOVAs of the SAS® approximation. Note that hand calculations used the LSD method while the REGWF method was used in SAS. I use REGWF when I can (i.i., in SAS) over the LSD because REGWF is more powerful than the LSD.

Essential features of the split-plot design

1. There are two sizes of experimental units and therefore two different estimates of variability associated with those units, 'Error a' and 'Error b'.
2. The main plots are randomly assigned as per the design they are in (usually an RCBD) and the sub-plots are randomly assigned within each main plot.
3. If possible, the factor of greatest interest should be the sub-plot, as this is where the most statistical information is.
4. The sub-plot main effects and sub-plot by main plot interactions are estimated with more precision than the main plot main effects.
5. Any gains in sub or sub by main interactions are offset by a loss in the precision of main plot main effect estimates.
6. Since there are two sizes of experimental units and two different estimates of Error (a and b), there are different standard errors needed for main plot effect and sub-plot effect confidence limit estimation and mean separation.

The Split-split-plot

Suppose in our strawberry experiment that we also wanted to examine within-the-row spacing on raised and flat beds. We could split the cultivar plots into six sub-plots to test within-the-row spacing of 3, 6, 9, 12, 15 and 18 inches. This creates another split in our original split-plot experiment. Each of the six spacings are randomly assigned within each cultivar within each bed height within each block. This is called a split-split-plot design. We now have three different-sized plots and will therefore require three error terms for testing. To make things a little simpler, we are only going to evaluate three cultivars rather than six; however, there is absolutely no reason you couldn't evaluate six cultivars if the time and space were available to do so.

Our SAS® statements for treatment assignment using PROC PLAN are:

```
01 TITLE1 'Chapter 14 Example 6.sas';
02 TITLE2 'Split-split-plot Design';
03 TITLE3 'PROC PLAN Treatment Assignment';
04 TITLE4 '2 Main plots, 3 sub-plots 6 sub-sub-plots and 5
   blocks';
05 PROC PLAN;
06 FACTORS Block=5 ORDERED BedHeight=2 Cultivar=3 Spacing = 6;
07 RUN;
```

which produces the following output:

'Chapter 14 Example 6.sas'
Split-split-plot Design
'PROC PLAN Treatment Assignment'
'2 Main plots, 3 sub-plots 6 sub-sub-plots and 5 blocks'

The PLAN Procedure

Factor	Select	Levels	Order
Block	5	5	Ordered
BedHeight	2	2	Random
Cultivar	3	3	Random
Spacing	6	6	Random

Block	BedHeight	Cultivar	Spacing					
1	1		2	2	6	3	4	1	5
		1	2	6	3	1	4	5
		3	3	4	6	2	5	1
	2	1	5	4	6	3	1	2
		2	1	4	3	2	5	6
		3	1	5	3	2	6	4
2	2	1	4	5	1	6	2	3
		2	5	2	1	4	6	3
		3	1	3	4	2	6	5
	1	3	3	2	6	1	4	5
		2	4	6	5	2	3	1
		1	1	4	2	5	3	6
3	1	1	6	4	5	3	2	1
		2	3	5	6	2	4	1
		3	6	2	1	5	4	3
	2	2	6	1	2	3	5	4
		1	3	4	1	2	6	5
		3	6	5	1	3	2	4
4	1	3	1	2	3	5	4	6
		2	5	4	3	1	2	6
		1	1	3	2	4	5	6
	2	2	5	4	1	3	6	2
		3	1	2	3	6	5	4
		1	4	3	1	2	5	6
5	1	2	4	1	2	3	6	5
		3	3	4	6	1	2	5
		1	1	6	5	4	3	2
	2	2	5	4	3	2	1	6
		3	6	3	1	4	5	2
		1	4	5	2	6	3	1

The ANOVA table for this split-split-plot experiment is:

Source	df	
Total	180	
μ	1	
Block	4	Main plot analysis
Bed height (BH)	1	
Error a (BH × Block)	4	
Cultivar (CV)	2	Sub-plot analysis
BH × CV	2	
Error b (CV × Block) + *(BH × CV × Block)*	16	
Spacing (SP)	5	
CV × SP	10	
BH × SP	5	
CV × BH × SP	10	
Error c		Sub-sub-plot analysis
(SP × Block) + (CV × SP × Block) + (BH × SP × Block) + (CV × SP × BH × Block)	120	

Here's an augmented data set to now include the spacing factor:
```
01  TITLE1 'Chapter 14 Example 7.sas';
02  TITLE2 'Split-split-plot Design';
03  TITLE3 'ANOVA';
04  DATA one;
05  INPUT block bedheight $ cultivar $ spacing yield @@;
06  CARDS;
07  1 r 1 3 545 1 f 1 3 521 2 r 1 3 576 2 f 1 3 599
08  3 r 1 3 511 3 f 1 3 501 4 r 1 3 584 4 f 1 3 576
09  5 r 1 3 563 5 f 1 3 555 1 r 1 6 547 1 f 1 6 527
10  2 r 1 6 579 2 f 1 6 506 3 r 1 6 522 3 f 1 6 504
11  4 r 1 6 594 4 f 1 6 581 5 r 1 6 569 5 f 1 6 567
12  1 r 1 9 652 1 f 1 9 629 2 r 1 9 687 2 f 1 9 601
13  3 r 1 9 617 3 f 1 9 610 4 r 1 9 688 4 f 1 9 681
14  5 r 1 9 670 5 f 1 9 663 1 r 1 12 753 1 f 1 12 734
15  2 r 1 12 783 2 f 1 12 714 3 r 1 12 725 3 f 1 12 714
16  4 r 1 12 706 4 f 1 12 784 5 r 1 12 770 5 f 1 12 776
17  1 r 1 15 851 1 f 1 15 829 2 r 1 15 879 2 f 1 15 807
18  3 r 1 15 819 3 f 1 15 809 4 r 1 15 890 4 f 1 15 877
19  5 r 1 15 873 5 f 1 15 863 1 r 1 18 951 1 f 1 18 934
20  2 r 1 18 981 2 f 1 18 911 3 r 1 18 926 3 f 1 18 910
21  4 r 1 18 997 4 f 1 18 982 5 r 1 18 969 5 f 1 18 969
22  1 r 2 3 368 1 f 2 3 354 2 r 2 3 345 2 f 2 3 399
23  3 r 2 3 311 3 f 2 3 311 4 r 2 3 302 4 f 2 3 301
24  5 r 2 3 322 5 f 2 3 322 1 r 2 6 474 1 f 2 6 455
25  2 r 2 6 447 2 f 2 6 403 3 r 2 6 417 3 f 2 6 416
26  4 r 2 6 407 4 f 2 6 401 5 r 2 6 424 5 f 2 6 430
27  1 r 2 9 674 1 f 2 9 655 2 r 2 9 654 2 f 2 9 600
28  3 r 2 9 615 3 f 2 9 617 4 r 2 9 602 4 f 2 9 602
29  5 r 2 9 623 5 f 2 9 632 1 r 2 12 783 1 f 2 12 759
30  2 r 2 12 757 2 f 2 12 704 3 r 2 12 719 3 f 2 12 724
31  4 r 2 12 712 4 f 2 12 710 5 r 2 12 731 5 f 2 12 737
32  1 r 2 15 773 1 f 2 15 755 2 r 2 15 750 2 f 2 15 703
33  3 r 2 15 714 3 f 2 15 722 4 r 2 15 704 4 f 2 15 707
34  5 r 2 15 723 5 f 2 15 732 1 r 2 18 685 1 f 2 18 662
35  2 r 2 18 652 2 f 2 18 615 3 r 2 18 627 3 f 2 18 617
36  4 r 2 18 619 4 f 2 18 610 5 r 2 18 628 5 f 2 18 637
37  1 r 3 3 621 1 f 3 3 644 2 r 3 3 601 2 f 3 3 611
38  3 r 3 3 699 3 f 3 3 699 4 r 3 3 663 4 f 3 3 655
39  5 r 3 3 676 5 f 3 3 676 1 r 3 6 630 1 f 3 6 655
40  2 r 3 6 608 2 f 3 6 623 3 r 3 6 604 3 f 3 6 605
41  4 r 3 6 668 4 f 3 6 666 5 r 3 6 684 5 f 3 6 686
42  1 r 3 9 724 1 f 3 9 748 2 r 3 9 713 2 f 3 9 713
43  3 r 3 9 702 3 f 3 9 700 4 r 3 9 766 4 f 3 9 756
44  5 r 3 9 781 5 f 3 9 779 1 r 3 12 838 1 f 3 12 860
45  2 r 3 12 813 2 f 3 12 826 3 r 3 12 806 3 f 3 12 816
46  4 r 3 12 869 4 f 3 12 867 5 r 3 12 889 5 f 3 12 892
47  1 r 3 15 931 1 f 3 15 951 2 r 3 15 912 2 f 3 15 919
```

```
48 3 r 3 15 900 3 f 3 15 903 4 r 3 15 972 4 f 3 15 964
49 5 r 3 15 981 5 f 3 15 984 1 r 3 18 1031 1 f 3 18 1066
50 2 r 3 18 1011 2 f 3 18 1033 3 r 3 18 1017 3 f 3 18 1016
51 4 r 3 18 1076 4 f 3 18 1076 5 r 3 18 1091 5 f 3 18 1098
52 RUN;
53 PROC PRINT;
54 RUN;
55 PROC ANOVA DATA = one;
56 CLASSES block bedheight cultivar spacing;
57 MODEL yield = block bedheight block*bedheight
58    cultivar cultivar*bedheight cultivar*bedheight(block)
59    spacing spacing*bedheight spacing*cultivar spacing*
   cultivar*bedheight spacing*cultivar*bedheight(block);
60 TEST H = bedheight E = block*bedheight;
61 TEST H = cultivar cultivar*bedheight E =
   cultivar*bedheight*spacing(block);
62 TEST H = spacing spacing*cultivar spacing*bedheight
   spacing*cultivar*bedheight E = spacing*cultivar*bedheight
   (block);
63 RUN;
64 MEANS bedheight / LSD E = block*bedheight PLOTS = none;
65 MEANS cultivar cultivar*bedheight / REGWQ E =
   cultivar*bedheight(block) PLOTS = none;
66 MEANS spacing spacing*cultivar spacing*bedheight
   spacing*cultivar*bedheight;
67 RUN;
```

Again, the 'MODEL' statement is a bit different than for simpler designs such as the CRD, RCBD and LS. Just like we did for the split-plot model, we include everything except 'Total' and 'μ' from our 'Source' column in our model statement. This means including the error terms. Note that by including the error terms in the model, SAS® has no 'ERROR' with which to perform any testing. This will lead to many '.' in our output. But this is a good thing, for it automatically reminds us that we need to tell SAS® the correct denominator for the tests we want to perform in the ANOVA.

Here are the ANOVA results:

Chapter 14 Example 7.sas
Split-split-plot Design
ANOVA

The ANOVA Procedure

Class Level Information		
Class	Levels	Values
block	5	1 2 3 4 5
bedheight	2	f r
cultivar	3	1 2 3
spacing	6	3 6 9 12 15 18

Number of Observations Read	180
Number of Observations Used	180

Chapter 14 Example 7.sas
Split-split-plot Design
ANOVA

The ANOVA Procedure

Dependent Variable: yield

Source	DF	Sum of Squares	Mean Square	F Value	Pr > F
Model	179	5806694.994	32439.637	.	.
Error	0	0.000	.		
Corrected Total	179	5806694.994			

R-Square	Coeff Var	Root MSE	yield Mean
1.000000	.	.	703.4944

Source	DF	Anova SS	Mean Square	F Value	Pr > F
block	4	39214.800	9803.700	.	.
bedheight	1	2020.050	2020.050	.	.
block*bedheight	4	4255.200	1063.800	.	.
cultivar	2	1628454.211	814227.106	.	.
bedheight*cultivar	2	4844.100	2422.050	.	.
bedhei*cultiv(block)	16	64246.800	4015.425	.	.
spacing	5	3561205.228	712241.046	.	.
bedheight*spacing	5	1156.517	231.303	.	.
cultivar*spacing	10	466394.256	46639.426	.	.
bedhei*cultiv*spacin	10	1370.233	137.023	.	.
bedh*cult*spac(bloc)	120	33533.600	279.447	.	.

Tests of Hypotheses Using the Anova MS for block*bedheight as an Error Term					
Source	DF	Anova SS	Mean Square	F Value	Pr > F
bedheight	1	2020.050000	2020.050000	1.90	0.2403

Tests of Hypotheses Using the Anova MS for bedh*cult*spac(bloc) as an Error Term					
Source	DF	Anova SS	Mean Square	F Value	Pr > F
cultivar	2	1628454.211	814227.106	2913.71	<.0001
bedheight*cultivar	2	4844.100	2422.050	8.67	0.0003
spacing	5	3561205.228	712241.046	2548.75	<.0001
cultivar*spacing	10	466394.256	46639.426	166.90	<.0001
bedheight*spacing	5	1156.517	231.303	0.83	0.5324
bedhei*cultiv*spacin	10	1370.233	137.023	0.49	0.8936

Examination of the ANOVA table reveals a significant interaction between cultivar and spacing (F-value = 166.90, $\alpha < 0.0001$) as well as a significant interaction between bed height and cultivar (F-value 8.67, $\alpha = 0.0003$). Let's examine the cultivar × spacing interaction first. The most appropriate way to examine the cultivar × spacing interaction is to develop regression equations for each cultivar and then compare the resulting three equations using dummy variables as outlined in Chapter 9.

A less desirable approach would be to present interaction means in a table along with appropriate LSD values for various mean comparisons. Just as with the split-plot, interaction mean separation is somewhat complicated and LSD values must be calculated by hand. The comparisons you would probably want to make are presented in the table that follows along with the formulas for the LSD values.

Comparison	LSD formula
Compare sub-sub-plot means	$t_c \sqrt{\dfrac{2(Ec)}{rab}}$
Compare sub-sub-plot means for the same main plot	$t_c \sqrt{\dfrac{2(Ec)}{rb}}$
Compare sub-sub-plot means for the same sub-plot	$t_c \sqrt{\dfrac{2(Ec)}{ra}}$
Compare sub-sub-plot means for the same sub- and main plot	$t_c \sqrt{\dfrac{2(Ec)}{r}}$
Compare sub-plot means for the same or different sub-sub-plot	$t_{bc} \sqrt{\dfrac{2((c-1)(Ec)+Eb)}{rac}}$
Compare sub-plot means for the same main plot and same or different sub-sub-plot	$t_{bc} \sqrt{\dfrac{2((c-1)(Ec)+Eb)}{rc}}$
Compare main plot means for the same or different sub-plot	$t_{ac} \sqrt{\dfrac{2((c-1)(Ec)+Ea)}{rbc}}$
Compare main plot means for the same or different sub- and sub-sub-plot	$t_{abc} \sqrt{\dfrac{2(b(c-1)(Ec)+(b-1)Eb+Ea)}{rbc}}$

where r = number of replications, a = number of levels of main plot factor, b = number of levels of sub-plot factor, c = number of levels of sub-sub-plot, Ea is MS Error for main plots, Eb is MS Error for sub-plots, Ec is MS Error for sub-sub-plots, t_a is t-value for df Ea, t_b is t-value for df Eb, t_c is t-value for df Ec. The values for t_{bc}, t_{ac} and t_{abc} are calculated from the following formulas:

$$t_{bc} = \frac{(c-1)Ec(t_c) + Eb(t_b)}{(c-1)Ec + Eb}$$

$$t_{ac} = \frac{(c-1)Ec(t_c) + Ea(t_a)}{(c-1)Ec + Ea}$$

$$t_{abc} = \frac{b(c-1)t_c + (b-1)Eb(t_b) + Ea(t_a)}{b(c-1) + (b-1)Eb + Ea}$$

To illustrate the comparisons suggested in the previous table, let's replace main, sub and sub-sub with bed height, cultivar and spacing so that the comparisons make a little more sense.

Comparison	LSD formula
Compare spacing means	$t_c \sqrt{\dfrac{2(Ec)}{rab}}$
Compare spacing means within the same bed height	$t_c \sqrt{\dfrac{2(Ec)}{rb}}$
Compare spacing means within the same cultivar	$t_c \sqrt{\dfrac{2(Ec)}{ra}}$
Compare spacing means within the same cultivar and bed height	$t_c \sqrt{\dfrac{2(Ec)}{r}}$
Compare cultivar means for the same or different spacing	$t_{bc} \sqrt{\dfrac{2((c-1)(Ec)+Eb)}{rac}}$
Compare cultivar means for the same bed height and the same or different spacing	$t_{bc} \sqrt{\dfrac{2((c-1)(Ec)+Eb)}{rc}}$
Compare bed height means for the same or different cultivar	$t_{ac} \sqrt{\dfrac{2((c-1)(Ec)+Ea)}{rbc}}$
Compare bed height means for the same or different cultivar and spacing	$t_{abc} \sqrt{\dfrac{2(b(c-1)(Ec)+(b-1)Eb+Ea)}{rbc}}$

But let's examine the interaction in the more desirable fashion by using dummy variables to determine if the lines have common intercepts or common slopes as outlined in Chapter 9. First, we can run a separate ANOVA for each cultivar to determine if spacing has an effect on yield.

The SAS statements to generate the ANOVAs for this approach are:

```
01 TITLE1 'Chapter 14 Example 8.sas';
02 TITLE2 'Split-split-plot Design';
03 TITLE3 'ANOVA';
04 DATA one;
05 INPUT block bedheight $ cultivar $ spacing yield @@;
06 CARDS;
07 1 r 1 3 545 1 f 1 3 521 2 r 1 3 576 2 f 1 3 599
08 3 r 1 3 511 3 f 1 3 501 4 r 1 3 584 4 f 1 3 576
09 5 r 1 3 563 5 f 1 3 555 1 r 1 6 547 1 f 1 6 527
10 2 r 1 6 579 2 f 1 6 506 3 r 1 6 522 3 f 1 6 504
11 4 r 1 6 594 4 f 1 6 581 5 r 1 6 569 5 f 1 6 567
12 1 r 1 9 652 1 f 1 9 629 2 r 1 9 687 2 f 1 9 601
13 3 r 1 9 617 3 f 1 9 610 4 r 1 9 688 4 f 1 9 681
14 5 r 1 9 670 5 f 1 9 663 1 r 1 12 753 1 f 1 12 734
15 2 r 1 12 783 2 f 1 12 714 3 r 1 12 725 3 f 1 12 714
16 4 r 1 12 706 4 f 1 12 784 5 r 1 12 770 5 f 1 12 776
17 1 r 1 15 851 1 f 1 15 829 2 r 1 15 879 2 f 1 15 807
18 3 r 1 15 819 3 f 1 15 809 4 r 1 15 890 4 f 1 15 877
19 5 r 1 15 873 5 f 1 15 863 1 r 1 18 951 1 f 1 18 934
20 2 r 1 18 981 2 f 1 18 911 3 r 1 18 926 3 f 1 18 910
21 4 r 1 18 997 4 f 1 18 982 5 r 1 18 969 5 f 1 18 969
```

```
22  1  r  2  3   368  1  f  2  3   354  2  r  2  3   345  2  f  2  3   399
23  3  r  2  3   311  3  f  2  3   311  4  r  2  3   302  4  f  2  3   301
24  5  r  2  3   322  5  f  2  3   322  1  r  2  6   474  1  f  2  6   455
25  2  r  2  6   447  2  f  2  6   403  3  r  2  6   417  3  f  2  6   416
26  4  r  2  6   407  4  f  2  6   401  5  r  2  6   424  5  f  2  6   430
27  1  r  2  9   674  1  f  2  9   655  2  r  2  9   654  2  f  2  9   600
28  3  r  2  9   615  3  f  2  9   617  4  r  2  9   602  4  f  2  9   602
29  5  r  2  9   623  5  f  2  9   632  1  r  2  12  783  1  f  2  12  759
30  2  r  2  12  757  2  f  2  12  704  3  r  2  12  719  3  f  2  12  724
31  4  r  2  12  712  4  f  2  12  710  5  r  2  12  731  5  f  2  12  737
32  1  r  2  15  773  1  f  2  15  755  2  r  2  15  750  2  f  2  15  703
33  3  r  2  15  714  3  f  2  15  722  4  r  2  15  704  4  f  2  15  707
34  5  r  2  15  723  5  f  2  15  732  1  r  2  18  685  1  f  2  18  662
35  2  r  2  18  652  2  f  2  18  615  3  r  2  18  627  3  f  2  18  617
36  4  r  2  18  619  4  f  2  18  610  5  r  2  18  628  5  f  2  18  637
37  1  r  3  3   621  1  f  3  3   644  2  r  3  3   601  2  f  3  3   611
38  3  r  3  3   699  3  f  3  3   699  4  r  3  3   663  4  f  3  3   655
39  5  r  3  3   676  5  f  3  3   676  1  r  3  6   630  1  f  3  6   655
40  2  r  3  6   608  2  f  3  6   623  3  r  3  6   604  3  f  3  6   605
41  4  r  3  6   668  4  f  3  6   666  5  r  3  6   684  5  f  3  6   686
42  1  r  3  9   724  1  f  3  9   748  2  r  3  9   713  2  f  3  9   713
43  3  r  3  9   702  3  f  3  9   700  4  r  3  9   766  4  f  3  9   756
44  5  r  3  9   781  5  f  3  9   779  1  r  3  12  838  1  f  3  12  860
45  2  r  3  12  813  2  f  3  12  826  3  r  3  12  806  3  f  3  12  816
46  4  r  3  12  869  4  f  3  12  867  5  r  3  12  889  5  f  3  12  892
47  1  r  3  15  931  1  f  3  15  951  2  r  3  15  912  2  f  3  15  919
48  3  r  3  15  900  3  f  3  15  903  4  r  3  15  972  4  f  3  15  964
49  5  r  3  15  981  5  f  3  15  984  1  r  3  18  1031  1  f  3  18  1066
50  2  r  3  18  1011  2  f  3  18  1033  3  r  3  18  1017  3  f  3  18  1016
51  4  r  3  18  1076  4  f  3  18  1076  5  r  3  18  1091  5  f  3  18  1098
52  RUN;
53  PROC SORT DATA=one;
54  BY cultivar;
55  PROC ANOVA;
56  BY cultivar;
57  CLASSES block bedheight spacing;
58  MODEL yield = block bedheight block*bedheight
59     spacing spacing*bedheight spacing*bedheight(block);
60  TEST H = bedheight E = block*bedheight;
61  TEST H = spacing spacing*bedheight E = spacing*bedheight(block);
62  RUN;
```

The output from this run is:

Chapter 14 Example 8.sas
Split-split-plot Design
ANOVA

The ANOVA Procedure

cultivar=1

Class Level Information		
Class	Levels	Values
block	5	1 2 3 4 5
bedheight	2	f r
spacing	6	3 6 9 12 15 18

Number of Observations Read	60
Number of Observations Used	60

Chapter 14 Example 8.sas
Split-split-plot Design
ANOVA

The ANOVA Procedure

Dependent Variable: yield

cultivar=1

Source	DF	Sum of Squares	Mean Square	F Value	Pr > F
Model	59	1391005.650	23576.367	.	.
Error	0	0.000		.	
Corrected Total	59	1391005.650			

R-Square	Coeff Var	Root MSE	yield Mean
1.000000	.	.	716.8500

Source	DF	Anova SS	Mean Square	F Value	Pr > F
block	4	30142.233	7535.558	.	.
bedheight	1	4914.150	4914.150	.	.
block*bedheight	4	6931.100	1732.775	.	.
spacing	5	1335421.750	267084.350	.	.
bedheight*spacing	5	1464.150	292.830	.	.
bedhei*spacin(block)	40	12132.267	303.307	.	.

Tests of Hypotheses Using the Anova MS for block*bedheight as an Error Term

Source	DF	Anova SS	Mean Square	F Value	Pr > F
bedheight	1	4914.150000	4914.150000	2.84	0.1675

Tests of Hypotheses Using the Anova MS for bedhei*spacin(block) as an Error Term

Source	DF	Anova SS	Mean Square	F Value	Pr > F
spacing	5	1335421.750	267084.350	880.58	<.0001
bedheight*spacing	5	1464.150	292.830	0.97	0.4504

Chapter 14 Example 8.sas
Split-split-plot Design
ANOVA

The ANOVA Procedure

Dependent Variable: yield

cultivar=2

Source	DF	Sum of Squares	Mean Square	F Value	Pr > F
Model	59	1386111.400	23493.414	.	
Error	0	0.000	.		
Corrected Total	59	1386111.400			

R-Square	Coeff Var	Root MSE	yield Mean
1.000000	.	.	580.9000

Source	DF	Anova SS	Mean Square	F Value	Pr > F
block	4	24841.400	6210.350	.	.
bedheight	1	1215.000	1215.000	.	.
block*bedheight	4	2802.667	700.667	.	.
spacing	5	1349238.000	269847.600	.	.
bedheight*spacing	5	886.000	177.200	.	.
bedhei*spacin(block)	40	7128.333	178.208	.	.

Tests of Hypotheses Using the Anova MS for block*bedheight as an Error Term

Source	DF	Anova SS	Mean Square	F Value	Pr > F
bedheight	1	1215.000000	1215.000000	1.73	0.2583

Tests of Hypotheses Using the Anova MS for bedhei*spacin(block) as an Error Term

Source	DF	Anova SS	Mean Square	F Value	Pr > F
spacing	5	1349238.000	269847.600	1514.23	<.0001
bedheight*spacing	5	886.000	177.200	0.99	0.4335

Chapter 14 Example 8.sas
Split-split-plot Design
ANOVA

The ANOVA Procedure

Dependent Variable: yield

cultivar=3

Source	DF	Sum of Squares	Mean Square	F Value	Pr > F
Model	59	1401123.733	23747.860	.	.
Error	0	0.000	.		
Corrected Total	59	1401123.733			

R-Square	Coeff Var	Root MSE	yield Mean
1.000000	.	.	812.7333

Source	DF	Anova SS	Mean Square	F Value	Pr > F
block	4	41411.067	10352.767	.	.
bedheight	1	735.000	735.000	.	.
block*bedheight	4	1588.333	397.083	.	.
spacing	5	1342939.733	268587.947	.	.
bedheight*spacing	5	176.600	35.320	.	.
bedhei*spacin(block)	40	14273.000	356.825	.	.

Tests of Hypotheses Using the Anova MS for block*bedheight as an Error Term					
Source	DF	Anova SS	Mean Square	F Value	Pr > F
bedheight	1	735.0000000	735.0000000	1.85	0.2453

Tests of Hypotheses Using the Anova MS for bedhei*spacin(block) as an Error Term					
Source	DF	Anova SS	Mean Square	F Value	Pr > F
spacing	5	1342939.733	268587.947	752.72	<.0001
bedheight*spacing	5	176.600	35.320	0.10	0.9918

A significant spacing effect was detected for each cultivar, therefore a regression equation should be developed for each separately, then compared for coincidence (not likely since we detected a significant interaction), similar slopes and similar intercepts.

We limit our analysis to the linear model, even though the quadratic or even cubic model may produce a better fit. (Further analysis of the dataset indicated that a significant lack of fit existed up to the cubic model for cultivar 1 and to the quartic model for cultivars 2 and 3.)

Since we have three cultivars, thus three lines we want to compare, we need two dummy variables, defined as:

A1 = 1 if cultivar = 1, otherwise A1 = 0;
A2 = 1 if cultivar = 2, otherwise A2 = 0;

```
01 TITLE1 'Chapter 14 Example 9.sas';
02 TITLE2 'Split-split-plot Design';
03 TITLE3 'ANOVA';
04 DATA one;
05 INPUT block bedheight $ cultivar $ spacing yield @@;
06 CARDS;
07 1 r 1 3 545 1 f 1 3 521 2 r 1 3 576 2 f 1 3 599
08 3 r 1 3 511 3 f 1 3 501 4 r 1 3 584 4 f 1 3 576
09 5 r 1 3 563 5 f 1 3 555 1 r 1 6 547 1 f 1 6 527
10 2 r 1 6 579 2 f 1 6 506 3 r 1 6 522 3 f 1 6 504
11 4 r 1 6 594 4 f 1 6 581 5 r 1 6 569 5 f 1 6 567
12 1 r 1 9 652 1 f 1 9 629 2 r 1 9 687 2 f 1 9 601
13 3 r 1 9 617 3 f 1 9 610 4 r 1 9 688 4 f 1 9 681
14 5 r 1 9 670 5 f 1 9     663 1 r 1 12 753 1 f 1 12 734
15 2 r 1 12 783 2 f 1 12 714 3 r 1 12 725 3 f 1 12 714
16 4 r 1 12 706 4 f 1 12 784 5 r 1 12 770 5 f 1 12 776
17 1 r 1 15 851 1 f 1 15 829 2 r 1 15 879 2 f 1 15 807
18 3 r 1 15 819 3 f 1 15 809 4 r 1 15 890 4 f 1 15 877
19 5 r 1 15 873 5 f 1 15 863 1 r 1 18 951 1 f 1 18 934
20 2 r 1 18 981 2 f 1 18 911 3 r 1 18 926 3 f 1 18 910
21 4 r 1 18 997 4 f 1 18 982 5 r 1 18 969 5 f 1 18 969
22 1 r 2 3 368 1 f 2 3 354 2 r 2 3 345 2 f 2 3 399
23 3 r 2 3 311 3 f 2 3 311 4 r 2 3 302 4 f 2 3 301
24 5 r 2 3 322 5 f 2 3 322 1 r 2 6 474 1 f 2 6 455
25 2 r 2 6 447 2 f 2 6 403 3 r 2 6 417 3 f 2 6 416
26 4 r 2 6 407 4 f 2 6 401 5 r 2 6 424 5 f 2 6 430
27 1 r 2 9 674 1 f 2 9 655 2 r 2 9 654 2 f 2 9 600
28 3 r 2 9 615 3 f 2 9 617 4 r 2 9 602 4 f 2 9 602
29 5 r 2 9 623 5 f 2 9 632 1 r 2 12 783 1 f 2 12 759
30 2 r 2 12 757 2 f 2 12 704 3 r 2 12 719 3 f 2 12 724
31 4 r 2 12 712 4 f 2 12 710 5 r 2 12 731 5 f 2 12 737
32 1 r 2 15 773 1 f 2 15 755 2 r 2 15 750 2 f 2 15 703
33 3 r 2 15 714 3 f 2 15 722 4 r 2 15 704 4 f 2 15 707
34 5 r 2 15 723 5 f 2 15 732 1 r 2 18 685 1 f 2 18 662
35 2 r 2 18 652 2 f 2 18 615 3 r 2 18 627 3 f 2 18 617
36 4 r 2 18 619 4 f 2 18 610 5 r 2 18 628 5 f 2 18 637
37 1 r 3 3 621 1 f 3 3 644 2 r 3 3 601 2 f 3 3 611
38 3 r 3 3 699 3 f 3 3 699 4 r 3 3 663 4 f 3 3 655
39 5 r 3 3 676 5 f 3 3 676 1 r 3 6 630 1 f 3 6 655
40 2 r 3 6 608 2 f 3 6 623 3 r 3 6 604 3 f 3 6 605
41 4 r 3 6 668 4 f 3 6 666 5 r 3 6 684 5 f 3 6 686
42 1 r 3 9 724 1 f 3 9 748 2 r 3 9 713 2 f 3 9 713
43 3 r 3 9 702 3 f 3 9 700 4 r 3 9 766 4 f 3 9 756
44 5 r 3 9 781 5 f 3 9 779 1 r 3 12 838 1 f 3 12 860
45 2 r 3 12 813 2 f 3 12 826 3 r 3 12 806 3 f 3 12 816
```

```
46 4 r 3 12 869 4 f 3 12 867 5 r 3 12 889 5 f 3 12 892
47 1 r 3 15 931 1 f 3 15 951 2 r 3 15 912 2 f 3 15 919
48 3 r 3 15 900 3 f 3 15 903 4 r 3 15 972 4 f 3 15 964
49 5 r 3 15 981 5 f 3 15 984 1 r 3 18 1031 1 f 3 18 1066
50 2 r 3 18 1011 2 f 3 18 1033 3 r 3 18 1017 3 f 3 18 1016
51 4 r 3 18 1076 4 f 3 18 1076 5 r 3 18 1091 5 f 3 18 1098
52 RUN;
53 DATA two;
54 SET one;
55 A1 = . ;
56 A2 = . ;
57 IF cultivar = 1 then A1 = 1; ELSE A1 = 0;
58 IF cultivar = 2 then A2 = 1; ELSE A2 = 0;
59 A1spacing = A1*spacing;
60 A2spacing = A2*spacing;
61 RUN;
```

The model we are using to make the comparison among the three lines utilizing our dummy variables is:

$$Y = \beta_0 + \beta_1 * X_i + \beta_2 * A_1 + \beta_3 * A_2 + \beta_4 * X * A_1 + \beta_5 X * A_2 + \epsilon_i$$

We assume equal variances among the three cultivars. Since our examples will always use dummy variables of the form Ai = 1 if the observation is in a specific group, otherwise Ai = 0, the above model provides a model for observations in either of our two cultivar groups simply by letting Ai = 0 or 1 depending on group:

Cultivar 1 (A1 = 1, A2 = 0): $Y = (\beta_0 + \beta_2) + (\beta_1 + \beta_4) * X + \epsilon_i$
Cultivar 2 (A1 = 0, A2 = 1): $Y = (\beta_0 + \beta_3) + (\beta_1 + \beta_5) * X + \epsilon_i$
Cultivar 3 (A1 = 0, A2 = 0): $Y = \beta_0 + \beta_1 * X_i + \epsilon_i$
Using the model:

$$Y = \beta_0 + \beta_1 * X_i + \beta_2 * A_1 + \beta_3 * A_2 + \beta_4 X * A_1 + \beta_5 X * A_2 + \epsilon_i$$

we can substitute our notation and obtain:

$$Y = \beta_0 + \beta_{1 * spacing} + \beta_2 * A_1 + \beta_3 * A_2 + \beta_{4 * spacing} * A_1 + \beta_{5 * spacing} * A_2 + \epsilon_i$$

1. The test for coincidence is obtained with the test:

H_0: $\beta2 = \beta3 = \beta4 = \beta5 = 0$

If we fail to reject H_0, then:

The model that corresponds to all cultivars is $Y = \beta_0 + \beta_1 X + \epsilon$.

2. The test for equal slopes (parallelism) is obtained with the test:

H_0: $\beta4 = \beta5 = 0$

If we fail to reject H_0:

The model for cultivar 1 is: $Y = (\beta_0 + \beta_2) + \beta_1 * X + \epsilon_i$
The model for cultivar 2 is: $Y = (\beta_0 + \beta_3) + \beta_1 * X + \epsilon_i$
The model for cultivar 3 is: $Y = \beta_0 + \beta_1 * X_i + \epsilon_i$

3. The hypothesis for equal intercepts is obtained with the test:

H_0: $\beta2 = \beta3 = 0$

If we fail to reject H_0:

The model for cultivar 1 is: $Y = \beta_0 + (\beta_1 + \beta_4) * X + \epsilon_i$

The model for cultivar 2 is: $Y = \beta_0 + (\beta_1 + \beta_5) * X + \epsilon_i$

The model for cultivar 3 is: $Y = \beta_0 + \beta_1 * X + \epsilon_i$

We now use 'A1', 'A2', 'spacing', 'A1spacing' and 'A2spacing' as predictors (independent variables) in a regression analysis using PROC REG to perform the above tests and get estimates of our β's.

We include TEST statements as well. The statement 'TEST A1 = 0, A2 = 0, A1spacing = 0, A2spacing = 0;' tests the H_0: $\beta_2 = \beta_3 = \beta_4 = \beta_5 = 0$. The statement 'TEST A1spacing = 0, A2spacing = 0;' tests the H_0: $\beta_4 = \beta_5 = 0$. The statement 'TEST A1 = 0, A2 = 0;' tests the H_0: $\beta_2 = \beta_3 = 0$.

```
62  PROC REG DATA = two PLOTS = none;
63  MODEL yield = A1 A2 spacing A1spacing A2spacing;
64  TEST A1 = 0, A2 = 0, A1spacing = 0, A2spacing = 0;
65  TEST A1spacing = 0, A2spacing = 0;
66  TEST A1=0, A2=0;
67  RUN;
```

This produces the following output:

Chapter 14 Example 9.sas
Split-split-plot Design
ANOVA

The REG Procedure
Model: MODEL1
Dependent Variable: yield

Number of Observations Read	180
Number of Observations Used	180

Analysis of Variance

Source	DF	Sum of Squares	Mean Square	F Value	Pr > F
Model	5	5093826	1018765	248.66	<.0001
Error	174	712869	4096.94686		
Corrected Total	179	5806695			

Root MSE	64.00740	R-Square	0.8772
Dependent Mean	703.49444	Adj R-Sq	0.8737
Coeff Var	9.09849		

Parameter Estimates

Variable	DF	Parameter Estimate	Standard Error	t Value	Pr > |t|
Intercept	1	513.65333	18.84327	27.26	<.0001
A1	1	-96.39333	26.64840	-3.62	0.0004
A2	1	-184.49333	26.64840	-6.92	<.0001
spacing	1	28.48381	1.61283	17.66	<.0001
A1spacing	1	0.04857	2.28089	0.02	0.9830
A2spacing	1	-4.50857	2.28089	-1.98	0.0497

The REG Procedure
Model: MODEL1

Test 1 Results for Dependent Variable yield				
Source	DF	Mean Square	F Value	Pr > F
Numerator	4	412508	100.69	<.0001
Denominator	174	4096.94686		

Chapter 14 Example 9.sas
Split-split-plot Design
ANOVA

The REG Procedure
Model: MODEL1

Test 2 Results for Dependent Variable yield				
Source	DF	Mean Square	F Value	Pr > F
Numerator	2	10788	2.63	0.0747
Denominator	174	4096.94686		

Chapter 14 Example 9.sas
Split-split-plot Design
ANOVA

The REG Procedure
Model: MODEL1

Test 3 Results for Dependent Variable yield				
Source	DF	Mean Square	F Value	Pr > F
Numerator	2	98252	23.98	<.0001
Denominator	174	4096.94686		

The test for coincidence (H_0: $\beta_2 = \beta_3 = \beta_4 = \beta_5 = 0$) produced an F-value of 100.69 with $\alpha < 0.0001$, therefore we reject the hypothesis of coincidence.

The test for equal slopes (H_0: $\beta_4 = \beta_5 = 0$) produced an F-value of 2.63 with $\alpha < 0.0747$, therefore we fail to reject the hypothesis of equal slopes.

The test for intercepts (H_0: $\beta_2 = \beta_3 = 0$) produced an F-value of 23.89 with $\alpha = 0.0001$, thus reject the hypothesis of equal intercepts.

Our β estimates are as follows:

$\beta 0 = 513.65$

$\beta 1 = 28.48$

$\beta 2 = -96.39$

$\beta 3 = -184.49$

$\beta 4 = 0.05$

$\beta 5 = -4.51$

Thus our models are:

Cultivar 1 is: $Y = (\beta_0 + \beta_2) + \beta_1 * X + \epsilon_i$
Cultivar 2 is: $Y = (\beta_0 + \beta_3) + \beta_1 * X + \epsilon_i$
Cultivar 3 is: $Y = \beta_0 + \beta_1 * X + \epsilon_i$

Thus:

Cultivar 1 is: $Y = (513.65 - 96.39) + 28.48 * spacing + \epsilon_i$
Cultivar 2 is: $Y = (513.65 - 184.49) + 28.48 * spacing + \epsilon_i$
Cultivar 3 is: $Y = 513.65 + 28.48 * spacing + \epsilon_i$

Thus:

Cultivar 1 is: $Y = 417.26 + 28.48 * spacing + \epsilon_i$
Cultivar 2 is: $Y = 329.16 + 28.48 * spacing + \epsilon_i$
Cultivar 3 is: $Y = 513.65 + 28.48 * spacing + \epsilon_i$

The cultivars respond similarly to spacing as indicated by the lack of different slopes. However, each cultivar has an inherently different yield potential, as indicated by the different intercepts, cultivar 2 being the least productive while cultivar 3 is the most productive.

Now let's evaluate the bed height × cultivar interaction to make our analysis complete. The approach you take depends on if you want to know which cultivar to use on raised or flat beds, or whether to use raised or flat beds for a particular cultivar. The structure of the experiment also deserves consideration when making this decision. Since cultivars were sub-plots within bed height, it intuitively makes sense to evaluate cultivars within bed height. Also, since there are more replications of cultivar compared with bed height (ten replications of cultivar vs five replications of bed height) the mean separation of cultivars within bed height is more powerful than the separation of bed heights within cultivar. We will proceed to evaluate cultivar within bed height; however, if you needed to compare bed height for each cultivar based on research or client need, that would be acceptable.

The SAS® statements for the analysis of cultivar within bed height are:

```
01 TITLE1 'Chapter 14 Example 10.sas';
02 TITLE2 'Split-split-plot Design';
03 TITLE3 'ANOVA';
04 DATA one;
05 INPUT block bedheight $ cultivar $ spacing yield @@;
06 CARDS;
07 1 r 1 3 545 1 f 1 3 521 2 r 1 3 576 2 f 1 3 599
08 3 r 1 3 511 3 f 1 3 501 4 r 1 3 584 4 f 1 3 576
09 5 r 1 3 563 5 f 1 3 555 1 r 1 6 547 1 f 1 6 527
10 2 r 1 6 579 2 f 1 6 506 3 r 1 6 522 3 f 1 6 504
11 4 r 1 6 594 4 f 1 6 581 5 r 1 6 569 5 f 1 6 567
12 1 r 1 9 652 1 f 1 9 629 2 r 1 9 687 2 f 1 9 601
13 3 r 1 9 617 3 f 1 9 610 4 r 1 9 688 4 f 1 9 681
14 5 r 1 9 670 5 f 1 9 663 1 r 1 12 753 1 f 1 12 734
15 2 r 1 12 783 2 f 1 12 714 3 r 1 12 725 3 f 1 12 714
16 4 r 1 12 706 4 f 1 12 784 5 r 1 12 770 5 f 1 12 776
17 1 r 1 15 851 1 f 1 15 829 2 r 1 15 879 2 f 1 15 807
```

```
18 3 r 1 15 819 3 f 1 15 809 4 r 1 15 890 4 f 1 15 877
19 5 r 1 15 873 5 f 1 15 863 1 r 1 18 951 1 f 1 18 934
20 2 r 1 18 981 2 f 1 18 911 3 r 1 18 926 3 f 1 18 910
21 4 r 1 18 997 4 f 1 18 982 5 r 1 18 969 5 f 1 18 969
22 1 r 2 3 368 1 f 2 3 354 2 r 2 3 345 2 f 2 3 399
23 3 r 2 3 311 3 f 2 3 311 4 r 2 3 302 4 f 2 3 301
24 5 r 2 3 322 5 f 2 3 322 1 r 2 6 474 1 f 2 6 455
25 2 r 2 6 447 2 f 2 6 403 3 r 2 6 417 3 f 2 6 416
26 4 r 2 6 407 4 f 2 6 401 5 r 2 6 424 5 f 2 6 430
27 1 r 2 9 674 1 f 2 9 655 2 r 2 9 654 2 f 2 9 600
28 3 r 2 9 615 3 f 2 9 617 4 r 2 9 602 4 f 2 9 602
29 5 r 2 9 623 5 f 2 9 632 1 r 2 12 783 1 f 2 12 759
30 2 r 2 12 757 2 f 2 12 704 3 r 2 12 719 3 f 2 12 724
31 4 r 2 12 712 4 f 2 12 710 5 r 2 12 731 5 f 2 12 737
32 1 r 2 15 773 1 f 2 15 755 2 r 2 15 750 2 f 2 15 703
33 3 r 2 15 714 3 f 2 15 722 4 r 2 15 704 4 f 2 15 707
34 5 r 2 15 723 5 f 2 15 732 1 r 2 18 685 1 f 2 18 662
35 2 r 2 18 652 2 f 2 18 615 3 r 2 18 627 3 f 2 18 617
36 4 r 2 18 619 4 f 2 18 610 5 r 2 18 628 5 f 2 18 637
37 1 r 3 3 621 1 f 3 3 644 2 r 3 3 601 2 f 3 3 611
38 3 r 3 3 699 3 f 3 3 699 4 r 3 3 663 4 f 3 3 655
39 5 r 3 3 676 5 f 3 3 676 1 r 3 6 630 1 f 3 6 655
40 2 r 3 6 608 2 f 3 6 623 3 r 3 6 604 3 f 3 6 605
41 4 r 3 6 668 4 f 3 6 666 5 r 3 6 684 5 f 3 6 686
42 1 r 3 9 724 1 f 3 9 748 2 r 3 9 713 2 f 3 9 713
43 3 r 3 9 702 3 f 3 9 700 4 r 3 9 766 4 f 3 9 756
44 5 r 3 9 781 5 f 3 9 779 1 r 3 12 838 1 f 3 12 860
45 2 r 3 12 813 2 f 3 12 826 3 r 3 12 806 3 f 3 12 816
46 4 r 3 12 869 4 f 3 12 867 5 r 3 12 889 5 f 3 12 892
47 1 r 3 15 931 1 f 3 15 951 2 r 3 15 912 2 f 3 15 919
48 3 r 3 15 900 3 f 3 15 903 4 r 3 15 972 4 f 3 15 964
49 5 r 3 15 981 5 f 3 15 984 1 r 3 18 1031 1 f 3 18 1066
50 2 r 3 18 1011 2 f 3 18 1033 3 r 3 18 1017 3 f 3 18 1016
51 4 r 3 18 1076 4 f 3 18 1076 5 r 3 18 1091 5 f 3 18 1098
52 RUN;
53 PROC PRINT;
54 RUN;
55 PROC SORT DATA = one; BY bedheight;
56 PROC ANOVA DATA = one; BY bedheight;
57 CLASSES block cultivar spacing;
58 MODEL yield = block cultivar cultivar*block
59    spacing spacing*cultivar spacing*cultivar(block);
60 TEST H = cultivar E = cultivar*block;
61 TEST H = spacing spacing*cultivar E = spacing*cultivar(block);
62 MEANS cultivar / REGWQ E = cultivar*block PLOTS = none;
63 RUN;
```
The results:

'Chapter 14 Example 10.sas'
Split-split-plot Design
ANOVA

The ANOVA Procedure

bedheight=f

Class Level Information		
Class	Levels	Values
block	5	1 2 3 4 5
cultivar	3	1 2 3
spacing	6	3 6 9 12 15 18

Number of Observations Read	90
Number of Observations Used	90

'Chapter 14 Example 10.sas'
Split-split-plot Design
ANOVA

The ANOVA Procedure

Dependent Variable: yield

bedheight=f

Source	DF	Sum of Squares	Mean Square	F Value	Pr > F
Model	89	2924601.122	32860.687	.	.
Error	0	0.000	.		
Corrected Total	89	2924601.122			

R-Square	Coeff Var	Root MSE	yield Mean
1.000000	.	.	700.1444

Source	DF	Anova SS	Mean Square	F Value	Pr > F
block	4	26590.844	6647.711	.	.
cultivar	2	865437.756	432718.878	.	.
block*cultivar	8	22956.689	2869.586	.	.
spacing	5	1759518.589	351903.718	.	.
cultivar*spacing	10	229263.178	22926.318	.	.
cultiv*spacin(block)	60	20834.067	347.234	.	.

Tests of Hypotheses Using the Anova MS for block*cultivar as an Error Term					
Source	DF	Anova SS	Mean Square	F Value	Pr > F
cultivar	2	865437.7556	432718.8778	150.79	<.0001

Tests of Hypotheses Using the Anova MS for cultiv*spacin(block) as an Error Term					
Source	DF	Anova SS	Mean Square	F Value	Pr > F
spacing	5	1759518.589	351903.718	1013.45	<.0001
cultivar*spacing	10	229263.178	22926.318	66.03	<.0001

The ANOVA Procedure

Ryan-Einot-Gabriel-Welsch Multiple Range Test for yield

bedheight=f

Note: This test controls the Type I experimentwise error rate.

Alpha	0.05
Error Degrees of Freedom	8
Error Mean Square	2869.586

Number of Means	2	3
Critical Range	31.894944	39.521421

yield REGWQ Grouping for Means of cultivar (Alpha = 0.05)

Means covered by the same bar are not significantly different.

cultivar	Estimate
3	816.23
1	707.80
2	576.40

'Chapter 14 Example 10.sas'
Split-split-plot Design
ANOVA

The ANOVA Procedure

bedheight=r

Class Level Information		
Class	Levels	Values
block	5	1 2 3 4 5
cultivar	3	1 2 3
spacing	6	3 6 9 12 15 18

Number of Observations Read	90
Number of Observations Used	90

The ANOVA Procedure

Dependent Variable: yield

bedheight=r

Source	DF	Sum of Squares	Mean Square	F Value	Pr > F
Model	89	2880073.822	32360.380	.	.
Error	0	0.000	.		
Corrected Total	89	2880073.822			

R-Square	Coeff Var	Root MSE	yield Mean
1.000000	.	.	706.8444

Source	DF	Anova SS	Mean Square	F Value	Pr > F
block	4	16879.156	4219.789	.	.
cultivar	2	767860.556	383930.278	.	.
block*cultivar	8	41290.111	5161.264	.	.
spacing	5	1802843.156	360568.631	.	.
cultivar*spacing	10	238501.311	23850.131	.	.
cultiv*spacin(block)	60	12699.533	211.659	.	.

Tests of Hypotheses Using the Anova MS for block*cultivar as an Error Term					
Source	DF	Anova SS	Mean Square	F Value	Pr > F
cultivar	2	767860.5556	383930.2778	74.39	<.0001

Tests of Hypotheses Using the Anova MS for cultiv*spacin(block) as an Error Term					
Source	DF	Anova SS	Mean Square	F Value	Pr > F
spacing	5	1802843.156	360568.631	1703.54	<.0001
cultivar*spacing	10	238501.311	23850.131	112.68	<.0001

'Chapter 14 Example 10.sas'
Split-split-plot Design
ANOVA

The ANOVA Procedure

Ryan-Einot-Gabriel-Welsch Multiple Range Test for yield

bedheight=r

Note: This test controls the Type I experimentwise error rate.

Alpha	0.05
Error Degrees of Freedom	8
Error Mean Square	5161.264

Number of Means	2	3
Critical Range	42.775023	53.003061

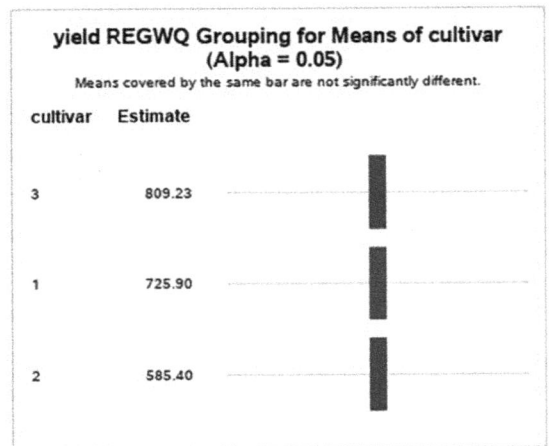

yield REGWQ Grouping for Means of cultivar
(Alpha = 0.05)

Means covered by the same bar are not significantly different.

cultivar	Estimate
3	809.23
1	725.90
2	585.40

The first interesting observation with these results is that significant cultivar × spacing interactions were detected on both raised and flat beds. You might think that this means we should look at spacing effects separately for each cultivar × bed height combination. However, check the original ANOVA results and you will see that the bed height × cultivar × spacing interaction was not significant ($\alpha = 0.89$).

Since the purpose of the analysis above was to determine cultivar differences for the two different bed height treatments, that's what we should focus on. The results can be presented nicely in a small table:

Yield of three strawberry cultivars grown on raised or flat beds

	Bed height	
	Flat	Raised
Cultivar	Yield (g)	
1	708 b	726 b
2	576 c	585 c
3	816 a	809 a

zMean separation within column with REGWQ test, $\alpha = 0.05$.

Even though the cultivar × bed height interaction was significant, which indicates that cultivar performance depended on bed height, cultivar 3 produced the highest yield on both raised and flat beds, followed by cultivar 1, then cultivar 2. Normally one would expect to see a different order of cultivar productivity depending on bed height given that there was a significant interaction. Additionally, when the data is analyzed to compare bed height within cultivar, no bed height effects were observed for any cultivar.

These results indicate that, even though significant interactions were detected in the original ANOVA, the effects were too small to be detected with the reduced power of the analyses (since there were fewer df for testing) evaluating cultivar within bed height or bed height within cultivar. The magnitude of the difference in yield comparing raised beds to flat beds within cultivar ranged from 7 g for cultivar 3, 9 g for cultivar 2, and 18 g for cultivar 1. The average weight of a strawberry normally ranges from 10 g to 30 g, so the difference the original ANOVA detected was in the order of 1 berry per plant. While this may seem insignificant, consider that there may be 15,000 or more plants per acre which could translate into 15,000 additional or fewer berries per acre depending on bed height chosen for production. Fifteen thousand berries weigh around 225 kg or 495 lb. The researcher would need to determine whether or not this value was horticulturally significant.

In this case, I think it would be prudent to compare bed height within cultivar using the formulas provided. This would be useful to determine whether or not it is worth the expense of making raised beds for cultivars 1 and 2, i.e. is the slight increase in yield due to the raised beds worth the cost?

Comparison	LSD formula
Compare bed height means for the same or different cultivar	$t_{ac}\sqrt{\dfrac{2((c-1)(Ec)+Ea)}{rbc}}$

$$t_{ac} = \frac{(c-1)Ec(t_c) + Ea(t_a)}{(c-1)Ec + Ea}$$

where r = number of replications, a = number of levels of bed height, b = number of levels of cultivar, c = number of levels of spacing, Ea is MS Error for main plots, Eb is MS Error for sub-plots, and Ec is MS Error for sub-sub plots, t_a is t-value for df Ea, t_b is t-value for df Eb, t_c is t-value for df Ec.

The t-values can be retrieved from a table of t probabilities or the values can be obtained from SAS® by replacing lines 64 to 67 in Example 7 with the following lines:

```
64 MEANS bedheight / LSD E = block*bedheight PLOTS = none;
65 MEANS cultivar / LSD E = cultivar*bedheight(block) PLOTS = none;
66 MEANS spacing / LSD E = spacing*cultivar*bedheight(block) PLOTS
   = none;
67 RUN;
```

Once you run the program, look for the output below in the mean separation section of the results. These boxes contain the information you need:

Alpha	0.05
Error Degrees of Freedom	4
Error Mean Square	1063.8
Critical Value of t	2.77645
Least Significant Difference	13.499

Alpha	0.05
Error Degrees of Freedom	16
Error Mean Square	4015.425
Critical Value of t	2.11991
Least Significant Difference	24.526

Alpha	0.05
Error Degrees of Freedom	120
Error Mean Square	279.4467
Critical Value of t	1.97993
Least Significant Difference	8.5458

Thus $r = 5$, $a = 2$, $b = 3$, $c = 6$, $Ea = 1063.8$, $Eb = 4015.425$, $Ec = 279.4467$, $ta = 2.77645$, $tb = 2.11991$ and $tc = 1.97993$.

Now enter these values in the formulas, calculating the t_{ac} first:

$$tac = \frac{(c-1)Ec(tc) + Ea(ta)}{(c-1)Ec + Ea}$$

$$tac = \frac{(6-1)279.4467(1.97993) + 1063.8(2.77645)}{(6-1)279.4467 + 1063.8}$$

$$tac = \frac{5720.012034}{2461.0335} = 2.324$$

Now calculate the LSD value:

$$t_{ac}\sqrt{\frac{2((c-1)(Ec) + Ea)}{rbc}}$$

$$LSD = 2.324\sqrt{\frac{2(5)(279.4467) + 1063.8}{5*3*6}} = 15.2$$

Now we can compare bed heights within cultivar:

Yield of three strawberry cultivars grown on raised or flat beds

	Bed height	
	Flat	Raised
Cultivar	Yield (g)	
1	708 b	726 a
2	576 a	585 a
3	816 a	809 a

^{z}Mean separation within row using $LSD_{\alpha = 0.05} = 15.2$.

Now we have revealed the nature of the bed height × cultivar interaction. Bed height has no impact on yield for cultivars 2 and 3. Raised beds enhance production for cultivar 1, by approximately 595 lb per acre (667 kg/ha) (assuming 15,000 plants/acre and average berry size of 15 g). This was a good example of where some hand calculations are worth the effort. The difference between raised- and flat-bed production using cultivar 1 would likely have been revealed with the SAS® estimation approach if greater replication were used. If we had used the SAS® estimation approach to reach our conclusions, we would have made a Type II error by declaring that two means were not different (cultivar 1 raised bed vs cultivar 1 flat bed) when in fact they were different.

Variations of the Split-plot

Two widely used variations of the split-plot design are the split-block and the strip-plot designs. In the split-block design, the main unit treatment is applied as in the split-plot.

The subunit treatment is applied as a strip to the main unit plots. Harvest or observation time can be considered a stripped factor in a split block experiment. This creates an additional error term and an additional standard error for mean separation calculations. In the strip-plot, the main unit treatment is applied as a strip within each block. The subunit treatment is applied as a strip to the main unit plots. Again, additional error terms must be calculated as well as additional standard errors for mean separation.

If you have a situation where either the main or the sub-plot treatments would more easily be applied as a strip across the field, one or both of these designs might be appropriate. There is simply not enough space in this text to cover these variations; however, two great references for them are *Agricultural Experimentation* by Little and Hills and *Statistical Procedures for Agricultural Research* by Gomez and Gomez.

References

Little, T.M. and Hills, F.J. (1978) *Agricultural Experimentation. Design and Analysis.* Wiley and Sons, New York. ISBN 0-471-02352-3.

Gomez, K.A. and Gomez, A.A. (1984) *Statistical Procedures for Agricultural Research.* Wiley and Sons, New York. ISBN 0-471-87092-7.

Expected Mean Squares

Expected mean squares are formulas based on statistical theory identifying the components of variability in sources of variation of an ANOVA table. Their theoretical derivation is beyond the scope of this text, thus they are presented in a simpler way here, providing a method for deriving expected mean squares without a background in statistical theory. If you know how to derive expected mean squares, you can determine appropriate denominators for any F-test in any ANOVA. This will become especially useful in the next chapter, where we cover experiments repeated over time and space.

Introduction

When we perform experiments, we gather data and calculate mean squares in order to estimate variability. These mean squares are estimates of variance associated with specific sources of variation in our ANOVA table. With variance estimates, we can perform tests to determine whether or not there are treatment effects.

Each mean square contains variance components or pieces of variability. Expected mean squares (EMS) are formulas, not numbers, which show us what pieces of variability are in our calculated mean square. In F-tests, we test these components to determine which account for the majority of the variability in our experiment. If a specific component contributes significant variability, we reject the null hypothesis that there is no effect in favor of the alternate hypothesis. The alternate is usually that there is a difference among the means of the levels of the specific component tested. The key in all F-tests is that you must know the correct denominator for your F-test to get correct results. In many designs, the correct denominator is 'built in', so there is no need to figure out the EMS.

What do we do if the EMS formulas are not 'built in'? We must derive our EMS formulas to determine the correct denominator(s) for our F-test(s) either by using mathematical proofs or a set of 'cookbook' rules.

Fixed vs Random Effects

First, we must briefly revisit fixed versus random effects. If the k-levels of interest for a particular factor exhaust all possible levels of interest for that factor, then we have a fixed effect. No inferences about the entire population of possible factor levels can be made. Only inferences about the levels investigated are valid. Most plant science experiments involve fixed effects. If k-levels of a factor constitute a random sample from a larger population of levels, then we have a random effect. Inferences about all of the possible levels of the factor can be made with a random effect.

For example, suppose we were interested in snap bean productivity. We determine that there are 100 cultivars of snap beans in existence (the population of snap bean cultivars).

DOI: 10.1079/9781789249927.0015

If we chose ten specific cultivars to trial based on their climate adaptation, we would say that cultivar is a fixed effect. We could not make any inferences about the productivity of the other 90 cultivars. Means of the levels are different by amounts that are distributed with a mean of 0 and no variance. They differ by the specific fixed amount denoted α_I.

But suppose we randomly picked ten cultivars to trial. We would then say that cultivar is random. We could make inferences about the potential performance of the other 90 cultivars (however, since cultivars are quite climate specific, it would not be a good idea to make such inferences even though they would be 'statistically legal'). Means of the levels differ by amounts that are distributed with a mean of 0 and variance σ^2_{effect}.

Another way to look at it is to consider a situation with factor A and its defined population of ten different levels. Also, let's consider two experiments. In experiment one, we investigate the fixed effects of factor A with the fixed levels of 1, 3 and 5. We repeat the experiment, again looking at the fixed levels of 1, 3 and 5. The fixed-effect model effect of factor A will be the same estimate in both experiments. It will differ only by σ^2 (random error) + α_I (the specific effect of the specific level of factor A under consideration). Now consider a second set of two experiments, but this time we consider the random effect of factor A. In experiment one, the randomly chosen levels of factor A are 1, 2 and 3. In the second experiment, the randomly chosen levels of factor A are 4, 6 and 8. The random-effects model effect of factor A will be different in the two different experiments. It differs by σ^2 (random error) + σ^2_A (error associated with random sampling of the levels of A).

The important thing is to know if you are considering fixed or random effects. If you are looking at a fixed effect, use α_I symbols in your EMS derivation table; and if you are looking at a random effect, use σ^2_A symbols in your table. A correct F-test depends on your EMS equations; and correct EMS equations depend on whether an effect is fixed or random.

In our source column in our ANOVA tables, we are interested in deriving EMS equations for main effects, interactions and error terms. We do not have to worry about the two sources, 'Total' and 'μ'. Suppose we had a single factor, A, at 'a' levels with 'n' replications in a CRD design. The sources in our ANOVA table we would be interested in are A and error. To generate the EMS equations, first set up a source column, then repeat it sideways as follows:

A		
Error		
	Error	A

Put an X in the error column, since every row or source has a component of variance due to error or natural variability, σ^2_E.

A	X	
Error	X	
	Error	A

Put an X on the diagonal, since each row will have its source as a component of the variance in its EMS.

A	X	X
Error	X	
	Error	A

For this example, the rest is easy. We need to convert our 'X' indicators to meaningful pieces of EMS equations. For the diagonally placed X indicating that each row has itself as component of variance in its EMS, we use the symbol σ^2_{Source} for a random effect and Σ^2_{Source} for fixed effect. In our example if A is random we indicate this via σ^2_A and if A is fixed we indicate it via Σ^2_A. For the component due to error we indicate it as σ^2_E. So far, our EMS equations for this simple example would look like:

Fixed:

A	$X = \sigma^2_E$	$X = \Sigma^2_A$
Error	$X = \sigma^2_E$	
	Error	A

Random:

A	$X = \sigma^2_E$	$X = \sigma^2_A$
Error	$X = \sigma^2_E$	
	Error	A

We add the columns together to produce the following:

Fixed:

Source	EMS
A	$\sigma^2_E + \Sigma^2_A$
Error	σ^2_E

Random:

Source	EMS
A	$\sigma^2_E + \sigma^2_A$
Error	σ^2_E

We're not quite done with the EMS equations yet. We need to add some coefficients to each piece of variation in the EMS equation to make them statistically legitimate. The rules are:

1. Each coefficient has an n in it (number of replications).
2. Any letter not in the name of the source (row) must be in the coefficient.

For example, in the piece in our EMS equation attributable to Error, the coefficient for σ^2_E would have 'n' (rule 1) and 'a' (rule 2). Thus the piece of our EMS equations attributable to σ^2_E would look like: $na\sigma^2_E$.

For the piece in our EMS equation attributable to factor A, the coefficient would be n. Thus our final EMS tables for this example would look like:

Fixed:

Source	EMS
A	$na\sigma^2_E + n\Sigma^2_A$
Error	$na\sigma^2_E$

Random:

Source	EMS
A	$na\sigma^2_E + n\sigma^2_A$
Error	$na\sigma^2_E$

In either case, to determine the correct denominator for an F-test, find a source of variation in the source column with everything in its EMS but the source to be tested. In this case for A, either fixed or random, the appropriate denominator contains only $na\sigma^2_E$ and the correspondingly appropriate source for the denominator is Error. More complicated cases are not much more difficult.

Let's consider an example. Two factors, 'A' and 'B' at 'a' and 'b' levels, respectively, replicated n times in a CRD. As before, set up a source column then repeat it sideways:

A				
B				
AB				
Error				
	Error	AB	B	A

Put an X in the error column, since every row or source has a component of variance due to error or natural variability, σ^2_E, and put an X on the diagonal, since each row will have its source as a component of the variance in its EMS.

A	X			X
B	X		X	
AB	X	X		
Error	X			
	Error	AB	B	A

Again we will use the symbols σ^2_{Source} for a random effect and Σ^2_{Source} for a fixed effect. With interactions, if one factor is random, then the entire interaction is random. Notice that we now have three empty boxes (we always only consider above the diagonal). We need to determine if there should be an X in each box to indicate that the column source of variation is in the row's EMS. To determine if there should be an X in the box, we examine each row. To put an X in a column, the column must pass two tests:

1. It must contain *all* the letters of the row.
2. All the letters in the column that are not in the row must correspond to random factors.

So for example, assume A and B are both random. Start with row B. Should column AB get an X? Does column contain all the letters of the row? Yes. Are all letters in the column (in this case, A) that are not in the row corresponding to random factors? Yes. Thus column AB gets an X in row B. Now what about row A? Does AB get an X? Does column contain all the letters of the row? Yes. Are all letters in the column (in this case, B) that are not in the row corresponding to random factors? Yes. Does column B get an X? No, column B does not contain A. Thus our table looks like:

A	X	X	–	X
B	X	X	X	
AB	X	X		
Error	X			
	Error	AB	B	A

Now we can complete our EMS table by deriving the equations from the previous table:

A	σ^2_E	σ^2_{AB}		σ^2_A
B	σ^2_E	σ^2_{AB}	σ^2_B	
AB	σ^2_E	σ^2_{AB}		
Error	σ^2_E			
	Error	AB	B	A

The coefficients for each piece of variation in the sources follow our previous rules: each coefficient has an 'n' in it (number of replications) and any letter not in the name of the source (row) must be in the coefficient. Thus for our two-factor example, both A and B random:

Source	EMS
A	$nab\sigma^2_E + n\sigma^2_{AB} + nb\sigma^2_A$
B	$nab\sigma^2_E + n\sigma^2_{AB} + na\sigma^2_B$
AB	$nab\sigma^2_E + n\sigma^2_{AB}$
Error	$nab\sigma^2_E$

Again, to test a source of variation for significance, look for a denominator that has an EMS with everything in it but the source you are testing. Thus, to test AB, look for a denominator with $nab\sigma^2_E$ in it. It's the Error row. To test both A and B, look for a denominator with $nab\sigma^2_E + n\sigma^2_{AB}$ in it. It's the AB row. But why have we never used the interaction to test for a main effect in any of our previous ANOVAs? Because we have always assumed we were dealing with fixed effects. This example assumed that both effects were random.

You should practice your EMS derivation skills and verify that the three EMS tables and associated F-tests which follow are correct:

A fixed, B random:

A	X	X	–	X
B	X	–	X	
AB	X	X		
Error	X			
	Error	AB	B	A

which yields:

Source	EMS
A	$nab\sigma^2_E + n\sigma^2_{AB} + nb\Sigma^2_A$
B	$nab\sigma^2_E + na\sigma^2_B$
AB	$nab\sigma^2_E + n\sigma^2_{AB}$
Error	$nab\sigma^2_E$

Test A with AB, test B and AB with Error.

A random, B fixed:

A	X	–	–	X
B	X	X	X	
AB	X	X		
Error	X			
	Error	AB	B	A

which yields:

Source	EMS
A	$nab\sigma^2_E + nb\sigma^2_A$
B	$nab\sigma^2_E + n\sigma^2_{AB} + na\Sigma^2_B$
AB	$nab\sigma^2_E + n\sigma^2_{AB}$
Error	$nab\sigma^2_E$

Test A with Error, test B with AB, test AB with Error.

Both A and B fixed:

A	X	–	–	X
B	X	–	X	
AB	X	X		
Error	X			
	Error	AB	B	A

which yields:

Source	EMS
A	$nab\sigma^2_E + nb\Sigma^2_A$
B	$nab\sigma^2_E + na\Sigma^2_B$
AB	$nab\sigma^2_E + n\Sigma^2_{AB}$
Error	$nab\sigma^2_E$

Test A, B and AB with Error.

Now suppose you have a three-factor example of A fixed, B and C random, at 'a', 'b' and 'c' levels respectively, replicated n times in a CRD. To make things a little easier to keep track of, add a column for F or R, to indicate fixed or random.

F or R	Source								
F	A	X							X
R	B	X						X	
R	C	X					X		
R	AB	X				X			
R	AC	X			X				
R	BC	X		X					
R	ABC	X	X						
	Error	X							
		Error	ABC	BC	AC	AB	C	B	A

Now test each column, row by row, to determine if it should be included in each row's EMS:

1. Column ABC
   a. In row BC? ABC contains BC. A random? No, thus no X.
   b. In row AC? ABC contains AC. B random? Yes, thus it gets an X.
   c. In row AB? ABC contains AB. C random? Yes, gets an X.
   d. In row C? ABC contains C. Both A and B random? No, no X.
   e. In row B? ABC contains B. Both A and C random? No, no X.
   f. In row A? ABC contains A. Both B and C random? Yes, gets an X

2. Column BC
   a. In row AC? No, BC does not contain A. No X.
   b. In row AB? No, BC does not contain A. No X.
   c. In row C? BC contains C. B random? Yes, gets an X.
   d. In row B? BC contains B. C random? Yes, gets an X.
   e. In row A? No, BC does not contain A. No X.

3. Column AC
   a. In row AB? No, AC does not contain B. No X.
   b. In row C? AC contains C. A random? No, no X.
   c. In row B? No, AC does not contain B. No X.
   d. In row A? AC contains A. C random? Yes, gets an X.

4. Column AB
   a. In row C? No, AB does not contain C. No X.
   b. In row B? AB contains B. A random? No, no X.
   c. In row A? AB contains A. B random? Yes, gets an X.

5. Column C
   a. In row B? No, C does not contain B. No X.
   b. In row A? No, C does not contain A. No X.

6. Column B
   a. In row A? No, B does not contain A. No X.

Now write out EMS based on the following:

F or R	Source								
F	A	X	X		X	X			X
R	B	X		X				X	
R	C	X		X			X		
R	AB	X	X			X			
R	AC	X	X		X				
R	BC	X		X					
R	ABC	X	X						
	Error	X							
		Error	ABC	BC	AC	AB	C	B	A

This yields (without coefficients):

Source	EMS							
A	σ^2	σ^2_{ABC}	—	σ^2_{AC}	σ^2_{AB}	—	—	Σ^2_A
B	σ^2	—	σ^2_{BC}	—	—	—	c^2_B	
C	σ^2	—	σ^2_{BC}	—	—	σ^2_C		
AB	σ^2	σ^2_{ABC}	—	—	σ^2_{AB}			
AC	σ^2	σ^2_{ABC}	—	σ^2_{AC}				
BC	σ^2	—	σ^2_{BC}					
ABC	σ^2	σ^2_{ABC}						
Error	σ^2							

Complete the EMS table by adding the coefficients:

Source	EMS	
A	$nabc\sigma^2 + n\sigma^2_{ABC}$	$+ nb\sigma^2_{AC} + nc\sigma^2_{AB} + nbc\Sigma^2_A$
B	$nabc\sigma^2 + na\sigma^2_{BC}$	$+ nac\sigma^2_B$
C	$nabc\sigma^2 + na\sigma^2_{BC}$	$+ nba\sigma^2_C$
AB	$nabc\sigma^2 + n\sigma^2_{ABC}$	$+ nc\sigma^2_{AB}$
AC	$nabc\sigma^2 + n\sigma^2_{ABC}$	$+ nb\sigma^2_{AC}$
BC	$nabc\sigma^2 + na\sigma^2_{BC}$	
ABC	$nabc\sigma^2 + n\sigma^2_{ABC}$	
Error	$nabc\sigma^2$	

Now determine correct tests:

1. Test ABC with error
2. Test BC with error
3. Test AC with ABC
4. Test AB with ABC
5. Test C with BC
6. Test B with BC
7. Test A with ?????

What do we do about testing A since there is no denominator in the EMS table that would be appropriate? We have to construct a denominator when an appropriate one is not in the EMS table. To test A, we need a denominator with $nabc\sigma^2 + n\sigma^2_{ABC} + nb\sigma^2_{AC} + nc\sigma^2_{AB}$ in it. Examine the EMS table carefully to determine if you could add and/or subtract EMS from different sources to get the one you need. This sometimes takes a little trial and error to find the right combination. In our case, AC + AB − ABC provides the needed denominator EMS. In your actual *F*-testing, you would add and subtract the actual numbers (mean squares) for these sources and the new mean square is designated MS* to remind you it was derived. Additionally, the df associated with it must be adjusted via the formula:

$$df^* = \frac{MS^{*2}}{\dfrac{MSAB^2}{dfAB} + \dfrac{MSAC^{*2}}{dfAC} + \dfrac{MSABC^2}{dfABC}}$$

A good exercise to test your EMS derivation skills is to derive the EMS for an RCB design to see why you don't test for blocks. In the RCBD, we assume blocks are random and treatments are fixed.

16 Analyzing a Series of Experiments

In many plant science experiments we normally can't control certain factors which may influence the outcome of our work. Such factors include weather and pest pressure. In order to get an accurate and reliable measure of crop performance we repeat the experiment over time, location or both. These repeated experiments should be analyzed in such a fashion that the data from each individual experiment are combined into one grand analysis. This chapter will cover these types of experiments and their analysis.

While there are other approaches to evaluating experiments over time and location using SAS® procedures such as PROC MIXED, I prefer to use the approach outlined in this chapter. The presented approach is a bit more involved, but I prefer it since it really shows you not only *how* to perform the analysis, but *why* to perform it in a specific manner.

Analysis Over Seasons

In many crops (mostly annuals) there may be several distinct crops produced in one calendar year. For example, broccoli can be produced in both the fall and spring growing season. Rice can be grown in the wet and the dry seasons. These planting seasons are fixed effects.

A randomized complete block design was used to study the influence of nitrogen fertilization on broccoli productivity during two seasons, spring (s) and fall (f). The data collected was for broccoli yields (kg) per 3 m plot.

Season	Block	Nitrogen	Yield
s	1	0	4
s	1	60	6
s	1	90	6
s	1	120	6
s	1	150	5
s	2	0	2
s	2	60	6
s	2	90	6
s	2	120	6
s	2	150	6
s	3	0	4
s	3	60	5
s	3	90	6
s	3	120	6
s	3	150	5
f	1	0	4
f	1	60	6

Continued

DOI: 10.1079/9781789249927.0016

Continued.

Season	Block	Nitrogen	Yield
f	1	90	6
f	1	120	4
f	1	150	3
f	2	0	3
f	2	60	6
f	2	90	5
f	2	120	4
f	2	150	4
f	3	0	5
f	3	60	6
f	3	90	5
f	3	120	5
f	3	150	3

The approach to analyzing an experiment over seasons is straightforward. The first step is to determine whether or not the variances of the seasons are homogeneous or not. This is important for determining the appropriate denominators for season, treatment and season × treatment F-tests.

STEP ONE

Test the homogeneity of variance assumption with Brown and Forsythe's (BF) test to determine if the variances of the two seasons are homogeneous. While there are a number of options for evaluating homogeneity of variances with the HOVTEST option of the MEANS statement, and all are fairly robust, the BF option is the one recommended by SAS® since it is powerful in detecting differences in variances and at the same time protects from Type I errors (saying variances are not homogeneous when in fact they are). A slight drawback to the BF option is that it requires significant computing power when there are many groups (seasons in this case) or when some groups (seasons) have many observations. The following statements will perform the test in SAS®:

```
TITLE1 'Chapter 16 Example 1.sas';
TITLE2 'Analysis of Data Over Seasons';
TITLE3 'Combined ANOVA';
TITLE4 'Test for Homogeneous Variances';
DATA one;
INPUT season $ block nitrogen yield;
nitrogen2 = nitrogen*nitrogen;
lof = nitrogen;
CARDS;
s 1 0 4
s 1 60 6
s 1 90 6
s 1 120     6
s 1 150     5
s 2 0 2
```

```
s 2 60 6
s 2 90 6
s 2 120        6
s 2 150        6
s 3 0 4
s 3 60 5
s 3 90 6
s 3 120        6
s 3 150        5
f 1 0 4
f 1 60 6
f 1 90 6
f 1 120        4
f 1 150        3
f 2 0 3
f 2 60 6
f 2 90 5
f 2 120        4
f 2 150        4
f 3 0 5
f 3 60 6
f 3 90 5
f 3 120        5
f 3 150        3
RUN;
PROC ANOVA DATA = one PLOTS = none;
CLASSES season;
MODEL yield = season;
MEANS season / HOVTEST = BF;
RUN;
```
The results:

Chapter 16 Example 1.sas
Analysis of Data Over Seasons
Combined ANOVA
Test for Homogeneous Variances

The ANOVA Procedure

Class Level Information		
Class	Levels	Values
season	2	f s

Number of Observations Read	30
Number of Observations Used	30

The ANOVA Procedure

Dependent Variable: yield

Source	DF	Sum of Squares	Mean Square	F Value	Pr > F
Model	1	3.33333333	3.33333333	2.55	0.1212
Error	28	36.53333333	1.30476190		
Corrected Total	29	39.86666667			

R-Square	Coeff Var	Root MSE	yield Mean
0.083612	23.15395	1.142262	4.933333

Source	DF	Anova SS	Mean Square	F Value	Pr > F
season	1	3.33333333	3.33333333	2.55	0.1212

The ANOVA Procedure

Brown and Forsythe's Test for Homogeneity of yield Variance ANOVA of Absolute Deviations from Group Medians					
Source	DF	Sum of Squares	Mean Square	F Value	Pr > F
season	1	0.3000	0.3000	0.32	0.5733
Error	28	25.8667	0.9238		

The ANOVA Procedure

Level of season	N	yield	
		Mean	Std Dev
f	15	4.60000000	1.12122382
s	15	5.26666667	1.16291915

To determine whether or not to reject the hypothesis of homogeneous variances between the seasons, find the output labeled 'Brown and Forsythe's Test for Homogeneity of yield Variance' and review the Pr > F for the 'season' source of variation in the ANOVA of absolute deviations from group medians. Reject the null hypothesis if Pr > F is less than or equal to 0.05. In our example the Pr > F = 0.5733, indicating that we should fail to reject the null hypothesis and that we most likely have homogeneous variances between the two seasons.

STEP TWO

Construct an ANOVA table for the combined analysis (s = seasons, r or b = replications or blocks, t = treatment):

Source	df
Total	sbt
μ	1
Seasons	s–1
Blocks(Season)	s(b–1)
Treatment	t–1
Season×Treatment	(s–1)(t–1)
Pooled error	s(b–1)(t–1)

Notice how we have a different notation for the 'Blocks' source of variation. While in other ANOVAs we have seen 'Blocks' source simply identified as 'blocks', in this situation we identify 'Blocks' as 'Blocks(season)' which is read as 'blocks within season'. This indicates that the blocks are randomly assigned, separately, within each season. The nice thing is that SAS® interprets this notation to correctly adjust ANOVA calculations. As long as you follow the above ANOVA table as a guideline for your analysis, you will get the correct results. With the ANOVA table in hand, we need to determine the correct denominators for F-tests by constructing an EMS table.

F or R						
F	Season					
R	Block(season)					
F	Treatment					
F	Season × treatment					
	Error					
		Error	Season × treatment	Treatment	Block(season)	Season

First place an 'X' in each cell of the 'Error' column, since all sources have a variance component attributed to error. Also place an 'X' in each diagonal cell, since each source has a component (variance for random effect and fixed component for fixed effect) attributed to itself. Your table should look like this:

F or R						
F	Season	X				X
R	Block(season)	X			X	
F	Treatment	X		X		
F	Season × treatment	X	X			
	Error	X				
		Error	Season × treatment	Treatment	Block(season)	Season

Now evaluate each column to determine if it should be included as a component of the EMS for each row (source of variation) following the rules set forth in Chapter 15.

As a refresher, those rules are as follows.

To determine if there should be an X in the cell of a column, the column must pass two tests:

1. It must contain *all* the components of the row.
2. All the letters in the column that are not in the row must correspond to random factors.

Then we add the coefficients to each piece of variation in the EMS equation using the rules:

1. Each coefficient has an *n* in it (number of replications).
2. Any letter not in the name of the source (row) must be in the coefficient.

- For the 'Season × treatment' column:
  Should the 'Treatment' cell get an 'X'? It contains all the components of the row. Is 'Season' random? No, thus do not put an 'X' in the cell.
  Should the 'Block(season)' cell get an 'X'? No, it doesn't contain all the components of the row.
  Should the 'Season' cell get an 'X'? It contains all the components of the row. Is 'Treatment' random? No, thus do not put an 'X' in the cell.
- For the 'Treatment' column:
  Should the 'Block(season)' cell get an 'X'? No, it doesn't contain all the components of the row. It doesn't contain 'Block(season)'.
  Should the 'Season' cell get an 'X'? No, it doesn't contain 'Block(season)'.
- For the 'Block(season)' column:
  Should the 'Season' cell get an 'X'? It contains all the components of the row (the 'Season' component). Is 'Block' random? Yes, thus put an 'X' in the cell.

Note that we don't pay attention to the way the factors such as block or season are represented in the Source column when determining if the column contains all of the components of the row. For example, we don't consider if blocks are represented as 'blocks × season' or 'blocks(season)'. We just consider whether or not the factor is present. By the way, the first representation 'blocks × season' is called 'crossed' while the second representation 'blocks(season)' is called 'nested'.

Your finished table should look like this:

F or R						
F	Season	X			X	X
R	Block(season)	X			X	
F	Treatment	X		X		
F	Season × treatment	X	X			
	Error	X				
		Error	Season × treatment	Treatment	Block(season)	Season

which translates into:

F or R						
F	Season	$nstb\sigma_E^2$			$nt\sigma_{B(S)}^2$	$nbt\sum_S^2$
R	Block(season)	$nstb\sigma_E^2$			$nt\sigma_{B(S)}^2$	
F	Treatment	$nstb\sigma_E^2$		$nbs\sum_T^2$		
F	Season × treatment	$nstb\sigma_E^2$	$nb\sum_{ST}^2$			
	Error	$nstb\sigma_E^2$				
		Error	Season × treatment	Treatment	Block(season)	Season

which is:

Season	$nstb\sigma_E^2 + nt\sigma_{B(S)}^2 + nbt\sum_S^2$
Block(season)	$nstb\sigma_E^2 + nt\sigma_{B(S)}^2$
Treatment	$nstb\sigma_E^2 + nbs\sum_T^2$
Season × treatment	$nstb\sigma_E^2 + nb\sum_{ST}^2$
Error	$nstb\sigma_E^2$

Thus if the variances are homogeneous, the correct tests are:

    Test for season: MS Season / MS Block(Season).

    Test for treatment: MS Treatment / MS Error.

    Test for interaction: MS Season*treatment / MS Error.

Note that the error in the above ANOVA and EMS tables is often called the 'pooled error', which is the error mean square for the combined analysis.

If error variances are not homogeneous as indicated by the BF test of STEP ONE above, the correct tests are:

    Test for season: MS Season / MS Rep(Season).

    Test for treatment: MS Treatment / MS Error.

The test for a season*treatment interaction becomes complicated due to the heterogeneous error variances. First of all, the interaction should be decomposed into a meaningful set of contrasts. The pooled error must then be decomposed into corresponding pieces for testing these contrasts. This can be quite cumbersome and is beyond the scope of this book. If the test for homogeneous variances indicates likely heterogeneous variances, an acceptable approach would be to analyze the seasons separately and evaluate whether or not the treatments performed differently over seasons.

For the combined analysis assuming homogeneous variances, the following program provides the ANOVA with appropriate *F*-tests:

```
TITLE1 'Chapter 16 Example 2.sas';
TITLE2 'Analysis of Data Over Seasons';
TITLE3 'Combined ANOVA';
```



```
DATA one;
INPUT season $ block nitrogen yield;
nitrogen2 = nitrogen*nitrogen;
lof = nitrogen;
CARDS;
s 1 0 4
s 1 60 6
s 1 90 6
s 1 120     6
s 1 150     5
s 2 0 2
s 2 60 6
s 2 90 6
s 2 120     6
s 2 150     6
s 3 0 4
s 3 60 5
s 3 90 6
s 3 120     6
s 3 150     5
f 1 0 4
f 1 60 6
f 1 90 6
f 1 120     4
f 1 150     3
f 2 0 3
f 2 60 6
f 2 90 5
f 2 120     4
f 2 150     4
f 3 0 5
f 3 60 6
f 3 90 5
f 3 120     5
f 3 150     3
RUN;
PROC ANOVA DATA = one PLOTS = none;
CLASSES block season nitrogen;
MODEL yield = season block(season) nitrogen nitrogen*season;
TEST H = season E = block(season);
RUN;
```
Results:

'Chapter 16 Example 2.sas'
Analysis of Data Over Seasons
'Combined ANOVA'

The ANOVA Procedure

Class Level Information		
Class	Levels	Values
block	3	1 2 3
season	2	f s
nitrogen	5	0 60 90 120 150

Number of Observations Read	30
Number of Observations Used	30

'Chapter 16 Example 2.sas'
Analysis of Data Over Seasons
'Combined ANOVA'

The ANOVA Procedure

Dependent Variable: yield

Source	DF	Sum of Squares	Mean Square	F Value	Pr > F
Model	13	32.40000000	2.49230769	5.34	0.0011
Error	16	7.46666667	0.46666667		
Corrected Total	29	39.86666667			

R-Square	Coeff Var	Root MSE	yield Mean
0.812709	13.84723	0.683130	4.933333

Source	DF	Anova SS	Mean Square	F Value	Pr > F
season	1	3.33333333	3.33333333	7.14	0.0167
block(season)	4	0.53333333	0.13333333	0.29	0.8829
nitrogen	4	20.20000000	5.05000000	10.82	0.0002
season*nitrogen	4	8.33333333	2.08333333	4.46	0.0130

Tests of Hypotheses Using the Anova MS for block(season) as an Error Term					
Source	DF	Anova SS	Mean Square	F Value	Pr > F
season	1	3.33333333	3.33333333	25.00	0.0075

STEP THREE

If the variances are not homogeneous, perform an ANOVA on each individual season. (The number of seasons is not limited to two, but in most cases would be no larger than three or four). In this example, the regression diagnostics have been omitted, since our focus is on the technique of combining seasons.

SAS® statements for the spring:

```
TITLE1 'Chapter 16 Example 3.sas';
TITLE2 'Analysis of Data Over Seasons';
TITLE3 'Season = s';
DATA one;
INPUT season $ block nitrogen yield;
nitrogen2 = nitrogen*nitrogen;
lof = nitrogen;
CARDS;
s 1 0 4
s 1 60 6
s 1 90 6
s 1 120    6
s 1 150    5
s 2 0 2
s 2 60 6
s 2 90 6
s 2 120    6
s 2 150    6
s 3 0 4
s 3 60 5
s 3 90 6
s 3 120    6
s 3 150    5
f 1 0 4
f 1 60 6
f 1 90 6
f 1 120    4
f 1 150    3
f 2 0 3
f 2 60 6
f 2 90 5
f 2 120    4
f 2 150    4
f 3 0 5
f 3 60 6
f 3 90 5
f 3 120    5
f 3 150    3
DATA two; SET one;
IF season = 's';
TITLE3 'Season = Spring';
PROC GLM DATA = two PLOTS = none;
CLASSES block lof;
MODEL yield = block nitrogen lof;
PROC GLM DATA = two PLOTS = none;
CLASSES block lof;
```

```
    MODEL yield = block nitrogen nitrogen2 lof;
    PROC REG DATA = two PLOTS = none;
    MODEL yield = nitrogen;
    MODEL yield = nitrogen nitrogen2;
    RUN;
Results:
```

Chapter 16 Example 3.sas
Analysis of Data Over Seasons
Season = s

The GLM Procedure

Class Level Information		
Class	Levels	Values
block	3	1 2 3
lof	5	0 60 90 120 150

Number of Observations Read	15
Number of Observations Used	15

Chapter 16 Example 3.sas
Analysis of Data Over Seasons
Season = s

The GLM Procedure

Dependent Variable: yield

Source	DF	Sum of Squares	Mean Square	F Value	Pr > F
Model	6	15.06666667	2.51111111	5.20	0.0183
Error	8	3.86666667	0.48333333		
Corrected Total	14	18.93333333			

R-Square	Coeff Var	Root MSE	yield Mean
0.795775	13.20041	0.695222	5.266667

Source	DF	Type I SS	Mean Square	F Value	Pr > F
block	2	0.13333333	0.06666667	0.14	0.8732
nitrogen	1	7.96036036	7.96036036	16.47	0.0036
lof	3	6.97297297	2.32432432	4.81	0.0337

Source	DF	Type III SS	Mean Square	F Value	Pr > F
block	2	0.13333333	0.06666667	0.14	0.8732
nitrogen	0	0.00000000	.	.	.
lof	3	6.97297297	2.32432432	4.81	0.0337

The GLM Procedure

Class Level Information		
Class	Levels	Values
block	3	1 2 3
lof	5	0 60 90 120 150

Number of Observations Read	15
Number of Observations Used	15

The GLM Procedure

Dependent Variable: yield

Source	DF	Sum of Squares	Mean Square	F Value	Pr > F
Model	6	15.06666667	2.51111111	5.20	0.0183
Error	8	3.86666667	0.48333333		
Corrected Total	14	18.93333333			

R-Square	Coeff Var	Root MSE	yield Mean
0.795775	13.20041	0.695222	5.266667

Source	DF	Type I SS	Mean Square	F Value	Pr > F
block	2	0.13333333	0.06666667	0.14	0.8732
nitrogen	1	7.96036036	7.96036036	16.47	0.0036
nitrogen2	1	6.95628176	6.95628176	14.39	0.0053
lof	2	0.01669121	0.00834561	0.02	0.9829

Source	DF	Type III SS	Mean Square	F Value	Pr > F
block	2	0.13333333	0.06666667	0.14	0.8732
nitrogen	0	0.00000000	.	.	.
nitrogen2	0	0.00000000	.	.	.
lof	2	0.01669121	0.00834561	0.02	0.9829

Chapter 16 Example 3.sas
Analysis of Data Over Seasons
Season = s

The REG Procedure
Model: MODEL1
Dependent Variable: yield

Number of Observations Read	15
Number of Observations Used	15

Analysis of Variance

Source	DF	Sum of Squares	Mean Square	F Value	Pr > F
Model	1	7.96036	7.96036	9.43	0.0089
Error	13	10.97297	0.84407		
Corrected Total	14	18.93333			

Root MSE	0.91874	R-Square	0.4204
Dependent Mean	5.26667	Adj R-Sq	0.3759
Coeff Var	17.44434		

Parameter Estimates

Variable	DF	Parameter Estimate	Standard Error	t Value	Pr > |t|
Intercept	1	4.08108	0.45312	9.01	<.0001
nitrogen	1	0.01411	0.00460	3.07	0.0089

Chapter 16 Example 3.sas
Analysis of Data Over Seasons
Season = s

The REG Procedure
Model: MODEL2
Dependent Variable: yield

Number of Observations Read	15
Number of Observations Used	15

Analysis of Variance

Source	DF	Sum of Squares	Mean Square	F Value	Pr > F
Model	2	14.91664	7.45832	22.28	<.0001
Error	12	4.01669	0.33472		
Corrected Total	14	18.93333			

Root MSE	0.57855	R-Square	0.7879
Dependent Mean	5.26667	Adj R-Sq	0.7525
Coeff Var	10.98519		

Parameter Estimates

Variable	DF	Parameter Estimate	Standard Error	t Value	Pr > |t|
Intercept	1	3.33382	0.32907	10.13	<.0001
nitrogen	1	0.05533	0.00949	5.83	<.0001
nitrogen2	1	-0.00027928	0.00006126	-4.56	0.0007

SAS® statements for the fall:

```
TITLE1 'Chapter 16 Example 4.sas';
TITLE2 'Analysis of Data Over Seasons';
TITLE3 'Season = f';
DATA one;
INPUT season $ block nitrogen yield;
nitrogen2 = nitrogen*nitrogen;
lof = nitrogen;
CARDS;
s 1 0 4
s 1 60 6
s 1 90 6
s 1 120     6
s 1 150     5
s 2 0 2
s 2 60 6
s 2 90 6
s 2 120     6
s 2 150     6
s 3 0 4
s 3 60 5
s 3 90 6
s 3 120     6
s 3 150     5
f 1 0 4
f 1 60 6
f 1 90 6
f 1 120     4
f 1 150     3
f 2 0 3
f 2 60 6
f 2 90 5
f 2 120     4
f 2 150     4
f 3 0 5
f 3 60 6
f 3 90 5
f 3 120     5
f 3 150     3
DATA two; SET one;
IF season = 'f';
TITLE3 'Season = Fall';
PROC GLM DATA = two PLOTS = none;
CLASSES block lof;
MODEL yield = block nitrogen lof;
PROC GLM DATA = two PLOTS = none;
CLASSES block lof;
```

```
MODEL yield = block nitrogen nitrogen2 lof;
PROC REG DATA = two PLOTS = none;
MODEL yield = nitrogen;
MODEL yield = nitrogen nitrogen2;
RUN;
```
Results:

Chapter 16 Example new 4.sas
Season = Fall

The GLM Procedure

Class Level Information		
Class	Levels	Values
block	3	1 2 3
lof	5	0 60 90 120 150

Number of Observations Read	15
Number of Observations Used	15

Chapter 16 Example new 4.sas
Season = Fall

The GLM Procedure

Dependent Variable: yield

Source	DF	Sum of Squares	Mean Square	F Value	Pr > F
Model	6	14.00000000	2.33333333	5.19	0.0184
Error	8	3.60000000	0.45000000		
Corrected Total	14	17.60000000			

R-Square	Coeff Var	Root MSE	yield Mean
0.795455	14.58305	0.670820	4.600000

Source	DF	Type I SS	Mean Square	F Value	Pr > F
block	2	0.40000000	0.20000000	0.44	0.6561
nitrogen	1	1.16756757	1.16756757	2.59	0.1459
lof	3	12.43243243	4.14414414	9.21	0.0057

Source	DF	Type III SS	Mean Square	F Value	Pr > F
block	2	0.40000000	0.20000000	0.44	0.6561
nitrogen	0	0.00000000	.	.	.
lof	3	12.43243243	4.14414414	9.21	0.0057

The GLM Procedure

Class Level Information		
Class	Levels	Values
block	3	1 2 3
lof	5	0 60 90 120 150

Number of Observations Read	15
Number of Observations Used	15

Chapter 16 Example new 4.sas
Season = Fall

The GLM Procedure

Dependent Variable: yield

Source	DF	Sum of Squares	Mean Square	F Value	Pr > F
Model	6	14.00000000	2.33333333	5.19	0.0184
Error	8	3.60000000	0.45000000		
Corrected Total	14	17.60000000			

R-Square	Coeff Var	Root MSE	yield Mean
0.795455	14.58305	0.670820	4.600000

Source	DF	Type I SS	Mean Square	F Value	Pr > F
block	2	0.40000000	0.20000000	0.44	0.6561
nitrogen	1	1.16756757	1.16756757	2.59	0.1459
nitrogen2	1	11.62929056	11.62929056	25.84	0.0009
lof	2	0.80314188	0.40157094	0.89	0.4468

Source	DF	Type III SS	Mean Square	F Value	Pr > F
block	2	0.40000000	0.20000000	0.44	0.6561
nitrogen	0	0.00000000	.	.	.
nitrogen2	0	0.00000000	.	.	.
lof	2	0.80314188	0.40157094	0.89	0.4468

Chapter 16 Example new 4.sas
Season = Fall

The REG Procedure
Model: MODEL1
Dependent Variable: yield

Number of Observations Read	15
Number of Observations Used	15

Analysis of Variance					
Source	DF	Sum of Squares	Mean Square	F Value	Pr > F
Model	1	1.16757	1.16757	0.92	0.3540
Error	13	16.43243	1.26403		
Corrected Total	14	17.60000			

Root MSE	1.12429	R-Square	0.0663
Dependent Mean	4.60000	Adj R-Sq	-0.0055
Coeff Var	24.44114		

Parameter Estimates					
Variable	DF	Parameter Estimate	Standard Error	t Value	Pr > |t|
Intercept	1	5.05405	0.55450	9.11	<.0001
nitrogen	1	-0.00541	0.00562	-0.96	0.3540

Chapter 16 Example new 4.sas
Season = Fall

The REG Procedure
Model: MODEL2
Dependent Variable: yield

Number of Observations Read	15
Number of Observations Used	15

Analysis of Variance					
Source	DF	Sum of Squares	Mean Square	F Value	Pr > F
Model	2	12.79686	6.39843	15.99	0.0004
Error	12	4.80314	0.40026		
Corrected Total	14	17.60000			

Root MSE	0.63266	R-Square	0.7271
Dependent Mean	4.60000	Adj R-Sq	0.6816
Coeff Var	13.75353		

Parameter Estimates					
Variable	DF	Parameter Estimate	Standard Error	t Value	Pr > |t|
Intercept	1	4.08787	0.35985	11.36	<.0001
nitrogen	1	0.04788	0.01038	4.61	0.0006
nitrogen2	1	-0.00036110	0.00006699	-5.39	0.0002

Interpretation of the results

Let's look at the results assuming homogeneous variances as indicated by our BF test of STEP ONE.

'Chapter 16 Example 2.sas'
Analysis of Data Over Seasons
'Combined ANOVA'

The ANOVA Procedure

Dependent Variable: yield

Source	DF	Sum of Squares	Mean Square	F Value	Pr > F
Model	13	32.40000000	2.49230769	5.34	0.0011
Error	16	7.46666667	0.46666667		
Corrected Total	29	39.86666667			

R-Square	Coeff Var	Root MSE	yield Mean
0.812709	13.84723	0.683130	4.933333

Source	DF	Anova SS	Mean Square	F Value	Pr > F
season	1	3.33333333	3.33333333	7.14	0.0167
block(season)	4	0.53333333	0.13333333	0.29	0.8829
nitrogen	4	20.20000000	5.05000000	10.82	0.0002
season*nitrogen	4	8.33333333	2.08333333	4.46	0.0130

Tests of Hypotheses Using the Anova MS for block(season) as an Error Term					
Source	DF	Anova SS	Mean Square	F Value	Pr > F
season	1	3.33333333	3.33333333	25.00	0.0075

Our ANOVA indicates that we have a significant season × nitrogen interaction (α = 0.013) meaning that, depending on the season under consideration, the yield response to nitrogen is different. To further analyze this response, we would develop separate regression lines for the spring and fall responses to nitrogen, which we have already done in STEP THREE. The equations from STEP THREE are:

Spring: $3.33 + 0.055*N - 0.00028*N^2$
Fall: $\quad 4.09 + 0.048*N - 0.00036*N^2$

We would then compare the lines from the two seasons using dummy variables as described in Chapter 9.

The question is whether these two equations represent different responses of each cultivar to nitrogen in two different seasons. The two null hypotheses you need to test to determine whether or not these equations are different are:

(1) $H_0: \beta_0 1 = \beta_0 2$

and

(2) $H_0: \beta_1 1 = \beta_1 2$

Are the slopes for the two lines the same (1) and are the intercepts for the two lines the same (2)?

We need to create one dummy variable, since there are two seasons. We define our dummy variable (A1) as:

A1 = 1 if season = 'spring', otherwise A1 = 0;

We do not have to define a dummy variable for the fall season since A1 = 0 is associated with any season other than spring, and we only have two seasons in this example.

In addition to the dummy variable $A1$, we must create the dummy variable indicating the interaction of $A1$ with the independent variable used in the regression analysis, in our case 'nitrogen'.

The following SAS® program generates the dummy variables for us by adding a few lines of code to the data step:

```
TITLE1 'Chapter 16 Example 5.sas';
TITLE2 'Analysis of Data Over Seasons';
TITLE3 'Combined ANOVA';
TITLE4 'Comparing Spring vs Fall Nitrogen Response With Dummy
Variables';
DATA one;
INPUT season $ block nitrogen yield;
nitrogen2 = nitrogen*nitrogen;
lof = nitrogen;
A1 = . ;
IF season = "s" then A1 = 1; ELSE A1 = 0;
A1nit = A1*nitrogen;
CARDS;
s 1 0 4
s 1 60 6
s 1 90 6
s 1 120      6
s 1 150      5
s 2 0 2
s 2 60 6
s 2 90 6
s 2 120      6
s 2 150      6
s 3 0 4
s 3 60 5
s 3 90 6
s 3 120      6
s 3 150      5
f 1 0 4
f 1 60 6
f 1 90 6
f 1 120      4
f 1 150      3
f 2 0 3
f 2 60 6
f 2 90 5
```

```
f 2 120     4
f 2 150     4
f 3 0 5
f 3 60 6
f 3 90 5
f 3 120     5
f 3 150     3
RUN;
PROC PRINT;
RUN;
```

We initially define our dummy variable 'A1' as a missing value (line 09) and give A1 a value of 1 if the season is "s" (note the quotation marks for the value of season) and 0 if the season is any other value, in this case "f" (line 10). We define our second dummy variable 'A1nit' as the interaction of A1 and nitrogen by giving it the value of 'A1*nitrogen' (line 11).

The model we are using to make the comparison between the two lines utilizing a dummy variable is:

$$Y = \beta_0 + \beta_1 X_i + \beta_2 A_i + \beta_3 X_i * A + \varepsilon_i$$

We assume that the variances from the two seasons are equal. Since our examples will always use dummy variables of the form $A_i = 1$ if the observation is in a specific group, otherwise $A_i = 0$, the above model provides a model for observation in either of our two season groups simply by letting $Ai = 0$ or 1, depending on group:

$$Season"s"(A1 = 1): Y = \beta_0 + \beta_1 X_i + \beta_2 A_i + \beta_3 X_i * A + \varepsilon_i$$

becomes:

$$Season"s"(A1 = 1): Y = \beta_0 + \beta_1 X_i + \beta_2 * 1 + \beta_3 X_i * 1 + \varepsilon_i$$

which is:

$$Season"s"(A1 = 1): Y = \beta_0 + \beta_1 X_i + \beta_2 + \beta_3 X_i + \varepsilon_i$$

and moving a few components of the equation around a bit:

$$Season"s"(A1 = 1): Y = \beta_0 + \beta_2 + \beta_1 X_i + \beta_3 X_i + \varepsilon_i$$

and finally:

$$Season"s" (A1 = 1): Y = \beta_0 + \beta_2 + (\beta_1 + \beta_3) * X + \varepsilon_i$$

Similarly, for season "f":

$$Season"f" (A1 = 0): Y = \beta_0 + \beta_1 X_i + \beta_2 A_i + \beta_3 X_i * A + \varepsilon_i$$

becomes:

$$Season"f"(A1 = 0): Y = \beta_0 + \beta_1 X_i + \beta_2 * 0 + \beta_3 X_i * 0 + \varepsilon_i$$

simplifying and moving a few components around:

$$Season"f"(A1=0): Y = \beta_0 + \beta_1 X_i + \varepsilon_i$$

Using the model:

$$Y = \beta_0 + \beta_1 X_i + \beta_2 A_i + \beta_3 X_i * A + \varepsilon_i$$

we can substitute our notation and obtain:

$$Y = \beta_0 + \beta_1 nitrogen + \beta_2 A + \beta_3 nitrogen * A + \varepsilon_i$$

1. The test for coincidence is obtained with the test

   $$H_0 : \beta_2 = \beta_3 = 0$$

   If we fail to reject H_0, then the model that corresponds to both seasons is:

   $$Y = \beta_0 + \beta_1 nitrogen + \varepsilon_i$$

2. The test for equal slopes (parallelism) is obtained with the test

   $$H_0 : \beta_3 = 0$$

   If we fail to reject H_0, the model for season "s" (where $A = 1$) is:

   $$Y = (\beta_0 + \beta_2) + \beta_1 nitrogen + \varepsilon_i$$

   and the model for season "f" is:

   $$Y = \beta_0 + \beta_1 nitrogen + \varepsilon_i$$

3. The hypothesis for equal intercepts is obtained with the test

   $$H_0: \beta_2 = 0$$

   If we fail to reject H_0, the model for season "s" is

   $$Y = \beta_0 + (\beta_1 + \beta_3) * nitrogen + \varepsilon_i$$

   and the model for season "f" is:

   $$Y = \beta_0 + \beta_1 nitrogen + \varepsilon_i$$

4. If we reject all three hypotheses, then both lines have different β_0's and β_1's.

We now use A1, *nitrogen* and A1*nit* as predictors (independent variables) in a regression analysis using PROC REG to perform the above tests and get estimates of our β's.

We will include a TEST statement (TEST A1nit = 0, nit=0;) in our program to test for coincidence, the hypothesis of $H_0: \beta_2 = \beta_3 = 0$. Since we observed a significant interaction between season and nitrogen in our overall ANOVA, we already know that the lines are not coincident, but we will run the test anyway just to illustrate how it's done:

```
PROC REG PLOTS = none DATA = one;
MODEL yield = A1 nitrogen A1nit;
TEST A1nit=0, nitrogen=0;
RUN;
```
produces the following output:

'Chapter 16 Example 5.sas'
Analysis of Data Over Seasons
'Combined ANOVA'
Comparing Spring vs Fall Nitrogen Response With Dummy Variables

The REG Procedure
Model: MODEL1
Dependent Variable: yield

Number of Observations Read	30
Number of Observations Used	30

Analysis of Variance					
Source	DF	Sum of Squares	Mean Square	F Value	Pr > F
Model	3	12.46126	4.15375	3.94	0.0192
Error	26	27.40541	1.05405		
Corrected Total	29	39.86667			

Root MSE	1.02667	R-Square	0.3126
Dependent Mean	4.93333	Adj R-Sq	0.2333
Coeff Var	20.81091		

Parameter Estimates					
Variable	DF	Parameter Estimate	Standard Error	t Value	Pr > |t|
Intercept	1	5.05405	0.50635	9.98	<.0001
A1	1	-0.97297	0.71609	-1.36	0.1859
nitrogen	1	-0.00541	0.00514	-1.05	0.3023
A1nit	1	0.01952	0.00726	2.69	0.0124

'Chapter 16 Example 5.sas'
Analysis of Data Over Seasons
'Combined ANOVA'
Comparing Spring vs Fall Nitrogen Response With Dummy Variables

The REG Procedure
Model: MODEL1

Test 1 Results for Dependent Variable yield				
Source	DF	Mean Square	F Value	Pr > F
Numerator	2	4.56396	4.33	0.0238
Denominator	26	1.05405		

The test for coincidence (H_0: $\beta_2 = \beta_3 = 0$) produces an F-value of 4.33 with $\alpha < 0.0238$, which is fairly significant. We would therefore reject the null hypothesis that the two lines are coincident.

From the above printout, note that our β estimates are as follows:

$\beta_0 = 5.05405, \beta_1 = -0.00541, \beta_2 = -0.97297$ and $\beta_3 = 0.01952$.

The test for equal slopes (parallelism) (H_0: $\beta_3 = 0$) produces a t-value of 2.69 with $\alpha < 0.0124$. We would therefore reject the hypothesis of equal slopes.

The hypothesis for equal intercepts (H_0: $\beta_2 = 0$) produces a t-value of -1.36 and $\alpha = 0.1859$, thus we fail to reject the hypothesis of equal intercepts.

The model for season "s" is:

$$Y = \beta_0 + (\beta_1 + \beta_3) * nitrogen + \varepsilon_i =$$

$$Y = 5.05405 + (-0.00541 - 0.97297) * nitrogen + \varepsilon_i$$

$$Y = 5.05405 - 0.97838 * nitrogen + \varepsilon_i$$

The model for season "f" is:

$$Y = \beta_0 + \beta_1 nitrogen + \varepsilon_i$$

$$Y = 5.05405 - 0.00541 * nitrogen + \varepsilon_i$$

The response to nitrogen is a negative one, and the response is more pronounced in the spring season compared with the fall. And there you have it, a nice experiment conducted over two seasons summarized with a single statement.

Analysis Over Years

Experiments are often repeated over time to verify results of initial trials. In lab experiments the time frame may be weeks. In field trials, the time frame may be years. Regardless of the time frame, the general methodology for combining a series of experiments repeated over time is the same.

The basic setup is that you have the same experiment and you have repeated the experiment one or more times for a total of n repeats. Each repeat should be re-randomized independently. The basic combined ANOVA can be modified to accommodate the design of the basic experiment which has been repeated. For our example, we will look at the analysis of an RCB repeated over time (years).

In experiments repeated over time, the time factor is considered random. Think about why seasons are fixed but time is random. When you evaluate a season, the inferences you make about a particular season are specific to that season only, say the rainy season or the fall season. However, when you make inferences about time, you want to make inferences for any given year that might be considered, not a specific year. The main difference in experiments repeated over season or over time is the denominator used for testing the treatment effect. The difference due to season is fixed, whereas time is random.

STEP ONE

Test homogeneity of variances from individual analysis to determine if pooling error terms is acceptable. If variances are not homogeneous, then variance partitioning should be performed.

Our experiment evaluated four lettuce cultivars for winter high-tunnel production and the experiment was performed over 3 years. The data collected was kg per 3 m plot. There were three blocks, each block a separate high tunnel. Since high tunnels are expensive, they limited the number of blocks possible.

```
TITLE1 'Chapter 16 Example 6.sas';
TITLE2 'Analysis of Data Over Years';
TITLE3 'Combined ANOVA';
TITLE4 'Test for Homogeneous Variances';
DATA one;
INPUT year block cultivar $ yield;
CARDS;
1 1 a 5
1 1 b 6
1 1 c 4
1 1 d 6
1 2 a 2
1 2 b 3
1 2 c 6
1 2 d 8
1 3 a 4
1 3 b 6
1 3 c 6
1 3 d 7
2 1 a 4
2 1 b 6
2 1 c 5
2 1 d 8
2 2 a 3
2 2 b 5
2 2 c 7
2 2 d 4
2 3 a 4
2 3 b 6
2 3 c 8
2 3 d 6
3 1 a 3
3 1 b 6
3 1 c 5
3 1 d 6
3 2 a 2
3 2 b 6
3 2 c 6
```

```
3 2 d 6
3 3 a 4
3 3 b 5
3 3 c 6
3 3 d 4
RUN;
PROC ANOVA DATA = one PLOTS = none;
CLASSES year;
MODEL yield = year;
MEANS year / HOVTEST = BF;
RUN;
```

The results:

Chapter 16 Example 6.sas
Analysis of Data Over Years
Combined ANOVA
Test for Homogeneous Variances

The ANOVA Procedure

Class Level Information		
Class	Levels	Values
year	3	1 2 3

Number of Observations Read	36
Number of Observations Used	36

Chapter 16 Example 6.sas
Analysis of Data Over Years
Combined ANOVA
Test for Homogeneous Variances

The ANOVA Procedure

Dependent Variable: yield

Source	DF	Sum of Squares	Mean Square	F Value	Pr > F
Model	2	2.05555556	1.02777778	0.41	0.6652
Error	33	82.16666667	2.48989899		
Corrected Total	35	84.22222222			

R-Square	Coeff Var	Root MSE	yield Mean
0.024406	30.21590	1.577941	5.222222

Source	DF	Anova SS	Mean Square	F Value	Pr > F
year	2	2.05555556	1.02777778	0.41	0.6652

The ANOVA Procedure

Brown and Forsythe's Test for Homogeneity of yield Variance ANOVA of Absolute Deviations from Group Medians					
Source	DF	Sum of Squares	Mean Square	F Value	Pr > F
year	2	0.3889	0.1944	0.17	0.8434
Error	33	38.8333	1.1768		

The ANOVA Procedure

Level of year	N	yield	
		Mean	Std Dev
1	12	5.25000000	1.71225529
2	12	5.50000000	1.62368828
3	12	4.91666667	1.37895437

To determine whether or not to reject the hypothesis of homogeneous variances between the years, find the output labeled 'Brown and Forsythe's Test for Homogeneity of yield Variance' and review the Pr > F for the 'years' source of variation in the ANOVA of absolute deviations from group medians. Reject the null hypothesis if Pr > F is less than or equal to 0.05. In our example the Pr > F = 0.8484, indicating that we should fail to reject the null hypothesis and that we most likely have homogeneous variances among the three seasons.

STEP TWO

Construct an ANOVA table for the combined analysis (y = years, r or b = replications or blocks, t = treatment):

Source	df
Total	ybt
μ	1
Year	y–1
Blocks(Year)	y(b–1)
Treatment	t–1
Year × Treatment	(y–1)(t–1)
Pooled error	y(b–1)(t–1)

Notice again we have a different notation for the 'Blocks' source of variation. While in other ANOVAs we have seen 'Blocks' source simply identified as 'blocks', in this situation we identify 'Blocks' as 'Blocks(Year)' which is read as 'blocks within year'. This indicates that the blocks are randomly assigned, separately, within each year. SAS® will interpret this notation to correctly adjust ANOVA calculations. As long as you follow the above ANOVA table as a guideline for your analysis, you will get the correct results. With the ANOVA table in hand, we need to determine the correct denominators for F-tests by constructing an EMS table:

F or R						
R	Year					
R	Block(year)					
F	Treatment					
F	Year × treatment					
	Error					
		Error	Year × treatment	Treatment	Block(year)	Year

First place an 'X' in each cell of the 'Error' column, since all sources have a variance component attributed to error. Also place an 'X' in each diagonal cell, since each source has a component (variance for random effect and fixed component for fixed effect) attributed to itself. Your table should look like this:

F or R						
R	Year	X				X
R	Block(year)	X			X	
F	Treatment	X	X	X		
F	Year × treatment	X	X			
	Error	X				
		Error	Year × treatment	Treatment	Block(year)	Year

Now evaluate each column to determine if it should be included as a component of the EMS for each row (source of variation).

- For the 'Year × treatment' column:
  Should the 'Treatment' cell get an 'X'? It contains all the components of the row. Is 'Year' random? Yes, thus put an 'X' in the cell.
  Should the 'Block(year)' cell get an 'X'? No, it doesn't contain all the components of the row.
  Should the 'Year' cell get an 'X'? It contains all the components of the row. Is 'Treatment' random? No, thus do not put an 'X' in the cell.
- For the 'Treatment' column:
  Should the 'Block(year)' cell get an 'X'? No, it doesn't contain all the components of the row. It doesn't contain 'Block(year)'.
  Should the 'Year' cell get an 'X'? No, it doesn't contain 'Block(year)'.
- For the 'Block(year)' column:
  Should the 'Year' cell get an 'X'? It contains all the components of the row (the 'Year' component). Is 'Block' random? Yes, thus put an 'X' in the cell.

Your finished table should look like this:

F or R						
R	Year	X			X	X
R	Block(year)	X			X	
F	Treatment	X	X	X		
F	Year × treatment	X	X			
	Error	X				
		Error	Year × treatment	Treatment	Block(year)	Year

which translates into:

F or R						
R	Year	$nytb\sigma_E^2$			$nt\sigma_{B(Y)}^2$	$nbt\sigma_Y^2$
R	Block(year)	$nytb\sigma_E^2$			$nt\sigma_{B(Y)}^2$	
F	Treatment	$nytb\sigma_E^2$	$nb\sum_{YT}^2$	$nby\sum_T^2$		
F	Year × treatment	$nytb\sigma_E^2$	$nb\sum_{YT}^2$			
	Error	$nytb\sigma_E^2$				
		Error	Year × treatment	Treatment	Block(year)	Year

which is:

Year	$nytb\sigma_E^2 + nt\sigma_{B(Y)}^2 + nbt\sigma_Y^2$
Block(year)	$nytb\sigma_E^2 + nt\sigma_{B(Y)}^2$
Treatment	$nytb\sigma_E^2 + nb\sum_{YT}^2 + nby\sum_T^2$
Year × treatment	$nytb\sigma_E^2 + nb\sum_{YT}^2$
Error	$nytb\sigma_E^2$

Thus if the variances are homogeneous, the correct tests are:

Test for year: MS Year / MS Block(Year).

Test for treatment: MS Treatment / MS Year × treatment.

Test for interaction: MS Year*treatment / MS Error.

If error variances are not homogeneous as indicated by the BF test of STEP ONE above, the correct tests are:

Test for season: MS Year / MS Block(Year).

Test for treatment: MS Treatment / MS Year × treatment.

The test for a year*treatment interaction is complicated with heterogeneous error variances. The interaction should be decomposed into a meaningful set of contrasts and the pooled error then decomposed into corresponding pieces for testing these contrasts. This can be quite cumbersome and is beyond the scope of this book. If the test for homogeneous variances indicates likely heterogeneous variances, an acceptable approach would be to analyze the years separately and evaluate whether or not the treatments performed differently over years.

For the combined analysis assuming homogeneous variances, the following program provides the ANOVA with appropriate *F*-tests:

```
TITLE1 'Chapter 16 Example 7.sas';
TITLE2 'Analysis of Data Over Seasons';
TITLE3 'Combined ANOVA';
DATA one;
INPUT year block cultivar $ yield;
CARDS;
1 1 a 5
1 1 b 6
1 1 c 4
1 d 6
1 2 a 2
1 2 b 3
1 2 c 6
1 2 d 8
1 3 a 4
1 3 b 6
1 3 c 6
1 3 d 7
2 1 a 4
2 1 b 6
2 1 c 5
2 1 d 8
2 2 a 3
2 2 b 5
2 2 c 7
2 2 d 4
2 3 a 4
2 3 b 6
2 3 c 8
2 3 d 6
3 1 a 3
3 1 b 6
3 1 c 5
3 1 d 6
3 2 a 2
3 2 b 6
3 2 c 6
3 2 d 6
3 3 a 4
3 3 b 5
3 3 c 6
3 3 d 4
RUN;
PROC ANOVA DATA = one PLOTS = none;
CLASSES block year cultivar;
MODEL yield = year block(year) cultivar cultivar*year;
```

```
TEST H = year E = block(year);
TEST H = cultivar E = cultivar*year;
MEANS year / REGWQ E = block(year);
MEANS cultivar / REGWQ E = cultivar*year;
MEANS year*cultivar;
RUN;
```
Results:

'Chapter 16 Example 7.sas'
Analysis of Data Over Seasons
'Combined ANOVA'

The ANOVA Procedure

Class Level Information		
Class	Levels	Values
block	3	1 2 3
year	3	1 2 3
cultivar	4	a b c d

Number of Observations Read	36
Number of Observations Used	36

'Chapter 16 Example 7.sas'
Analysis of Data Over Seasons
'Combined ANOVA'

The ANOVA Procedure

Dependent Variable: yield

Source	DF	Sum of Squares	Mean Square	F Value	Pr > F
Model	17	54.55555556	3.20915033	1.95	0.0853
Error	18	29.66666667	1.64814815		
Corrected Total	35	84.22222222			

R-Square	Coeff Var	Root MSE	yield Mean
0.647757	24.58345	1.283802	5.222222

Source	DF	Anova SS	Mean Square	F Value	Pr > F
year	2	2.05555556	1.02777778	0.62	0.5472
block(year)	6	5.66666667	0.94444444	0.57	0.7467
cultivar	3	40.00000000	13.33333333	8.09	0.0013
year*cultivar	6	6.83333333	1.13888889	0.69	0.6598

Tests of Hypotheses Using the Anova MS for block(year) as an Error Term					
Source	DF	Anova SS	Mean Square	F Value	Pr > F
year	2	2.05555556	1.02777778	1.09	0.3951

Tests of Hypotheses Using the Anova MS for year*cultivar as an Error Term					
Source	DF	Anova SS	Mean Square	F Value	Pr > F
cultivar	3	40.00000000	13.33333333	11.71	0.0064

'Chapter 16 Example 7.sas'
Analysis of Data Over Seasons
'Combined ANOVA'

The ANOVA Procedure

Ryan-Einot-Gabriel-Welsch Multiple Range Test for yield

Note: This test controls the Type I experimentwise error rate.

Alpha	0.05
Error Degrees of Freedom	6
Error Mean Square	0.944444

Number of Means	2	3
Critical Range	0.9707897	1.2173184

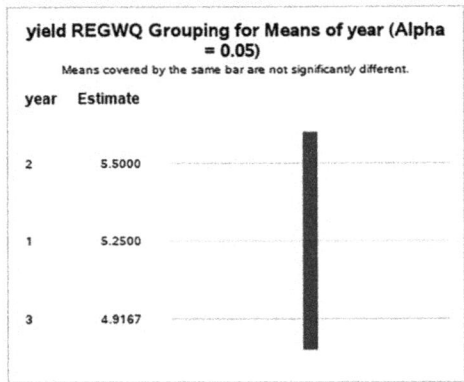

yield REGWQ Grouping for Means of year (Alpha = 0.05)

Means covered by the same bar are not significantly different.

year	Estimate
2	5.5000
1	5.2500
3	4.9167

'Chapter 16 Example 7.sas'
Analysis of Data Over Seasons
'Combined ANOVA'

The ANOVA Procedure

Ryan-Einot-Gabriel-Welsch Multiple Range Test for yield

Note: This test controls the Type I experimentwise error rate.

Alpha	0.05
Error Degrees of Freedom	6
Error Mean Square	1.138889

Number of Means	2	3	4
Critical Range	1.4885096	1.5435689	1.7415045

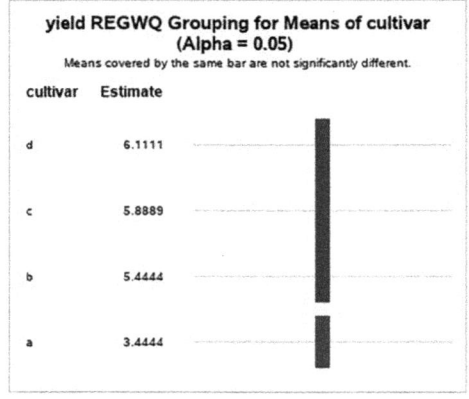

yield REGWQ Grouping for Means of cultivar (Alpha = 0.05)

Means covered by the same bar are not significantly different.

cultivar	Estimate
d	6.1111
c	5.8889
b	5.4444
a	3.4444

Level of year	Level of cultivar	N	yield	
			Mean	Std Dev
1	a	3	3.66666667	1.52752523
1	b	3	5.00000000	1.73205081
1	c	3	5.33333333	1.15470054
1	d	3	7.00000000	1.00000000
2	a	3	3.66666667	0.57735027
2	b	3	5.66666667	0.57735027
2	c	3	6.66666667	1.52752523
2	d	3	6.00000000	2.00000000
3	a	3	3.00000000	1.00000000
3	b	3	5.66666667	0.57735027
3	c	3	5.66666667	0.57735027
3	d	3	5.33333333	1.15470054

Interpretation of the results

Interpretation of these results is straightforward and we evaluate the interaction between year and treatment first. The F-value for the year × cultivar interaction is 0.69 and is significant at $\alpha = 0.6598$, indicating that there is no difference in cultivar performance over time. This is important when selecting cultivar adaptation, since you would want production-stable cultivars. The F-values for year and cultivar are 1.09 and 11.71, respectively, which are significant at $\alpha = 0.3951$ and 0.0064, respectively. The cultivar means are:

Cultivar	Yield (kg/plot)
A	3.4 b[z]
B	5.4 a
C	5.9 a
D	6.1 a

[z]Mean separation with REGWQ $\alpha = 0.05$.

Our cultivar trial over 3 years has indicated that of the four tested cultivars, 'a' had the poorest performance based on yield. Cultivars b, c and d were similar and not statistically different from each other. Be careful that you don't always judge a cultivar by its yield alone, since quality is also a very important consideration.

Analysis Over Locations (Sites)

This situation arises when you are interested in testing a new technology (e.g. cultivar, rootstock, etc.) over a number of locations. You want to learn if the new technology is really better than the current and where this new technology might be adapted. Sites are normally selected and are therefore fixed effects. Treatments are also fixed. This situation is similar to an analysis over seasons. The approach is exactly the same, just substitute 'site' for 'season' in your approach.

Analysis Over Locations (Sites) and Time (Years)

The general procedures outlined above can be extended for many variations of a series of experiments over time, location and seasons and combinations thereof. Additionally, instead of a single factor at several levels, a factorial treatment structure can be analyzed by extending the basic concepts already presented. Probably the most important things to remember are: (1) which factors are random and which are fixed; and (2) appropriate F-test denominators as determined via construction of an EMS table.

Although we have stated that in an experiment over locations, location is a fixed effect, there may be times when location should be considered random. For example, you only have five research farms available to you in New Jersey, but you would like to make inferences about your treatments for the entire state. Here you would probably consider location as random. Appropriate modifications to the EMS table and subsequent F-testing denominator selection should be made in these cases.

17 The Analysis of Covariance

The goal of nearly all statistical procedures is to account for variability among experimental units. By accounting for variability, we can determine important sources of it and how much of an impact each has on the characteristics we measure. In addition to accounting for variability, we attempt to control it as much as possible.

The first way to control or account for variability in an experiment is by selecting an appropriate experimental design. This often includes blocking experimental units according to known or assumed sources of variability. These may include soil moisture or fertility gradients in the field, temperature or light gradients in a greenhouse, plant vigor, etc.

A second valuable method for evaluating variability is the analysis of covariance. If we can account for more variability in an experiment, we remove that variability from our estimate of error (estimated random variability) and thus increase the precision of our test for a treatment effect.

In the analysis of covariance, there is some measurable characteristic associated with experimental units which seems to be contributing significant variability to an experiment. This variability inflates the error term and makes it harder to reject the null hypothesis of no treatment effect. If you could pull this variability out of the error term, you would get a larger F-value, and therefore possibly reject the null hypothesis. This is increasing precision.

Realize that just because you think the characteristic is causing variability in your experiment doesn't necessarily mean that it is. If it really isn't, then your treatment F-value may not change very much with the analysis of covariance and the original failure to reject the null hypothesis stands. Having the test of the null hypothesis remain the same after a covariance analysis is just as informative as if it had changed. I have often encountered researchers who 'think the experiment didn't work' if they fail to reject the null hypothesis. Finding no treatment effect is still as informative as detecting a treatment effect; it just might not be as exciting.

The measurable characteristic that seems to be attributing to variability is called the covariate and is often noted as 'X'. There are two important considerations regarding X: (1) the variable (an independent variable) must not be affected by the treatment; and (2) the relationship between X and the dependent variable (yield for example) must be linear. The assumption that the X variable is not affected by treatment is automatically taken care of if the characteristic can be measured before applying experimental treatments. Thus it is always a good idea to think about covariance in the planning stages of an experiment. If you decide that you need to perform an analysis of covariance after you start your experiment, and must therefore measure the covariate after treatment application, you must give some thought as to whether or not assumption (1) is true. The assumption of linearity between X and Y is usually valid for much plant science work.

In certain crops, experiments called 'uniformity trials' are often conducted. These are essentially experiments where many variables of interest or potential interest are studied without applying any treatments. In this way sources of variability and relationships among these variables can be established for future reference. Uniformity trials are time consuming, but can be very valuable in the long run. They often provide great insight into the nature of a specific crop and set a strong foundation for future work.

Let's briefly explore five uses of the analysis of covariance in plant science work: (1) increasing precision in an experiment by reducing variability associated with the error term in an analysis of variance; (2) accounting for sources of bias in an experiment; (3) shedding light on the nature of treatment effects; (4) adjusting treatment means to a common covariate level; and (5) estimating missing values.

Increasing Precision

The covariate is measured before treatment application. This variable is often associated with plant productivity or vigor prior to the experiment. It is often used with perennial crops, but can be used with annual crops as well. Examples of covariates include prior yield, plant size, plant vigor (could be a rating or in tree fruit is often a measure of trunk cross-sectional area), plant stand, etc.

For example, suppose you were interested in yield response to some treatment. Some of the plants are inherently more productive than others, but you don't know this by looking at the plants, i.e. there is no obvious difference in vigor by which you could block. What if, even with proper randomization, some treatment was applied primarily to more productive plants? You might attribute this greater productivity to the treatment. But now suppose that you had records of prior yield for these plants and performed an analysis of covariance on the data, using prior yields as the covariate. This type of data would also be useful for crops where high productivity is often followed by a period of low productivity, as in the biennial bearing of some fruit crops. Either type of variability associated with productivity can be accounted for by the analysis of covariance. By accounting for this variability, you have prevented it from either: (1) being dumped into error, thereby inflating it and making it harder to reject the null hypothesis; or (2) being incorrectly attributed to some treatment effect.

Adjusting for Sources of Bias in an Experiment

Suppose that you were studying obesity as it is related to work required (physical activity) in some occupation. You have data of weight, age and occupation. Suppose age and obesity are linearly related (we get heavier as we get older) and some particular occupation has primarily one age group in it. You could perform an analysis of covariance using weight as your dependent variable, occupation as your treatment and age as a covariate. If age is not accounted for (i.e. no analysis of covariance), some weight differences ascribed to occupation may actually be attributable to age differences. (I realize this is not a plant science example, but it serves the purpose.)

Shedding Light on the Nature of Treatment Effects

Suppose you are studying treatment effects on nematode populations in the soil and yield. Your data shows variation in both the number of nematodes and yield per plot. Can yield differences be attributed to differences in nematode populations or are they due to direct treatment effects? One way to shed light on this question is to see if treatment differences disappear after adjusting for nematode numbers in an analysis of covariance using nematode population as a covariate. This is a good example of the importance of measuring the covariate before the experiment. If you measured nematodes before the treatment, there's no problem with assuming no treatment effect on covariate. But if you measured the nematode population after applying some treatment, this assumption may not be valid. If treatment differences disappear after adjusting for nematode number, then nematodes directly affected yield. If not, then treatments really affected yield. Let's look at this example with real numbers.

The experiment was a completely random design with three treatments and ten replications. Our covariate (X) is the number of nematodes per quantity of soil before treatment, and our dependent variable (Y) is yield after treatment. The data and the initial analysis of variance are as follows.

```
TITLE1 'Chapter 17, Example 1.sas';
TITLE2 'Analysis of Covariance';
DATA one;
INPUT treatment $ nematodes yield;
CARDS;
a 11 6
a 6 4
d 6 0
d 8 4
f 16 13
f 16 12
a 8 0
a 10 13
d 6 2
d 19 14
f 13 10
f 12 5
a 5 2
a 6 1
d 7 3
d 8 9
f 11 18
f 12 16
a 14 8
a 11 8
d 8 1
d 5 1
f 9 5
f 7 1
a 19 11
a 3 0
d 18 18
d 15 9
```

```
f 21 23
f 12 20
RUN;
PROC ANOVA PLOTS = none;
CLASSES treatment;
MODEL yield = treatment;
MEANS treatment / REGWQ;
RUN;
```
SAS® output:

'Chapter 17 Example 1.sas'
'Analysis of Covariance'

The ANOVA Procedure

Class Level Information		
Class	Levels	Values
treatment	3	a d f

Number of Observations Read	30
Number of Observations Used	30

'Chapter 17 Example 1.sas'
'Analysis of Covariance'

The ANOVA Procedure

Dependent Variable: yield

Source	DF	Sum of Squares	Mean Square	F Value	Pr > F
Model	2	293.600000	146.800000	3.98	0.0305
Error	27	995.100000	36.855556		
Corrected Total	29	1288.700000			

R-Square	Coeff Var	Root MSE	yield Mean
0.227826	76.84655	6.070878	7.900000

Source	DF	Anova SS	Mean Square	F Value	Pr > F
treatment	2	293.6000000	146.8000000	3.98	0.0305

The ANOVA Procedure

Ryan-Einot-Gabriel-Welsch Multiple Range Test for yield

Note: This test controls the Type I experimentwise error rate.

Alpha	0.05
Error Degrees of Freedom	27
Error Mean Square	36.85556

Number of Means	2	3
Critical Range	5.5706499	6.7313738

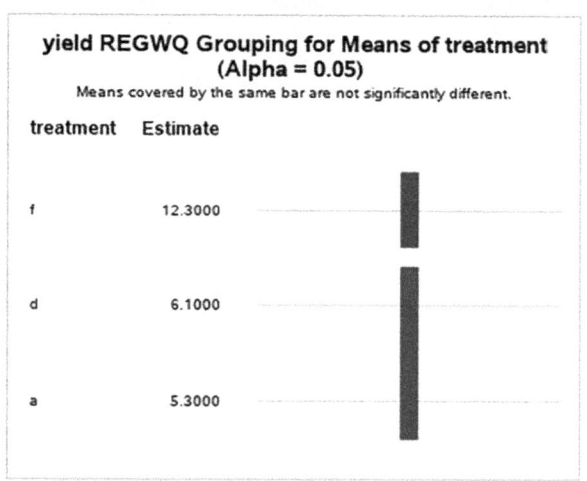

yield REGWQ Grouping for Means of treatment (Alpha = 0.05)

Means covered by the same bar are not significantly different.

treatment	Estimate
f	12.3000
d	6.1000
a	5.3000

Now we perform the analysis of covariance with the following SAS® statements:

```
TITLE1 'Chapter 17 Example 2.sas';
TITLE2 'Analysis of Covariance';
DATA one;
INPUT treatment $ nematodes yield;
CARDS;
a 11 6
a 6 4
d 6 0
d 8 4
f 16 13
f 16 12
a 8 0
a 10 13
d 6 2
d 19 14
f 13 10
f 12 5
a 5 2
```

```
a 6 1
d 7 3
d 8 9
f 11 18
f 12 16
a 14 8
a 11 8
d 8 1
d 5 1
f 9 5
f 7 1
a 19 11
a 3 0
d 18 18
d 15 9
f 21 23
f 12 20
RUN;
PROC GLM DATA = one PLOTS = none;
CLASSES treatment;
MODEL yield = treatment nematodes / SOLUTION;
MEANS treatment / REGWQ;
RUN;
```

Note that 'X' (nematodes) is not in the classes statement. The ' / SOLUTION' option at the end of the MODEL statement tells SAS® that you want the β estimates printed out. Of the β estimates, you are only interested in the estimate for 'nematodes'. The output is:

'Chapter 17 Example 2.sas'
'Analysis of Covariance'

The GLM Procedure

Class Level Information		
Class	Levels	Values
treatment	3	a d f

Number of Observations Read	30
Number of Observations Used	30

'Chapter 17 Example 2.sas'
'Analysis of Covariance'

The GLM Procedure

Dependent Variable: yield

Source	DF	Sum of Squares	Mean Square	F Value	Pr > F
Model	3	871.497403	290.499134	18.10	<.0001
Error	26	417.202597	16.046254		
Corrected Total	29	1288.700000			

R-Square	Coeff Var	Root MSE	yield Mean
0.676261	50.70604	4.005778	7.900000

Source	DF	Type I SS	Mean Square	F Value	Pr > F
treatment	2	293.6000000	146.8000000	9.15	0.0010
nematodes	1	577.8974030	577.8974030	36.01	<.0001

Source	DF	Type III SS	Mean Square	F Value	Pr > F
treatment	2	68.5537106	34.2768553	2.14	0.1384
nematodes	1	577.8974030	577.8974030	36.01	<.0001

Parameter	Estimate		Standard Error	t Value	Pr > |t|
Intercept	-0.434671164	B	2.47135356	-0.18	0.8617
treatment a	-3.446138280	B	1.88678065	-1.83	0.0793
treatment d	-3.337166948	B	1.85386642	-1.80	0.0835
treatment f	0.000000000	B	.	.	.
nematodes	0.987183811		0.16449757	6.00	<.0001

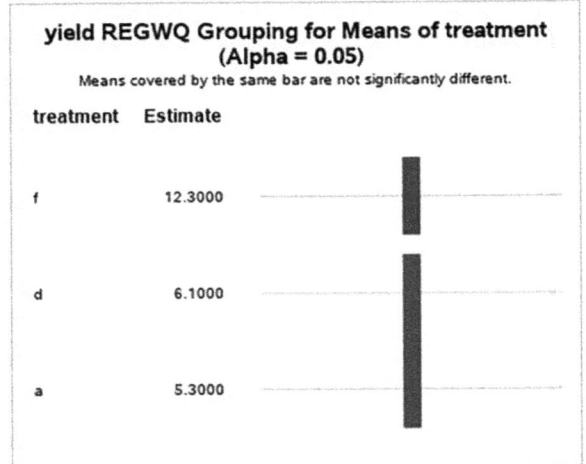

'Chapter 17 Example 2.sas'
'Analysis of Covariance'

The GLM Procedure

Ryan-Einot-Gabriel-Welsch Multiple Range Test for yield

Note: This test controls the Type I experimentwise error rate.

Alpha	0.05
Error Degrees of Freedom	26
Error Mean Square	16.04825

Number of Means	2	3
Critical Range	3.6823413	4.4513769

yield REGWQ Grouping for Means of treatment
(Alpha = 0.05)

Means covered by the same bar are not significantly different.

treatment	Estimate
f	12.3000
d	6.1000
a	5.3000

Notice that in the initial analysis of variance there were 27 df for error and a significant treatment effect at $\alpha = 0.0305$. When you examine the analysis of covariance, notice that there are now only 26 df for error. By performing the covariance, you had to 'pay' for it with 1 df from error. Now look at the Type I SS for treatment. It is exactly the same as the SS treatment in the initial analysis of variance. In the covariance analysis we have an additional Type I SS, namely that attributable to X (nematodes) printed out. This is the variability that is accounted for *after* the variability due to treatment. Note that the treatment Type I SS is *not* adjusted for nematodes (X). Thus the important SS we need to look at are the Type III SS for both treatment and X (nematodes). Notice when the variability due to treatment is estimated while accounting for variability due to 'X' concomitantly, the treatment effect is no longer significant ($P > F = 0.1384$). Also note that the variability associated with 'X' is significant at the 0.0001 level. This means that variability in yield was significantly linearly associated with varying levels of nematodes in the soil before treatment. Note that the parameter estimate for X is 0.987 and is significant at the 0.0001 level.

If we had not performed the covariance analysis we would have incorrectly rejected the null hypothesis and said that there was a significant treatment effect on yield, when in fact the treatment differences were really due to differences in pre-treatment nematode populations.

Adjusting Treatment Means to a Common Covariate Level

If the covariate has a significant influence on our dependent variable, it doesn't really make sense to look at regular treatment means provided by SAS® because each treatment may have a different mean level of the covariate associated with it, thus the mean reported differs by both the treatment and covariate effects. We can, however, ask SAS® to provide treatment means that have been adjusted for a common level of the covariate. This common level is the overall mean of the covariate. We get these means using the following SAS® statement:

```
LSMEANS treatment / STDERR PDIFF;
```

The STDERR option tells SAS® to print out the standard errors of each mean. The PDIFF option tells SAS® to give *P*-values for all possible pairwise mean comparisons. Notice the note that SAS® prints out to warn about using the *P*-values.

The output is:

'Chapter 17 Example 3.sas'
'Analysis of Covariance'

The GLM Procedure
Least Squares Means

treatment	yield LSMEAN	Standard Error	Pr > |t|	LSMEAN Number
a	6.7149635	1.2884943	<.0001	1
d	6.8239348	1.2724690	<.0001	2
f	10.1611017	1.3159234	<.0001	3

Least Squares Means for effect treatment
Pr > \|t\| for H0: LSMean(i)=LSMean(j)

Dependent Variable: yield

i/j	1	2	3
1		0.9521	0.0793
2	0.9521		0.0835
3	0.0793	0.0835	

You would use this information to indicate which of the pairwise comparisons were significant. Notice that you only have to consider half of the table above, since it is mirrored along the diagonal comparing means of treatments 1 to 2, 1 to 3 and 2 to 3. All three comparisons indicate that there were no significant differences among treatments when adjusted for the mean nematode population.

Estimating Missing Values

In many experiments we run into the problem of missing data. Briefly, we deal with missing data by:

1. Using PROC GLM in SAS and cautiously interpreting output.
2. Calculating a missing value replacement via an analysis of covariance.

To use an analysis of covariance to estimate a missing value, follow this procedure:

1. For the missing value, enter a 0.
2. Set up a dummy X variable as the covariate. For the missing value, $X = 1$. For all observed values, $X = 0$.
3. Perform an analysis of covariance.
4. The missing value estimate is $-\beta$ (the parameter estimate for X).

An example:

```
TITLE1 'Chapter 17 Example 4.sas';
TITLE2 'Analysis of Covariance';
TITLE3 'Missing value estimation';
DATA one;
INPUT x treatment y;
CARDS;
1 1 0
0 1 21
0 1 9
0 1 21
0 2 24
0 2 30
0 2 11
0 2 31
0 3 15
0 3 22
0 3 13
0 3 20
0 4 24
0 4 27
0 4 17
0 4 18
RUN;
PROC GLM PLOTS = none;
CLASSES treatment;
MODEL y = treatment x / SOLUTION;
RUN;
```

The results:

'Chapter 17 Example 4.sas'
'Analysis of Covariance'
'Missing value estimation'

The GLM Procedure

Class Level Information		
Class	Levels	Values
treatment	4	1 2 3 4

Number of Observations Read	16
Number of Observations Used	16

'Chapter 17 Example 4.sas'
'Analysis of Covariance'
'Missing value estimation'

The GLM Procedure

Dependent Variable: y

Source	DF	Sum of Squares	Mean Square	F Value	Pr > F
Model	4	506.9375000	126.7343750	2.95	0.0696
Error	11	472.0000000	42.9090909		
Corrected Total	15	978.9375000			

R-Square	Coeff Var	Root MSE	y Mean
0.517845	34.59012	6.550503	18.93750

Source	DF	Type I SS	Mean Square	F Value	Pr > F
treatment	3	290.1875000	96.7291667	2.25	0.1391
x	1	216.7500000	216.7500000	5.05	0.0461

Source	DF	Type III SS	Mean Square	F Value	Pr > F
treatment	3	124.4000000	41.4666667	0.97	0.4430
x	1	216.7500000	216.7500000	5.05	0.0461

Parameter	Estimate		Standard Error	t Value	Pr > |t|
Intercept	21.50000000	B	3.27525155	6.56	<.0001
treatment 1	-4.50000000	B	5.00302939	-0.90	0.3877
treatment 2	2.50000000	B	4.63190516	0.54	0.6001
treatment 3	-4.00000000	B	4.63190516	-0.86	0.4063
treatment 4	0.00000000	B	.	.	.
x	-17.00000000		7.56386946	-2.25	0.0461

The missing value estimate is $-\beta_X$ (17). You would then plug in 17 for the missing value and perform an appropriate analysis of variance. In theory you should subtract 1 df from the error you get in this analysis and calculate a new MSE and F-value for treatment by hand since you must pay for the missing value with 1 df from error.

For two missing values, you would need two dummy covariates (X1 and X2). The rest of the procedure is similar to the one used for calculating one missing value. Here's an example with two missing values: (lines 7 and 18 contain the missing value, with both y values set to 0)

```
TITLE1 'Chapter 17 Example 5.sas';
TITLE2 'Analysis of Covariance';
TITLE3 'Estimation of 2 missing values';
DATA one;
INPUT x2 x1 treatment y;
CARDS;
0 1 1 0
0 0 1 21
0 0 1 9
0 0 1 21
0 0 2 24
0 0 2 30
0 0 2 11
0 0 2 31
0 0 3 15
0 0 3 22
0 0 3 13
1 0 3 0
0 0 4 24
0 0 4 27
0 0 4 17
0 0 4 18
RUN;
PROC GLM;
CLASSES treatment;
MODEL y = treatment x1 x2 / SOLUTION;
RUN;
```

'Chapter 17 Example 5.sas'
'Analysis of Covariance'
'Estimation of 2 missing values'

The GLM Procedure

Class Level Information		
Class	Levels	Values
treatment	4	1 2 3 4

Number of Observations Read	16
Number of Observations Used	16

The GLM Procedure

Dependent Variable: y

Source	DF	Sum of Squares	Mean Square	F Value	Pr > F
Model	5	847.770833	169.554167	3.66	0.0385
Error	10	463.666667	46.366667		
Corrected Total	15	1311.437500			

R-Square	Coeff Var	Root MSE	y Mean
0.646444	38.49785	6.809307	17.68750

Source	DF	Type I SS	Mean Square	F Value	Pr > F
treatment	3	422.6875000	140.8958333	3.04	0.0795
x1	1	216.7500000	216.7500000	4.67	0.0559
x2	1	208.3333333	208.3333333	4.49	0.0601

Source	DF	Type III SS	Mean Square	F Value	Pr > F
treatment	3	132.6904762	44.2301587	0.95	0.4514
x1	1	216.7500000	216.7500000	4.67	0.0559
x2	1	208.3333333	208.3333333	4.49	0.0601

Parameter	Estimate		Standard Error	t Value	Pr > |t|
Intercept	21.50000000	B	3.40465368	6.31	<.0001
treatment 1	-4.50000000	B	5.20069440	-0.87	0.4072
treatment 2	2.50000000	B	4.81490741	0.52	0.6149
treatment 3	-4.83333333	B	5.20069440	-0.93	0.3746
treatment 4	0.00000000	B	.	.	.
x1	-17.00000000		7.86271087	-2.16	0.0559
x2	-16.66666667		7.86271087	-2.12	0.0601

The missing value estimates are 17 and 16.7. You should adjust the error df by 2 (for the two missing values calculated) and perform hand calculations of an adjusted *F*-value as described for one missing value.

Note that you can use this procedure for any experimental design with any treatment structure. I used the CRD for simplicity.

18 Procedures for Non-normal Data

The majority of this text is devoted to parametric statistical procedures. These procedures are used because we assume that our data are from a particular probability distribution. While we assume that we know the distribution, we don't know the parameters of the distribution. Most of the time we assume a normal distribution, thus the parameters we don't know are μ and σ^2. We use parametric procedures to estimate these parameters, then perform statistical tests on them to test specific hypotheses we have posed (for example, the F-test of an ANOVA to evaluate the equality of treatment means). If our data truly come from the assumed distribution, we can be confident in the results we obtain and conclusion we draw regarding our data. While we never know for sure if our data are from a particular distribution, we can perform a statistical test on our data to give us a good idea if it does or does not fit our assumption (i.e. the Shapiro–Wilks test to test the normality assumption).

But what about when we are pretty sure that our data are not normally distributed or when we have no clue regarding the underlying distribution of our data? We need statistical techniques which we can use regardless of the data's true distribution. These statistical techniques are called nonparametric or distribution-free methods. This chapter will serve as an introduction to nonparametric statistics with an emphasis on several methods supported by SAS® along with some plant science examples. We will cover the methodology one would use when we would normally use an ANOVA if the data were normal.

We will also look at the approach that can be adopted for any of the parametric procedures we have learned that first requires transformation of our data using what is called the 'aligned rank transformation' (ART). Once transformed, the analysis of the transformed data proceeds just as if the data had not been transformed. This approach was recently reflected in an article in the journal *Horticulturae* (Durner, 2019) which highlighted the ART procedure for evaluating treatment interactions (most nonparametric tests are not suited to evaluating interactions). The section of this chapter discussing this methodology was adapted from the Durner paper. It is an extremely useful approach and once you understand how to use this methodology, it will become an invaluable tool in your statistical arsenal for evaluating experiments with non-normal data with and without interactions.

Introduction

Most statistical techniques commonly used in horticultural research, such as the ANOVA, *t*-tests and linear regression, are parametric techniques, valid only if the data in the analysis are normally and independently distributed with $\mu = 0$ and common variance, σ^2, i.e. normal data with homogeneous variances. Even though parametric tests are robust when data 'slightly' deviate from normality, significant departure can lead to incorrect conclusions.

DOI: 10.1079/9781789249927.0018

When parametric procedures are used on non-normal data, the probability of detecting a treatment effect when it does in fact exist (the power of the test) is greatly reduced and the probability of a Type I error, declaring a significant treatment effect when in fact there is none, greatly increases.

The first step most researchers take if they determine that their data are not normal is to transform them using one of many techniques. Subjecting transformed data to a parametric analysis is valid as long as the transformed data are normal. If transformed data are still not normal, a different transformation can be applied to the original data until a transformation which renders the data normal is revealed. There are many statistical texts that discuss this subject at great length, providing guidance indicating which transformations are usually most effective for different types of data. The process can be quite time consuming, cumbersome and confusing.

The alternative to parametric analysis of transformed data is a nonparametric or distribution-free test. Let's examine the general theory of how this works and look at several tests that are easily accomplished using SAS®.

How Do Nonparametric Tests Work?

Nonparametric tests do not rely on any assumptions regarding data distribution and are therefore often called distribution-free tests. The nice thing about most nonparametric tests is that they are almost as powerful as their parametric counterpart even when the data are normal and a nonparametric test is used where a parametric test would seem more appropriate.

Nonparametric tests are often based on ranked data rather than the raw data. The prominent procedure in SAS® which performs nonparametric testing, PROC NPAR1WAY, ranks the raw data so the user doesn't have to. The PROC RANK procedure of SAS® performs ranking based on user specifications in cases where the user needs ranked data for use in analyses other than those performed by PROC NPAR1WAY.

Choices If Data Is Not Normal

Certain data naturally lend themselves to nonparametric testing. Data based on counts or ranks are the two data types often subjected to nonparametric analysis. When you are unsure as to the nature of your data's distribution, the test for normality using PROC UNIVARIATE is appropriate. If the normality test indicates non-normal data, you can proceed in one of four directions:

1. Use a parametric procedure relying on the robust nature of parametric testing even with non-normal data. (Not recommended.)
2. Evaluate transformations which make the data 'more normal' (see Chapter 20) followed by parametric analysis of transformed data. (Recommended, may be time consuming and transformations may not make data 'more normal'.)
3. Use a nonparametric procedure. (Recommended, especially if transformations do not render data more normal. However, if treatments are of a factorial nature and interactions are possible, use option 4.)
4. Use the ARTool app to align rank transform data, re-test transformed data for normality, then proceed with parametric testing. (Recommended.)

Option 1 is not recommended, since confidence in results generated by such an approach will always be questionable. Option 2 is recommended if appropriate transformations are available which render the data more normal. The transformations to utilize if this option is chosen are presented in Chapter 20. There are many possible transformations to increase normality, and the same transformations are used for other 'data problems'.

If your data have no interactions (thus not a factorial treatment structure, only main effects) then option 3 is a good choice, but option 4 is also a good choice. If there are interactions (thus a factorial treatment structure) option 4 is the best choice. Note that option 4 can be used in both situations, with or without interactions, but must be used when there are interactions, since option 3 does not work for evaluating interactions (see below).

OPTION 3: Several Nonparametric Tests

There are many nonparametric options available for testing non-normal data and the choice of which to use in different situations can be bewildering, to say the least. To that end, the discussion of nonparametric methods from this point forward will be limited to specific examples of which procedure to use to accomplish some very specific tasks. Hopefully this makes the use and implementation of nonparametric testing easier.

Example 1: Where is my data centered?

In most situations we measure the center of our data with the mean. In the nonparametric world we measure it with the median. The median is a value where half of the data lies below it and half lies above it. If you are interested in making an inference about the central location of your non-normal data, you can use the Sign test in PROC UNIVARIATE. This test makes no assumptions about your data. It simply tests the hypothesis that the median of the population from which your sample is drawn is equal to some specific value (by default, 0).

The following sample of yield (dozens of ears per hectare) from a new sweetcorn cultivar was gathered. Since a test that the median is equal to 0 is really not very meaningful, let's make a more meaningful test that the median yield is equal to the average yield for sweetcorn in the region where the new cultivar was tested. That yield is 2965. To perform this test, use the following code:

```
TITLE1 'Chapter 18 Example 1.sas';
TITLE2 'Sweetcorn Yield';
TITLE3 'dozens of ears per hectare';
TITLE4 'New cultivar X';
DATA one;
INPUT sample yield;
CARDS;
1 2533
2 2762
3 3028
4 3101
5 2800
```

```
6 2965
7 3003
8 2902
9 2893
10 2348
RUN;
PROC UNIVARIATE DATA = one NORMAL LOCATION = 2965;
VAR yield;
RUN;
```

The results:

'Chapter 18 Example 1.sas'
'Sweetcorn Yield'
'dozens of ears per hectare'
'New cultivar X'

The UNIVARIATE Procedure
Variable: yield

Moments			
N	10	Sum Weights	10
Mean	2833.5	Sum Observations	28335
Std Deviation	234.730697	Variance	55098.5
Skewness	-1.1576549	Kurtosis	0.79751346
Uncorrected SS	80783109	Corrected SS	495886.5
Coeff Variation	8.28412553	Std Error Mean	74.2283639

Basic Statistical Measures			
Location		Variability	
Mean	2833.500	Std Deviation	234.73070
Median	2897.500	Variance	55099
Mode	.	Range	753.00000
		Interquartile Range	241.00000

Tests for Location: Mu0=2965				
Test		Statistic	p Value	
Student's t	t	-1.77156	Pr > |t|	0.1102
Sign	M	-1.5	Pr >= |M|	0.5078
Signed Rank	S	-14	Pr >= |S|	0.1055

Tests for Normality				
Test	Statistic		p Value	
Shapiro-Wilk	W	0.8975	Pr < W	0.2056
Kolmogorov-Smirnov	D	0.200052	Pr > D	>0.1500
Cramer-von Mises	W-Sq	0.077099	Pr > W-Sq	0.2091
Anderson-Darling	A-Sq	0.463192	Pr > A-Sq	0.2079

To evaluate the hypothesis that the median is equal to 2965, we include the 'LOCATION = 2965' option on line 19. For any other value of the median you would like to test, simply replace the '2965' with that number. The test we want to look at is the 'Sign' test, whose value is −1.5 with Pr >= |M| = 0.5078, indicating that we should fail to reject the hypothesis that the calculated median of the data 2897.5 is not statistically different from 2965.

In addition, the Shapiro–Wilks test indicates that we should fail to reject the assumption of normality at $\alpha = 0.2056$, indicating that it appears that our data are normally distributed and we don't need to use a nonparametric measure of the data center (the median) and can instead look at the mean. The t-test of the null hypothesis that $\mu = 2965$ (since that is the value in our LOCATION option on line 19) reveals that we should fail to reject the null hypothesis at $\alpha = 0.1102$, indicating that our calculated mean of statistic 2833.5 is not statistically different from 2965.

Example 2: Comparing two independent samples

Suppose you had data from two different samples of sweetcorn yield that you wanted to compare. One sample was the original sample from the last section and the second sample was from another test where you had yield estimates for a cultivar that is the most widely grown cultivar in your area. You would like to know whether or not the two samples are different from each other. We identify our two different samples with 'X' and 'Y' to indicate which cultivar the observation is from.

```
TITLE1 'Chapter 18 Example 2.sas';
TITLE2 'Sweetcorn Yield';
TITLE3 'dozens of ears per hectare';
TITLE4 'New cultivar X vs standard';
DATA one;
INPUT sample cultivar $ yield;
CARDS;
1 X 2533
2 X 2762
3 X 3028
4 X 3101
5 X 2800
6 X 2965
7 X 3003
8 X 2902
9 X 2893
10 X 2348
```

```
11  Y  2433
12  Y  2562
13  Y  2728
14  Y  3001
15  Y  2600
16  Y  2765
17  Y  2803
18  Y  2702
19  Y  2693
20  Y  2448
RUN;
PROC NPAR1WAY DATA = one WILCOXON PLOTS = ncne;
CLASS cultivar;
VAR yield;
RUN;
```

To test if the medians of the two groups differ, we use PROC NPAR1WAY. The CLASS statement indicates the name of our variable identifying which group the data are in. We include the WILCOXON option on line 29 to request the Mann–Whitney test (Wilcoxon sign ranks, identified in the printout as the Kruskal–Wallis test) which is actually an ANOVA of the ranks (Wilcoxon scores). The results for our test are:

Kruskal-Wallis Test		
Chi-Square	DF	Pr > ChiSq
3.2914	1	0.0696

Our statistic is 3.2914 with 1 df (the number of class levels −1). The p-value (the probability of obtaining a test statistic as large or larger than that obtained, simply by chance) is 0.0696. Thus we would fail to reject the null hypothesis that the two groups did not differ.

Both the chi-square test statistic used and the calculated p-value are asymptotic values, meaning that it estimates the exact value. As the number of observations increases, the test statistic and therefore the p-value behave like the exact value. Exact values aren't routinely used for these procedures in SAS® because they often require excessive computing time. You really don't need to worry about this, as long as you know how to interpret the p-values.

Comparing more than two independent samples

When you want to compare more than two independent samples, you use exactly the same procedure as described for two samples. When the number of samples is greater than two, the test is known as the Kruskal–Wallis test.

Nonparametric multiple comparisons

If you are interested in multiple comparisons using nonparametric tests, please see a reference text which deals exclusively with nonparametric statistics. One excellent text is Conover (1999).

There are many more nonparametric methods available for evaluating research data, but there simply is not enough room to cover them in this text. Since we have already learned many parametric techniques applicable to normally distributed data, it makes sense to cover option 4 thoroughly.

OPTION 4: Analysis of Non-normal Data Using the Aligned Rank Transform, ARTool and SAS®

Option 4 teaches how to appropriately transform data via the aligned rank transformation using a readily downloadable app so that parametric procedures can be used on non-normal data. Once you understand how to use the methodology of option 4 you will have a vast arsenal of options to choose from when confronted with non-normal data.

The main problem with nonparametric tests is that they are not appropriate for evaluating interactions common in horticultural research. Many of the nonparametric tests are based on simple ranks with or without slight modifications. Because of this, non-significant interactions may be declared significant (Type I error) or significant interactions may not be detected (low power). Thus there is little confidence in such results.

The aligned rank transformation (ART) is a useful tool which allows nonparametric testing of interactions and main effects using standard ANOVA techniques. The ART transformation mathematically strips each tested effect of all other effects in the analysis while maintaining the underlying relationship among the factors. The two problems (Type I errors and low power) often associated with other tests based on ranking do not exist with ART.

The ART technique has not been widely adapted, due to the rigorous mathematical nature of the transformation, especially when two or more factors are tested. Computer code performing the aligned rank transformation is available, but many horticulture researchers are not familiar with programming and are generally unable to utilize such information. Luckily, Wobbrock *et al.* (2011–2018) developed a downloadable program (ARTool) which performs the mathematics needed for the transformation. The program, which was originally developed for those working in the field of computer–user interface research, is applicable to all fields generating data that may not be normally distributed. While correct use of the program is time consuming and meticulous, it is not difficult.

The program is freely downloadable as a Windows program, a package for R, or as source code. Using ARTool and SAS® together provides an easy method for testing normality, transforming data via the aligned rank procedure and data analysis using common methods such as ANOVA and linear regression.

First, download the operating system appropriate ART app using the link: http://depts. washington.edu/madlab/proj/art/. Double-click the file and extract the contents to the folder where the app will reside, such as a subfolder in the SAS® 'myfolders' folder called 'ARToolExe'. The two components that must be in this subdirectory are 'ARTool.exe' and 'WobbrockLib.dll'. The subfolder 'data' which is created upon file extraction contains supplementary material included in the download. Copy the two required files directly into a folder where data files will reside. This will be a folder on your computer. Once you have transformed the data, you will upload the transformed file to your SAS OnDemand for Academics folder as described in Chapter Two. Create a shortcut to the 'ARTool.exe' file and pin the shortcut to the taskbar for easy access. Change the icon of the shortcut to a meaningful one, in this case the letter 'A' to remember what the shortcut points to.

We'll examine three examples to illustrate how to integrate the ART app with SAS®. Each example is small for simplicity and each includes a brief description of the data, treatment structure and experimental design. With each sample, data will be tested for normality and the variances tested for homogeneity using SAS®, data will be transformed using the ART app, then the transformed data will be analyzed appropriately with SAS®.

Example 3: Completely Random Design

A completely random design was used to examine the effects of nitrogen source and cultivar on strawberry yield using five single plant replicates. Three sources of nitrogen (urea, calcium nitrate and potassium nitrate) and four cultivars ('Chandler', 'Earliglow', 'Jewel' and 'Flavorfest') were evaluated for their effects on productivity (yield, g/plant).

Enter the following program into the SAS® editor window:

```
1   TITLE1 'Chapter 18 Example 3a.sas';
2   TITLE2 'ART of CRD';
3   TITLE3 'Strawberry and Nitrogen Study';
4   DATA one;
5   INPUT rep nitrogen $ cultivar $ yield;
6   CARDS;
7   1 Urea Chand 375
8   2 Urea Chand 424
9   3 Urea Chand 368
10  4 Urea Chand 407
11  5 Urea Chand 372
12  1 Urea Eglow 349
13  2 Urea Eglow 339
14  3 Urea Eglow 367
15  4 Urea Eglow 365
16  5 Urea Eglow 374
17  1 Urea Jewel 407
18  2 Urea Jewel 335
19  3 Urea Jewel 382
20  4 Urea Jewel 415
21  5 Urea Jewel 352
22  1 Urea Ffest 418
23  2 Urea Ffest 366
24  3 Urea Ffest 365
25  4 Urea Ffest 355
26  5 Urea Ffest 364
27  1 Calnit Chand 406
28  2 Calnit Chand 410
29  3 Calnit Chand 333
30  4 Calnit Chand 416
31  5 Calnit Chand 354
32  1 Calnit Eglow 371
33  2 Calnit Eglow 337
```

```
34 3 Calnit Eglow 330
35 4 Calnit Eglow 405
36 5 Calnit Eglow 371
37 1 Calnit Jewel 364
38 2 Calnit Jewel 367
39 3 Calnit Jewel 391
40 4 Calnit Jewel 396
41 5 Calnit Jewel 388
42 1 Calnit Ffest 335
43 2 Calnit Ffest 389
44 3 Calnit Ffest 386
45 4 Calnit Ffest 380
46 5 Calnit Ffest 361
47 1 Potnit Chand 333
48 2 Potnit Chand 333
49 3 Potnit Chand 343
50 4 Potnit Chand 364
51 5 Potnit Chand 388
52 1 Potnit Eglow 424
53 2 Potnit Eglow 409
54 3 Potnit Eglow 361
55 4 Potnit Eglow 373
56 5 Potnit Eglow 364
57 1 Potnit Jewel 340
58 2 Potnit Jewel 333
59 3 Potnit Jewel 391
60 4 Potnit Jewel 396
61 5 Potnit Jewel 389
62 1 Potnit Ffest 341
63 2 Potnit Ffest 365
64 3 Potnit Ffest 331
65 4 Potnit Ffest 414
66 5 Potnit Ffest 410
67 RUN;
68 PROC PRINT;
69 PROC UNIVARIATE NORMAL;
70 VAR yield;
71 RUN;
72 PROC ANOVA;
73 CLASSES nitrogen cultivar;
74 MODEL yield = nitrogen;
75 MEANS nitrogen / HOVTEST = levene;
76 RUN;
77 PROC ANOVA;
78 CLASSES nitrogen cultivar;
79 MODEL yield = cultivar;
80 MEANS cultivar / HOVTEST = levene;
```

```
81 RUN;
82 PROC ANOVA;
83 CLASSES nitrogen cultivar;
84 MODEL yield = nitrogen*cultivar;
85 MEANS nitrogen*cultivar / HOVTEST= levene;
86 RUN;
```

Lines 1 through 67 generate the dataset, line 68 provides a printout of it, and lines 69 through 71 provide a test of normality for 'yield'.

Lines 72 through 86 provide the test for homogeneity of variances among the nitrogen groups, cultivar groups and nitrogen*cultivar groups, respectively.

We first check the assumption of normality by finding the results of the Shapiro–Wilks test in the results window:

Tests for Normality				
Test		Statistic	p Value	
Shapiro-Wilk	W	0.949734	Pr < W	0.0151
Kolmogorov-Smirnov	D	0.092079	Pr > D	>0.1500
Cramer-von Mises	W-Sq	0.100884	Pr > W-Sq	0.1088
Anderson-Darling	A-Sq	0.775608	Pr > A-Sq	0.0430

The Shapiro–Wilks normality test is used for datasets with seven or more but fewer than 2000 observations and tests the null hypothesis that the data is normally distributed. The test provides the probability of obtaining a W statistic less than the one calculated for the data in question, simply by chance, if the data is normally distributed. For our example, $Pr < W = 0.0151$, thus the evidence is that the data is not normally distributed and should be transformed to use parametric tests.

The tests for homogeneity of variances use Levene's test (there are other tests available). Three main groups are involved: the nitrogen groups (three groups), the cultivar groups (four groups) and the nitrogen*cultivar groups (12 groups), and we must test for heterogeneity within each main group. The three consecutive ANOVAs accomplish this.

'Chapter 18 Example 3.sas'
'ART of CRD'
'Strawberry and Nitrogen Study'

The ANOVA Procedure

Levene's Test for Homogeneity of yield Variance ANOVA of Squared Deviations from Group Means					
Source	DF	Sum of Squares	Mean Square	F Value	Pr > F
nitrogen	2	762144	381072	0.69	0.5036
Error	57	31284930	548858		

'Chapter 18 Example 3.sas'
'ART of CRD'
'Strawberry and Nitrogen Study'

The ANOVA Procedure

	Levene's Test for Homogeneity of yield Variance ANOVA of Squared Deviations from Group Means				
Source	DF	Sum of Squares	Mean Square	F Value	Pr > F
cultivar	3	971055	323685	0.53	0.6608
Error	56	33937912	606034		

'Chapter 18 Example 3.sas'
'ART of CRD'
'Strawberry and Nitrogen Study'

The ANOVA Procedure

	Levene's Test for Homogeneity of yield Variance ANOVA of Squared Deviations from Group Means				
Source	DF	Sum of Squares	Mean Square	F Value	Pr > F
nitrogen*cultivar	11	6148230	558930	1.71	0.0998
Error	48	15693982	326958		

All three tests have p-values > 0.05, which suggests that variances are homogeneous within each group. Overall, our tests reveal that our data are not normal and do not suffer from heterogeneous variances.

We now know that transforming our data is the next appropriate step. In order to use the ARTool app, data must be specifically formatted in a CSV (comma-separated values) file. Data in the CSV file are long-format data where each line represents one observation and the right-most column contains the dependent variable (Y), i.e. in this case, 'yield'. The first column represents the experimental unit or observation number (OBS) and is not used in the ARTool calculations. Each column between the OBS and Y columns represents one factor in the experiment. For this example, there are two columns between OBS and Y, nitrogen (N) and cultivar (CV). The ARTool program will generate aligned and ranked columns for each main effect and interaction and save them in the output data file which is saved to the default folder where the ARTool app is located (myfolders) with the name of the input file processed appended with '.art' (i.e. 'data.csv' would produce the output file 'data.art.csv'). When you start the ARTool app it should look like this:

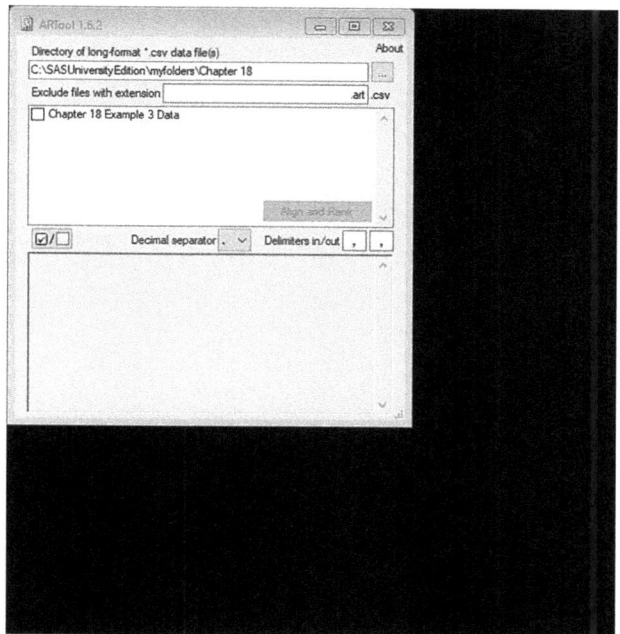

The CSV files in our 'myfolders' folder appear in the top window of the program. To process the data in the file and create an output data file, click the check-box next to the filename of the file for processing ('Chapter 18 Example 3 Data') and click the 'Align and Rank' button in the lower right of the same window. A message will appear in the bottom window indicating the file has been processed and the output file created:

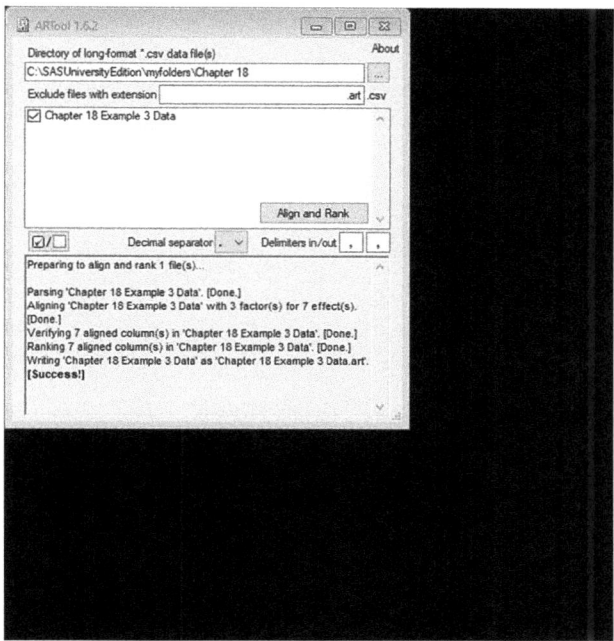

Problems will be indicated with an error message. Two most common errors are: (1) the CSV file is still being used by Excel; and (2) blank spaces or non-numeric 'Y' values have been detected. For problem 1, close the CSV file in Excel. Problem 2 may be a bit more complicated. First make sure there are no observations with missing values indicated by blank spaces or placeholders such as '.'. If there are, delete all rows in the original CSV file where this occurs. Sometimes the CSV file has blank columns to the right of the 'Y' column or blank rows past the final observation of the CSV file, either of which the ARTool program will detect as a blank space and produce an error. To fix this, simply delete all columns to the right of 'Y' and all rows past the final observation, then resave the CSV file.

With successful processing, a CSV file will be generated by ARTool. The file will contain variables in the first row which include ones we supplied in our CSV data file as well as variables created by ARTool. For this example the variables are Obs, Nitrogen, Cultivar, Yield, aligned(Yield) for Nitrogen, aligned(Yield) for Cultivar, aligned(Yield) for Nitrogen*Cultivar, ART(Yield) for Nitrogen, ART(Yield) for Cultivar and ART(Yield) for Nitrogen*Cultivar.

The ARTool-generated CSV file is used directly in the next SAS® program ('Chapter 18 Example 3b.sas') for an ANOVA, testing for nitrogen and cultivar main effects and an interaction between the two. To make things a bit easier we are going to access the data directly from this CSV file generated by the ARTool app. This is accomplished in lines 6 through 8.

```
1  TITLE1 'Chapter 18 Example 3b.sas';
2  TITLE2 'ART of CRD';
3  TITLE3 'Strawberry and Nitrogen Study';
4  TITLE4 'Analysis of transformed data';
5  DATA one;
6  INFILE '/folders/myfolders/CABI Text/Chapter 18/Chapter 18
   Example 3 Data.art.csv' DLM =',' FIRSTOBS=2;
7  INPUT obs nitrogen $ cultivar $ yield a b c Anit Acult
   Anitcult;
8  CARDS;
9  RUN;
10 PROC ANOVA PLOTS = none;
11 CLASSES nitrogen cultivar;
12 MODEL yield Anit Acult Anitcult =
   nitrogen cultivar nitrogen*cultivar;
13 MEANS nitrogen cultivar / REGWQ;
14 MEANS nitrogen*cultivar;
15 RUN;
```

Let's look at lines 6 and 7 a little more closely.

```
6  INFILE '/folders/myfolders/CABI Text/Chapter 18/Chapter 18
   Example 3 Data.art.csv' DLM =',' FIRSTOBS=2;
```

This indicates where the data for dataset 'one' is located. The 'INFILE' keyword indicates that the data is in a file, located in the folder 'myfolders/CABI Text/Chapter 18/'. The entire piece within the single quotes is the complete file name. The 'DLM' keyword indicates that the delimiter separating the data information described in line 7 is a comma (indicated by the ','). The 'FIRSTOBS' keyword indicates that the first line of data is the second line of the file.

```
7 INPUT obs nitrogen $ cultivar $ yield a b c Anit Acult Anitcult;
```

The 'INPUT' keyword indicates that the data is read from the file identified in line 6 as obs, nitrogen, cultivar and yield. Note that the ARTool app has added the variable 'obs' to our data as well as several other variables after yield. Note that the variable names have been modified from the ARTool .art.csv file to facilitate easier SAS® programing (for example, 'ART(Yield) for Nitrogen' is changed to 'Anit'). The variables we will analyze are the ART (aligned and rank transformed) variables generated by the ARTool program. Their names have been shortened to 'Anit', 'Acult' and 'Anitcult', which we use in the ANOVA. Note also that the 'aligned' variables are not used in the analysis so they are labeled 'a', 'b' and 'c'.

The ART value for each factor or interaction is the Y value stripped of all effects except the one under consideration during the aligning and ranking procedure. For example, 'Anit' is the value of Y (yield) stripped of all cultivar and nitrogen*cultivar effects, 'Acult' is Y (yield) stripped of all nitrogen and nitrogen*cultivar effects, and 'Anitcult' is Y (yield) stripped of all nitrogen and cultivar main effects. The significance value for each tested source of variation is obtained from the test for the effect indicated in the Anit, Acult or Anitcult variable name. Thus to obtain a significance value for nitrogen, examine the significance of the nitrogen effect on the Anit variable. For a cultivar effect, examine the significance of cultivar with Acult and for the interaction, nitrogen*cultivar, examine the nitrogen*cultivar effect with Anitcult. The full factorial analysis of the data is achieved with lines 10 through 15 of the program above.

Note that the model statement has, as dependent variables, yield (non-transformed yield values), 'Anit' (ART values for Y considering only nitrogen), 'Acult' (ART values for Y considering only cultivar, and 'Anitcult' (ART values for Y considering only nitrogen*cultivar).

Results and discussion

SAS® output:

'Chapter 18 Example 3b.sas'
'ART of CRD'
'Strawberry and Nitrogen Study'
Analysis of transformed data

The ANOVA Procedure

Dependent Variable: yield

Source	DF	Sum of Squares	Mean Square	F Value	Pr > F
Model	11	7022.98333	638.45303	0.81	0.6324
Error	48	37954.00000	790.70833		
Corrected Total	59	44976.98333			

R-Square	Coeff Var	Root MSE	yield Mean
0.156146	7.535046	28.11954	373.1833

Source	DF	Anova SS	Mean Square	F Value	Pr > F
nitrogen	2	287.233333	143.616667	0.18	0.8345
cultivar	3	459.516667	153.172222	0.19	0.9002
nitrogen*cultivar	6	6276.233333	1046.038889	1.32	0.2654

'Chapter 18 Example 3b.sas'
'ART of CRD'
'Strawberry and Nitrogen Study'
Analysis of transformed data

The ANOVA Procedure

Dependent Variable: Anit

Source	DF	Sum of Squares	Mean Square	F Value	Pr > F
Model	11	177.60000	16.14545	0.04	1.0000
Error	48	17814.90000	371.14375		
Corrected Total	59	17992.50000			

R-Square	Coeff Var	Root MSE	Anit Mean
0.009871	63.16423	19.26509	30.50000

Source	DF	Anova SS	Mean Square	F Value	Pr > F
nitrogen	2	136.8750000	68.4375000	0.18	0.8322
cultivar	3	26.4333333	8.8111111	0.02	0.9950
nitrogen*cultivar	6	14.2916667	2.3819444	0.01	1.0000

'Chapter 18 Example 3b.sas'
'ART of CRD'
'Strawberry and Nitrogen Study'
Analysis of transformed data

The ANOVA Procedure

Dependent Variable: Acult

Source	DF	Sum of Squares	Mean Square	F Value	Pr > F
Model	11	256.20000	23.29091	0.06	1.0000
Error	48	17736.30000	369.50625		
Corrected Total	59	17992.50000			

R-Square	Coeff Var	Root MSE	Acult Mean
0.014239	63.02474	19.22255	30.50000

Source	DF	Anova SS	Mean Square	F Value	Pr > F
nitrogen	2	0.3000000	0.1500000	0.00	0.9996
cultivar	3	251.4000000	83.8000000	0.23	0.8773
nitrogen*cultivar	6	4.5000000	0.7500000	0.00	1.0000

The ANOVA Procedure

Dependent Variable: Anitcult

Source	DF	Sum of Squares	Mean Square	F Value	Pr > F
Model	11	2250.70000	204.60909	0.62	0.7994
Error	48	15742.80000	327.97500		
Corrected Total	59	17993.50000			

R-Square	Coeff Var	Root MSE	Anitcult Mean
0.125084	59.37731	18.11008	30.50000

Source	DF	Anova SS	Mean Square	F Value	Pr > F
nitrogen	2	3.325000	1.662500	0.01	0.9949
cultivar	3	8.366667	2.788889	0.01	0.9989
nitrogen*cultivar	6	2239.008333	373.168056	1.14	0.3552

In the ANOVA of non-transformed data, the p-values for nitrogen, cultivar and nitrogen*cultivar are 0.8345, 0.9002 and 0.2654, respectively. However, these values are not valid, since the normality assumption of the data has been violated.

Considering the nitrogen effect for transformed data, the p-value for the significance of the nitrogen effect for the dependent variable Anit is 0.8322, which is not much different than the p-value produced for non-transformed data (0.8345). The difference between the two is the confidence one can have in asserting that there is not a significant effect of nitrogen on yield when considering results from the transformed analysis. One cannot be confident using the analysis of non-transformed data, since the data were identified as non-normal. The same is true for cultivar and the nitrogen*cultivar interaction.

With all three analyses of transformed data, the p-values for effects not considered in each analysis (i.e. for cultivar and nitrogen*cultivar in the analysis for a nitrogen main effect) is close to 1.00. This is a characteristic of the analysis which is a good check of the effectiveness of the ART procedure. If these p-values are far from 1.00, the ART procedure may not be adapted to your data and some other method should be employed. Alternatives to the ART procedure are given by Sawilowsky (1990) and Higgins (2004). Fortunately, most data are amenable to the ART procedure.

Example 4: Randomized Complete Block Design

Suppose the experiment in Example 1 was a randomized complete block design (RCBD) instead of a CRD. The ART process would be exactly like that used in Example 1. The difference between the two is the SAS model statement. With the RCBD, an effect attributable to blocking would be accounted for and the analysis would be like that shown below. Note that in the INPUT statement, 'rep' has been changed to 'block' to reflect the 'new' design and 'block' has been added to the model statement.

```
TITLE1 'Chapter 18 Example 4.sas';
TITLE2 'ART of RCBD';
TITLE3 'Strawberry and Nitrogen Study';
TITLE4 'Analysis of transformed data';
DATA one;
INFILE '/folders/myfolders/CABI Text/Chapter 18/Chapter 18
Example 3 Data.art.csv' DLM =',' FIRSTOBS=2;
INPUT block nitrogen $ cultivar $ yield a b c Anit Acult
Anitcult;
CARDS;
RUN;
PROC ANOVA PLOTS = none;
CLASSES block nitrogen cultivar;
MODEL yield Anit Acult Anitcult = block nitrogen cultivar
nitrogen*cultivar;
MEANS nitrogen cultivar / REGWQ;
MEANS nitrogen*cultivar;
RUN;
```

Results:

'Chapter 18 Example 4.sas'
'ART of RCBD'
'Strawberry and Nitrogen Study'
Analysis of transformed data

The ANOVA Procedure

Dependent Variable: yield

Source	DF	Sum of Squares	Mean Square	F Value	Pr > F
Model	15	12482.21667	832.14778	1.13	0.3624
Error	44	32494.76667	738.51742		
Corrected Total	59	44976.98333			

R-Square	Coeff Var	Root MSE	yield Mean
0.277525	7.282125	27.17568	373.1833

Source	DF	Anova SS	Mean Square	F Value	Pr > F
block	4	5459.233333	1364.808333	1.85	0.1367
nitrogen	2	287.233333	143.616667	0.19	0.8240
cultivar	3	459.516667	153.172222	0.21	0.8907
nitrogen*cultivar	6	6276.233333	1046.038889	1.42	0.2299

'Chapter 18 Example 4.sas'
'ART of RCBD'
'Strawberry and Nitrogen Study'
Analysis of transformed data

The ANOVA Procedure

Dependent Variable: Anit

Source	DF	Sum of Squares	Mean Square	F Value	Pr > F
Model	15	2433.68333	162.24556	0.46	0.9488
Error	44	15558.81667	353.60947		
Corrected Total	59	17992.50000			

R-Square	Coeff Var	Root MSE	Anit Mean
0.135261	61.65412	18.80451	30.50000

Source	DF	Anova SS	Mean Square	F Value	Pr > F
block	4	2256.083333	564.020833	1.60	0.1925
nitrogen	2	136.875000	68.437500	0.19	0.8247
cultivar	3	26.433333	8.811111	0.02	0.9945
nitrogen*cultivar	6	14.291667	2.381944	0.01	1.0000

'Chapter 18 Example 4.sas'
'ART of RCBD'
'Strawberry and Nitrogen Study'
Analysis of transformed data

The ANOVA Procedure

Dependent Variable: Acult

Source	DF	Sum of Squares	Mean Square	F Value	Pr > F
Model	15	2649.07500	176.60500	0 51	0.9238
Error	44	15343.42500	348.71420		
Corrected Total	59	17992.50000			

R-Square	Coeff Var	Root MSE	Acult Mean
0.147232	61.22587	18.67389	30.50000

Source	DF	Anova SS	Mean Square	F Value	Pr > F
block	4	2392.875000	598.218750	1.72	0.1636
nitrogen	2	0.300000	0.150000	0.00	0.9996
cultivar	3	251.400000	83.800000	0.24	0.8678
nitrogen*cultivar	6	4.500000	0.750000	0.00	1.0000

'Chapter 18 Example 4.sas'
'ART of RCBD'
'Strawberry and Nitrogen Study'
Analysis of transformed data

The ANOVA Procedure

Dependent Variable: Anitcult

Source	DF	Sum of Squares	Mean Square	F Value	Pr > F
Model	15	4563.03333	304.20222	1.00	0.4754
Error	44	13430.46667	305.23788		
Corrected Total	59	17993.50000			

R-Square	Coeff Var	Root MSE	Anitcult Mean
0.253593	57.28216	17.47106	30.50000

Source	DF	Anova SS	Mean Square	F Value	Pr > F
block	4	2312.333333	578.083333	1.89	0.1284
nitrogen	2	3.325000	1.662500	0.01	0.9946
cultivar	3	8.366667	2.788889	0.01	0.9988
nitrogen*cultivar	6	2239.008333	373.168056	1.22	0.3131

Interpreting the results is similar to Example 1. When interpreting the SAS® output, remember that p-values are only valid for testing a nitrogen effect using the dependent variable Anit, for a cultivar effect using Acult and an interaction using Anitcult.

Example 5: Split-plot Design

This final example is a bit more complicated and considers an experiment where six rates of three nitrogen sources were evaluated for strawberry yield (g/plant). The experimental design was a split-plot with nitrogen source as main plot and nitrogen rate as sub-plot. There were five replicates of the main plot and main plots were set in a randomized complete block design. Data are provided as the supplementary file 'Chapter 18 Example 5 Data.csv'.

The SAS® program for evaluating data normality and variance homogeneity is presented below.

```
TITLE1 'Chapter 18 Example 5.sas';
TITLE2 'ART of Split Plot';
TITLE3 'Strawberry and Nitrogen Study';
TITLE4 'Analysis of transformed data';
DATA one;
INFILE '/folders/myfolders/CABI Text/Chapter 18/Chapter 18
Example 5 Data.art.csv' DLM =',' FIRSTOBS=2;
INPUT block nitrogen $ rate yield a b c Anit Arate Anitrate;
CARDS;
```

```
RUN;
PROC PRINT;
PROC UNIVARIATE normal;
VAR yield;
RUN;
PROC ANOVA PLOTS = none;
CLASSES nitrogen rate;
MODEL yield = nitrogen;
MEANS nitrogen / HOVTEST= levene;
RUN;
PROC ANOVA PLOTS = none;
CLASSES nitrogen rate;
MODEL yield = rate;
MEANS rate / HOVTEST= levene;
RUN;
PROC ANOVA PLOTS = none;
CLASSES nitrogen rate;
MODEL yield = nitrogen*rate;
MEANS nitrogen*rate / HOVTEST= levene;
RUN;
PROC ANOVA PLOTS = none;
CLASSES block nitrogen rate;
MODEL yield = block nitrogen block*nitrogen rate
nitrogen*rate;
TEST h = nitrogen e = blk*nitrogen;
RUN;
```

Results:

Tests for Normality				
Test	Statistic		p Value	
Shapiro-Wilk	W	0.962988	Pr < W	0.0116
Kolmogorov-Smirnov	D	0.116968	Pr > D	<0.0100
Cramer-von Mises	W-Sq	0.307764	Pr > W-Sq	<0.0050
Anderson-Darling	A-Sq	1.574098	Pr > A-Sq	<0.0050

Levene's Test for Homogeneity of yield Variance ANOVA of Squared Deviations from Group Means					
Source	DF	Sum of Squares	Mean Square	F Value	Pr > F
nitrogen	2	21633744	10816872	1.16	0.3188
Error	87	8.1231E8	9336941		

Levene's Test for Homogeneity of yield Variance ANOVA of Squared Deviations from Group Means					
Source	DF	Sum of Squares	Mean Square	F Value	Pr > F
rate	5	1.2485E8	24969851	1.26	0.2903
Error	84	1.6688E9	19866384		

Levene's Test for Homogeneity of yield Variance ANOVA of Squared Deviations from Group Means					
Source	DF	Sum of Squares	Mean Square	F Value	Pr > F
nitrogen*rate	17	89510677	5265334	0.94	0.5331
Error	72	4.0359E8	5605385		

The Shapiro–Wilks normality test for this data produced a Shapiro-Wilk W statistic of 0.962988 with a Pr $< W = 0.0116$, indicating that the data are not normally distributed. The data appear to have homogeneous variances as determined by Levene's test, thus it is appropriate to use the ARTool app to transform the data.

The analysis of the transformed data is straightforward:

```
TITLE1 'Chapter 18 Example 5 ART.sas';
TITLE2 'ART of Split Plot';
TITLE3 'Strawberry and Nitrogen Study';
TITLE4 'Analysis of transformed data';
DATA one;
INFILE '/folders/myfolders/CABI Text/Chapter 18/Chapter 18
Example 5 Data.art.csv' DLM =',' FIRSTOBS=2;
INPUT block nitrogen $ rate yield a b c Anit Arate Anitrate;
CARDS;
RUN;
PROC ANOVA PLOTS = none;
CLASSES block nitrogen rate;
MODEL yield = block nitrogen block*nitrogen rate
nitrogen*rate;
TEST h = nitrogen e = block*nitrogen;
RUN;
```

Remember that for a split-plot experiment, the correct test for the main plot must be explicitly requested ('Test h = nitrogen e = block*nitrogen'). Also, remember that p-values are valid for testing a nitrogen effect using the dependent variable Anit, for a rate effect using Arate and an interaction using Anitrate.

Results:

Dependent Variable: yield

Source	DF	Sum of Squares	Mean Square	F Value	Pr > F
Model	29	197153.2333	6798.3874	3.91	<.0001
Error	60	104406.8667	1740.1144		
Corrected Total	89	301560.1000			

R-Square	Coeff Var	Root MSE	yield Mean
0.653778	11.23678	41.71468	371.2333

Source	DF	Anova SS	Mean Square	F Value	Pr > F
block	4	4246.1556	1061.5389	0.61	0.6570
nitrogen	2	140697.2667	70348.6333	40.43	<.0001
block*nitrogen	8	6518.1778	814.7722	0.47	0.8737
rate	5	14324.6333	2864.9267	1.65	0.1617
nitrogen*rate	10	31367.0000	3136.7000	1.80	0.0795

Tests of Hypotheses Using the Anova MS for block*nitrogen as an Error Term					
Source	DF	Anova SS	Mean Square	F Value	Pr > F
nitrogen	2	140697.2667	70348.6333	86.34	<.0001

The ANOVA Procedure

Dependent Variable: Anit

Source	DF	Sum of Squares	Mean Square	F Value	Pr > F
Model	29	37545.43333	1294.67011	3.35	<.0001
Error	60	23195.56667	386.59278		
Corrected Total	89	60741.00000			

R-Square	Coeff Var	Root MSE	Anit Mean
0.618123	43.21310	19.66196	45.50000

Source	DF	Anova SS	Mean Square	F Value	Pr > F
block	4	916.11111	229.02778	0.59	0.6694
nitrogen	2	33931.46667	16965.73333	43.89	<.0001
block*nitrogen	8	2014.42222	251.80278	0.65	0.7315
rate	5	105.70000	21.14000	0.05	0.9980
nitrogen*rate	10	577.73333	57.77333	0.15	0.9987

Tests of Hypotheses Using the Anova MS for block*nitrogen as an Error Term					
Source	DF	Anova SS	Mean Square	F Value	Pr > F
nitrogen	2	33931.46667	16965.73333	67.38	<.0001

The ANOVA Procedure

Dependent Variable: Arate

Source	DF	Sum of Squares	Mean Square	F Value	Pr > F
Model	29	19416.51667	669.53506	0.97	0.5201
Error	60	41322.98333	688.71639		
Corrected Total	89	60739.50000			

R-Square	Coeff Var	Root MSE	Arate Mean
0.319669	57.67782	26.24341	45.50000

Source	DF	Anova SS	Mean Square	F Value	Pr > F
block	4	2114.80556	528.70139	0.77	0.5505
nitrogen	2	198.71667	99.35833	0.14	0.8660
block*nitrogen	8	3662.31111	457.78889	0.66	0.7202
rate	5	11611.13333	2322.22667	3.37	0.0095
nitrogen*rate	10	1829.55000	182.95500	0.27	0.9863

Tests of Hypotheses Using the Anova MS for block*nitrogen as an Error Term					
Source	DF	Anova SS	Mean Square	F Value	Pr > F
nitrogen	2	198.7166667	99.3583333	0.22	0.8095

The ANOVA Procedure

Dependent Variable: Anitrate

Source	DF	Sum of Squares	Mean Square	F Value	Pr > F
Model	29	23188.25000	799.59483	1.28	0.2093
Error	60	37550.75000	625.84583		
Corrected Total	89	60739.00000			

R-Square	Coeff Var	Root MSE	Anitrate Mean
0.381769	54.98222	25.01691	45.50000

Source	DF	Anova SS	Mean Square	F Value	Pr > F
block	4	2411.02778	602.75694	0.96	0.4344
nitrogen	2	55.85000	27.92500	0.04	0.9564
block*nitrogen	8	3247.62222	405.95278	0.65	0.7337
rate	5	219.06667	43.81333	0.07	0.9964
nitrogen*rate	10	17254.68333	1725.46833	2.76	0.0073

Tests of Hypotheses Using the Anova MS for block*nitrogen as an Error Term					
Source	DF	Anova SS	Mean Square	F Value	Pr > F
nitrogen	2	55.85000000	27.92500000	0.07	0.9341

Chapter 18

With non-transformed data the ANOVA reveals a significant effect of nitrogen source (*p*-value < 0.0001), and non-significant effects of nitrogen rate ($\alpha = 0.1617$) and the interaction between the two (*p*-value = 0.0795).

Confidence in these assertions is not high, since with non-normal data the test power is not high and the chances of a Type I error are relatively high. When the data are appropriately transformed, all three sources of variation are significant (*p*-values = 0.0001, 0.0095 and 0.0073 for source, rate and interaction, respectively) and confidence is high since the ANOVA is legitimate with transformed data.

The significant interaction suggests the next appropriate step in the analysis: is there a rate effect for each separate source of nitrogen?

To do this, the data must be separated into three CSV files, one corresponding to each source of nitrogen. Each dataset is then separately ART transformed.

Urea

The SAS® code for evaluating Urea:

```
TITLE1 'Chapter 18 Example 5 Urea ART.sas';
TITLE2 'ART of Split Plot';
TITLE3 'Strawberry and Nitrogen Study';
TITLE3 'Analysis of transformed data UREA';
DATA one;
INFILE '/folders/myfolders/CABI Text/Chapter 18/Chapter 18
Example 5 Data Urea.art.csv' DLM =',' FIRSTOBS=2;
INPUT block rate yield a Arate;
CARDS;
RUN;
PROC PRINT;
PROC UNIVARIATE normal;
VAR yield;
RUN;
PROC ANOVA;
CLASSES block rate;
MODEL yield Arate = block rate;
RUN;
```

Results:

Tests for Normality				
Test	Statistic		p Value	
Shapiro-Wilk	W	0.910448	Pr < W	0.0153
Kolmogorov-Smirnov	D	0.204342	Pr > D	<0.0100
Cramer-von Mises	W-Sq	0.22421	Pr > W-Sq	<0.0050
Anderson-Darling	A-Sq	1.251637	Pr > A-Sq	<0.0050

Dependent Variable: Arate

Source	DF	Sum of Squares	Mean Square	F Value	Pr > F
Model	9	557.983333	61.998148	0.74	0.6733
Error	20	1687.016667	84.350833		
Corrected Total	29	2245.000000			

R-Square	Coeff Var	Root MSE	Arate Mean
0.248545	59.25336	9.184271	15.50000

Source	DF	Anova SS	Mean Square	F Value	Pr > F
block	4	291.5833333	72.8958333	0.86	0.5023
rate	5	266.4000000	53.2800000	0.63	0.6779

Results suggest that transformed data should be used in the analysis, since the data is not normally distributed. The rate effect for the source Urea has a p-value $= 0.6779$, thus rate does not seem to impact yield response when using urea as a nitrogen source.

Calcium nitrate

Evaluation of the calcium nitrate rate effect proceeds in a similar fashion:

```
TITLE1 'Chapter 18 Example 5 Calcium nitrate ART.sas';
TITLE2 'ART of Split Plot';
TITLE3 'Strawberry and Nitrogen Study';
TITLE3 'Analysis of transformed data CALCIUM NITRATE';
DATA one;
INFILE '/folders/myfolders/CABI Text/Chapter 18/Chapter 18
Example 5 Data Calcium Nitrate.art.csv' DLM =',' FIRSTOBS=2;
INPUT block rate yield a Arate;
CARDS;
RUN;
PROC PRINT;
PROC UNIVARIATE normal;
VAR yield;
RUN;
PROC ANOVA;
CLASSES block rate;
MODEL yield Arate = block rate;
RUN;
```

Results:

Tests for Normality				
Test	Statistic		p Value	
Shapiro-Wilk	W	0.962651	Pr < W	0.3614
Kolmogorov-Smirnov	D	0.129208	Pr > D	>0.1500
Cramer-von Mises	W-Sq	0.054486	Pr > W-Sq	>0.2500
Anderson-Darling	A-Sq	0.374638	Pr > A-Sq	>0.2500

The data appear to be normally distributed, thus non-transformed data can be used for the estimation of the calcium nitrate rate effect.

Dependent Variable: yield

Source	DF	Sum of Squares	Mean Square	F Value	Pr > F
Model	9	33137.53333	3681.94815	2.53	0.0403
Error	20	29124.33333	1456.21667		
Corrected Total	29	62261.86667			

R-Square	Coeff Var	Root MSE	yield Mean
0.532228	8.956440	38.16041	426.0667

Source	DF	Anova SS	Mean Square	F Value	Pr > F
block	4	3292.86667	823.21667	0.57	0.6906
rate	5	29844.66667	5968.93333	4.10	0.0100

There is a significant rate effect (p-value = 0.0100), thus a linear regression to estimate the relationship between yield and rate would be appropriate as the next step in the analysis. The reader is left to pursue this on their own.

Potassium nitrate

The SAS® code for analyzing potassium nitrate:

```
TITLE1 'Chapter 18 Example 5 Potassium nitrate ART.sas';
TITLE2 'ART of Split Plot';
TITLE3 'Strawberry and Nitrogen Study';
TITLE3 'Analysis of transformed data POTASSIUM NITRATE';
DATA one;
INFILE '/folders/myfolders/CABI Text/Chapter 18/Chapter 18
Example 5 Data Potassium Nitrate.art.csv' DLM =',' FIRSTOBS=2;
INPUT block rate yield a Arate;
CARDS;
RUN;
PROC PRINT;
PROC UNIVARIATE normal;
VAR yield;
RUN;
PROC ANOVA;
CLASSES block rate;
MODEL yield Arate = block rate;
RUN;
```
Results:

Tests for Normality				
Test	Statistic		p Value	
Shapiro-Wilk	W	0.842853	Pr < W	0.0004
Kolmogorov-Smirnov	D	0.202797	Pr > D	<0.0100
Cramer-von Mises	W-Sq	0.237843	Pr > W-Sq	<0.0050
Anderson-Darling	A-Sq	1.613322	Pr > A-Sq	<0.0050

The data are not normally distributed as suggested by the Shapiro–Wilk test ($\alpha = 0.0004$), thus transformed data should be used for estimating the rate effect.

The ANOVA Procedure

Dependent Variable: Arate

Source	DF	Sum of Squares	Mean Square	F Value	Pr > F
Model	9	1268.683333	140.964815	2.89	0.0230
Error	20	975.316667	48.765833		
Corrected Total	29	2244.000000			

R-Square	Coeff Var	Root MSE	Arate Mean
0.565367	45.05325	6.983254	15.50000

Source	DF	Anova SS	Mean Square	F Value	Pr > F
block	4	610.0833333	152.5208333	3.13	0.0376
rate	5	658.6000000	131.7200000	2.70	0.0506

A significant rate effect (p-value = 0.0506) suggests that a linear regression would be appropriate to examine the relationship between yield and rate.

Since the data for calcium nitrate were normally distributed, the reader was left to evaluate the linear relationship between rate and yield on their own. With potassium nitrate the data are not normally distributed, thus an illustration of the procedure with non-normal data is appropriate.

The p-values for the significance levels of the regression analysis of potassium nitrate and yield should be estimated using *transformed* data. The parameter estimates are obtained using *non-transformed* data. SAS® code below illustrates how to proceed.

```
TITLE1  'Chapter  18  Example  5  Potassium  nitrate  ART
regression.sas';
TITLE2 'ART of Split Plot';
TITLE3 'Strawberry and Nitrogen Study';
TITLE3 'Regression  analysis  of  transformed  data  POTASSIUM
NITRATE';
DATA one;
```

```
INFILE  '/folders/myfolders/CABI  Text/Chapter  18/Chapter  18
Example 5 Data Potassium Nitrate.art.csv' DLM =',' FIRSTOBS=2;
INPUT block rate yield a Arate;
rate2 = rate*rate;
CARDS;
RUN;
PROC REG Plots = none;
MODEL yield Arate = rate;
MODEL yield Arate = rate rate2;
RUN;
```
Results:

The REG Procedure
Model: MODEL1
Dependent Variable: yield

| Number of Observations Read | 30 |
| Number of Observations Used | 30 |

Analysis of Variance

Source	DF	Sum of Squares	Mean Square	F Value	Pr > F
Model	1	3192.14000	3192.14000	3.01	0.0939
Error	28	29722	1061.50571		
Corrected Total	29	32914			

Root MSE	32.58076	R-Square	0.0970
Dependent Mean	353.30000	Adj R-Sq	0.0647
Coeff Var	9.22184		

Parameter Estimates

Variable	DF	Parameter Estimate	Standard Error	t Value	Pr > |t|
Intercept	1	374.44000	13.56445	27.60	<.0001
rate	1	-6.04000	3.48303	-1.73	0.0939

The REG Procedure
Model: MODEL1
Dependent Variable: Arate

Number of Observations Read	30
Number of Observations Used	30

Analysis of Variance					
Source	DF	Sum of Squares	Mean Square	F Value	Pr > F
Model	1	139.54571	139.54571	1.86	0.1839
Error	28	2104.45429	75.15908		
Corrected Total	29	2244.00000			

Root MSE	8.66943	R-Square	0.0622
Dependent Mean	15.50000	Adj R-Sq	0.0287
Coeff Var	55.93183		

Parameter Estimates					
Variable	DF	Parameter Estimate	Standard Error	t Value	Pr > |t|
Intercept	1	19.92000	3.60937	5.52	<.0001
rate	1	-1.26286	0.92680	-1.36	0.1839

'Chapter 18 Example 5 Potassium nitrate ART regression.sas'
'ART of Split Plot'
'Regression analysis of transformed data POTASSIUM NITRATE'

The REG Procedure
Model: MODEL2
Dependent Variable: yield

Number of Observations Read	30
Number of Observations Used	30

Analysis of Variance

Source	DF	Sum of Squares	Mean Square	F Value	Pr > F
Model	2	5146.51143	2573.25571	2.50	0.1007
Error	27	27768	1028.43661		
Corrected Total	29	32914			

Root MSE	32.06925	R-Square	0.1564
Dependent Mean	353.30000	Adj R-Sq	0.0939
Coeff Var	9.07706		

Parameter Estimates

Variable	DF	Parameter Estimate	Standard Error	t Value	Pr > |t|
Intercept	1	344.24000	25.65540	13.42	<.0001
rate	1	16.61000	16.78446	0.99	0.3312
rate2	1	-3.23571	2.34723	-1.38	0.1794

'ART of Split Plot'
'Regression analysis of transformed data POTASSIUM NITRATE'

The REG Procedure
Model: MODEL2
Dependent Variable: Arate

Number of Observations Read	30
Number of Observations Used	30

Analysis of Variance

Source	DF	Sum of Squares	Mean Square	F Value	Pr > F
Model	2	355.97964	177.98982	2.55	0.0971
Error	27	1888.02036	69.92668		
Corrected Total	29	2244.00000			

Root MSE	8.36222	R-Square	0.1586
Dependent Mean	15.50000	Adj R-Sq	0.0963
Coeff Var	53.94979		

Parameter Estimates

Variable	DF	Parameter Estimate	Standard Error	t Value	Pr > |t|
Intercept	1	9.87000	6.68977	1.48	0.1517
rate	1	6.27464	4.37663	1.43	0.1631
rate2	1	-1.07679	0.61205	-1.76	0.0899

Both linear and quadratic components will be examined. Note the variable 'rate2' generated in line 8 of the SAS® code. This allows testing for the quadratic effect. The model statements in lines 12 and 13 used in the regression procedure of SAS® will test the linear component followed by the quadratic component. The p-values are obtained from the tests on Arate (transformed data) while the parameter estimates are extracted from the tests evaluating yield (non-transformed).

Neither linear nor quadratic components were significant. You might question why the rate was significant (p-value = 0.0506) but neither linear (p-value = 0.1839) nor quadratic (p-value = 0.0899) components were significant. This apparent contradiction suggests that, although there is a relationship between rate and yield, the relationship is not linear. Further analysis might investigate non-linear models. Additionally, the rate effect was marginally significant and one might argue that a p-value of 0.0899 for a quadratic response is also marginally significant.

Suppose that, for our purposes, we consider the significance level of the quadratic component (p-value = 0.0899) sufficient to derive the regression equation for a presentation.

The parameter estimates would be obtained from the analysis of non-transformed data. The regression equation would be:

$$Y = 344.2 + 16.6 * rate - 3.2 * rate^2$$

There seems to be a marginal relationship between rate and yield using potassium nitrate, thus further experiments should investigate a broader range of rates with a greater number of replications to determine if the quadratic component in the initial experiment was indeed relevant. Non-linear models might also be examined.

Other designs and experiments

The beauty of the ART procedure is that it is applicable to nearly any situation and easy to use. The key to remember is to use the correct effect when testing. For example, use Atreatment when testing for a treatment effect, etc.

References and Further Reading

Conover, W. J. (1999) *Practical Nonparametric Statistics*, 3rd edn. John Wiley & Sons, New York.

Conover, W.J. and Iman, R.L. (1981) Rank transformations as a bridge between parametric and nonparametric statistics. *The American Statistician* 35, 124–129.

Durner, E. (2019) Effective analysis of interactive effects with non-normal data using the aligned rank transform, ARTool and SAS® University Edition. *Horticulturae* 5, 57. doi:10.3390/horticulturae5030057

Gomez, K.A. and Gomez, A.A. (1984) *Statistical Procedures in Agricultural Research*, 2nd edn. John Wiley and Sons, New York.

Higgins, J.J. (2004) *Introduction to Modern Nonparametric Statistics*. Duxbury Press, Pacific Grove, California.

Higgins, J.J. and Tashtoush, S. (1994) An aligned rank transform test for interaction. *Nonlinear World* 1, 201–211.

Higgins, J.J., Blair, R.C. and Tashtoush, S. (1990) The aligned rank transform procedure. In: *Proceedings, Conference on Applied Statistics in Agriculture*. Kansas State, Kansas, pp. 185–195.

Richter, S.J. and Payne, M.E. (2002) SAS Program to perform analysis of factorial experiments using aligned ranks. *Journal of Statistical Computation and Simulation* 72, 14–17.

Salter, K.C. and Fawcett, R.F. (1993) The art test of interaction: A robust and powerful rank test of interaction in factorial models. *Communications in Statistics: Simulation and Computation* 22, 137–153.

Sawilowsky, S.S. (1990) Nonparametric tests of interaction in experimental design. *Review of Educational Research* 60, 91–126.

Snedecor, G.W. and Cochran, W.G. (1989) *Statistical Methods*, 8th edn. Iowa State University Press, Ames, Iowa.

Wobbrock, J.O., Findlater, L., Gergle, D. and Higgins, J.J. (2011) The aligned rank transform for nonparametric factorial analyses using only ANOVA procedures. In: *Proceedings of the SIGCHI Conference on Human Factors in Computing Systems*, Vancouver, British Columbia, pp. 143–146

Wobbrock, J.O., Findlater, L., Gergle, D., Higgins, J.J. and Kay, M. (2011–2018) ARTool Align-and-rank data for a nonparametric ANOVA. Available at: http://depts.washington.edu/madlab/proj/art/ (accessed 1 May 2019).

19 Sampling

Introduction

In research, we depend on parameter estimates (such as means and variances) to determine the effects different factors might have on the individuals in those populations. We normally can't measure all of the individuals of a population, therefore we measure only a fraction of them by taking a sample. In statistics, sample theory explores the methods of sampling and making inferences regarding populations. We generally obtain one of two types of samples in plant science research: (1) samples from populations where there is no imposed experimental design; or (2) samples from a designed experiment.

When parameter estimates are desired but no experimental design has been established, samples are drawn from a population to estimate them. Appropriate sampling includes random selection of individuals, which helps prevent biasing measurements towards or away from experimental units with or without specific attributes. We also take samples from designed agricultural experiments. When we take measurements during the implementation of a designed experiment, the desired random aspect of our sampling is 'built in' if we follow accepted design and implementation protocols.

This chapter covers sampling techniques for plant science research. Even though there are whole texts written regarding such techniques, the relatively limited number presented here are usually sufficient for most applied plant science work. We will examine some of the protocols for sampling from populations without an imposed experimental design as well as from designed experiments.

Design- vs Model-based Approaches to Sampling

There are two fundamentally different approaches to sampling, both providing estimates of parameters from which inferences about a population can be made. The two methodologies include design-based approaches and model-based approaches.

Design-based approaches are the 'classical' sampling procedures which provide unbiased estimators of the parameters we are interested in, such as means and variances. The design-based approach uses a well-defined random selection process for implementing a survey and well-defined methods for estimating parameters based on the sampling design. *Model-based* approaches have no requirements on the method for selecting samples and the samples are typically selected by targeted (non-random) sampling and the estimation methods are well defined.

We will investigate design-based approaches. They are nearly universally accepted for plant science research and require no specialized theoretical statistical knowledge as is often the case with model-based methods. In addition, design-based methods provide the reliable estimates we need in order to have confidence in the inferences we make with our results.

10.1079/9781789249927.0019

Populations and Samples

A *population* is simply a defined group of individuals (also called elements) meeting preset characteristics for inclusion into the group. Individuals can be plants, animals, human beings or other experimental units, such as field plots. If an individual has all of the characteristics describing the group, then he/she/it is considered a member of the population. For example, if the population is defined as all students attending college, then any student in college would be a member of the population. If the population is defined as all tomato plants growing in the state of New Jersey during the 2020 growing season, then all tomato plants in the state during the 2020 growing season would be members of the population. The key is the definition of membership for inclusion in a population. The list of all individuals in the population to be sampled is called the *sampling frame*.

Populations can be *homogeneous* (members very similar to each other) or *heterogeneous* (members very different from each other). We can define what variables we are considering to make the distinction between homogeneous and heterogeneous. We might consider variables such as age, vigor or cultivar to decide what type of population we have. Which variables are chosen depends on the nature of the research.

Since we often want to estimate something about the members of a population which leads to inferences about our population, we must consider how to best measure the characteristic of interest. Characteristics might be exam scores for students or yield for tomato plants. It's important to clearly identify the characteristics you are interested in so that the population and subsequent sample will be appropriate. It's also important to limit the number of characteristics you are interested in by limiting the objectives of your sampling so that the process does not become overly complex.

We normally cannot measure all members of the defined population so we only measure a portion of it. That portion of a population we measure to obtain our estimate of interest is called a *sample*. The method of selecting members of a population for inclusion in a sample followed by how to estimate our characteristic of interest from the sample data is what sampling is all about.

We use methods for sampling which prevent bias as much as possible. *Bias* is simply a skewing of our estimate one way or another which leads to erroneous conclusions and inferences about our population. For example, if we are measuring yield of broccoli and some plots seem vigorous while others seem weak, measuring only vigorous plots or only weak plots would lead to bias. In order to prevent bias, our sampling protocol should be one that includes individuals from both vigorous and weak groups.

Sampling Methods

Non-random sampling

When the selection of individuals in a population for inclusion in a sample is based on subjective judgement of the researcher and does not include any randomness in the selection, we have what is called *non-probability* (or *purposive*) sampling. In non-probability sampling, every individual in the population does not have an equal chance of being included in any sample. Essentially the researcher handpicks which individuals are in the sample. While this type of sampling might be suitable for some research needs, it is not suited to plant science research for the most part.

While non-random sampling methods generally require less time and effort than random methods, they are prone to systematic errors and biases and inferences based on data from non-random samples are not acceptable, since they are not based on the population as a whole but rather a select portion of it.

A few types of non-random sampling include snowball sampling, convenience sampling and quota sampling. *Snowball sampling* usually involves samples of people. Samples consist of individuals who are sampled that were recommended for sampling by others in a population. This often occurs when the population is difficult to sample, such as members of a club or isolated and remote populations. *Convenience sampling* is used when individuals are easy to access for evaluation. Again, this normally involves sampling of people. For example, a professor might use their class of students to evaluate the taste of a new variety of lettuce simply because the students are readily available to give their opinion.

Quota sampling involves non-randomly selecting individuals for inclusion in a sample until a quota has been met, for example if you want to obtain the opinions of 50 men and 50 women for a particular study and simply select folks entering a farmer's market until your quotas have been met. The problem with all types of non-random sampling is that all of them can be extremely biased and do not represent the population well.

Probability (random) sampling

Most plant science research relies on one or more forms of probability or random sampling. Simply put, *random sampling* occurs when individuals are randomly selected from a defined population for inclusion in a sample and all individuals in the population have a known, non-zero, usually equal, objectively assigned chance of being included in any sample drawn. While random sampling requires significant planning and effort on the researcher's part, it is highly desirable for several reasons. Random sampling greatly reduces the chance of systematic errors or biasing. It also produces an accurate representation of the populations as long as a suitable sampling protocol has been employed. (We will cover sampling protocols in the next section.) Finally, random sampling is desirable since accurate and reliable inferences regarding the population can be made from correctly sampled estimates.

Some preliminary sampling concepts

Before we explore different sampling methods, let's examine a sampling example closely to illustrate some of the complexities often encountered in nearly all types of sampling, especially from a statistical perspective.

Suppose we had a population of five plants growing in a greenhouse from which we have yield data as number of fruits per plant produced from October through March. For the sake of illustrating some points, let's suppose that the data from the five plants were: a = 143, b = 245, c = 194, d = 222 and e = 224. Since there are five plants, using simple random sampling, each plant would have a $1/5 = 0.20$ chance of being selected for any sample taken. Our task is to estimate the average yield per plant from the greenhouse by measuring only two of the plants.

Now to illustrate some important concepts regarding sampling, let's establish that we really do know that the average production per plant (since the data is presented above) is:

$$\frac{(143+245+194+222+224)}{5} = \frac{1028}{5} = 205.6$$

Also recall from high school math class that the number of samples of n objects taken r at a time with no replacement and order is not important is calculated with the following combination equation:

$$nCr = \frac{n(n-1)(n-2)...(n-r+1)}{r!} = \frac{n!}{r!(n-r)!}$$

Thus for our example:

$$5C2 = \frac{5(4)(3)(2)}{2(1)} = \frac{5!}{2!(3)!} = \frac{5*4*3*2*1}{2*1(3*2*1)} = \frac{120}{12} = 10$$

There are ten possible samples of size 2 from a population of size 5. The 10 possible samples along with the estimate of the mean derived from each are presented in the table below along with the estimate of sample accuracy (mean square error), standard error (s) and coefficient of variation (cv).

An estimate of accuracy in our sample is the mean square error (MSE) or sample variance, which is calculated as:

$$MSE = \hat{\sigma}^2 = s^2 = \sum_{i=1}^{n} \frac{(Y_i - \bar{Y})^2}{(n)}$$

Note that we use n rather than n-1 since we 'know' the population mean, thus we don't have to sample it. The estimates of the mean obtained with our sampling scheme are presented in the table below.

The standard error of our sampling scheme is the square root of 466.62 or 21.6 which is approximately 10% of the size of our sample estimate. We want our standard error to be as small as possible, and 10% of the sample estimate is not too bad. The coefficient of variation (CV, expressed as a percentage) is a measure of the relative variability in a population or sample which allows general comparisons between different samples:

$$Population \; CV(\%) = \frac{\sigma}{|\mu|} * 100$$

$$Sample \; CV(\%) = \frac{s}{|\bar{Y}|} * 100$$

A CV of 10% or lower is desirable for 'good' sampling.

Sample #	Sample drawn	Sample mean	s^2(MSE)	S	CV
1	143, 245	194	5202.0	72.1	37.2
2	143, 194	168.5	1300.5	36.1	21.4
3	143, 222	182.5	3120.5	55.9	30.6
4	143, 224	183.5	3280.5	57.3	31.2
5	245, 194	219.5	1300.5	36.1	16.4
6	245, 222	233.5	264.5	16.3	6.9
7	245, 224	234.5	220.5	14.9	6.3
8	194, 222	208	392.0	19.8	9.5
9	194, 224	209	450.0	21.2	10.1
10	222, 224	223	2.0	1.4	0.6

Notice that the CV values range from 0.6 to 37.2. If the random samples we happen to randomly choose are 6, 7, 8, 9 or 10, our sample size of 2 is adequate using the CV rule. However, if our samples happen to be 1, 2, 3, 4 or 5 a sample of size of 2 is inadequate. This illustrates how important the selection of an adequate sample size is if one wants to have faith in the estimates made and inferences drawn. It all comes down to variability in our data. Notice the range in s^2 within the samples that seemed to be adequate (6 through 10). The values for s^2 range from 2 to 450. Also note that s^2 for samples 6 through 9 have much closer values, ranging from 220 to 450. If you had only randomly chosen sample 10, you wouldn't know that 2 is a 'small' value for s^2. Similarly, if you had only chosen sample 1, you wouldn't know that 5202 is a 'large' value for s^2.

Obtaining a reasonable estimate of s^2

One way to get the best estimate of s^2 so that we can determine an acceptable sample size is to preliminarily sample as many individuals as possible to obtain an estimate of s^2. This may be 3 or 4 or it might be 30 or 40, all depending on your particular situation. One approach to this preliminary sampling is, once at least two individuals have been sampled, to calculate s^2. Sample a third individual and recalculate s^2. Compare the new estimate to the original estimate. Did it increase, decrease or stay about the same? Continue sampling and calculating s^2 so that you can plot sample size against s^2. Once your estimate of s^2 becomes stable and levels off, you can consider this a reasonable stopping point for your preliminary estimate of overall variance. If there is little variability among members of your population, the variance estimate will level off quickly. If the variability is great, the estimate will change significantly and take much longer to level out. Once you have an estimate of s^2 that you are reasonably comfortable with, you can estimate the sample size needed using one of the techniques that follows.

Sample size

This brings up the question of 'How large a sample do I need?' There are many factors that enter into this decision, such as precision required, cost of sampling and population variability. Several approaches help to answer this often-asked question.

THE CV METHOD   The general rule of thumb is that the CV on a mean basis should be < 10%. Suppose you had to determine the number of fruit samples needed to estimate the fruit soluble solids for an experiment you were performing with peaches. Should you take 10? 30? 100? The first thing you should do is to take a preliminary sample of a relatively small number as described in the previous section. From that sample, calculate the mean and standard deviation for your preliminary sample.
Recall:

$$Sample\,CV_{obs}\,(\%) = \frac{s}{|\bar{Y}|} * 100$$

On a mean basis this becomes:

$$Sample\,CV_{mean} = \frac{\frac{s}{\sqrt{n}}}{\bar{Y}} * 100$$

You have an estimate of s and $\bar{Y}$ so solve for n in the following equation:

$$\frac{\frac{s}{\sqrt{n}}}{\bar{Y}} * 100 = 10\%$$

Suppose $s = 50$ and $\bar{Y} = 120$:

$$\frac{\frac{50}{\sqrt{n}}}{120} * 100 = 10\%$$

$$\frac{\frac{50}{\sqrt{n}}}{120} = \frac{10}{100}$$

$$\frac{50}{\sqrt{n}} = \frac{1200}{100}$$

$$\frac{50}{\sqrt{n}} = 12$$

$$50 = 12\sqrt{n}$$

$$4.17 = \sqrt{n}$$

$$17.36 = n$$

Thus an appropriate sample size would be 18 individuals.

THE CONFIDENCE LIMIT METHOD   Suppose we have a situation where we want to sample such that our estimate of the mean is within a particular range of allowable error we'll call 'E'. We are also willing to assume a 5% chance that our E will fall outside the range we desire.

The 95% confidence limits for a sample mean assuming we are dealing with a normal distribution are given as (from Chapter 3):

$$CL = \bar{Y} \pm ts_{\bar{Y}}$$

where t is a tabular value with $(n-1)$ df at the $1-(\alpha/2)$ significance level.

To determine the sample size needed for estimating a mean within allowable error of E and estimated variance s^2, solve:

$$n = \frac{4s^2}{E^2}$$

We obtain s^2 from a previous similar experiment or we estimate it with our preliminary small sample. We determine what E should be by thinking about our experiment and how large the error could be yet still be acceptable for our work.

If our calculated n is > 10% of our population size (N), then adjust n to n' with the formula:

$$n' = \frac{n}{\left(1 + \dfrac{n}{N}\right)}$$

For example, we want to sample one pint of fruit from raspberry growers to estimate average fruit size for each grower, but we don't know how many growers we should contact to obtain a sample from. We know from previous work that the variance of the average fruit size calculated as *(fruit weight)/(fruit number)* is 0.84. We are setting our error range to 0.4 g. We chose this value since the average fruit weight, in general, of raspberries grown in our region is 3 to 5 g per berry, thus we're ok with a 10% error.

Our estimate for n is:

$$n = \frac{4s^2}{E^2}$$

$$n = \frac{4 * 0.84}{(0.4)^2}$$

$$n = \frac{3.36}{0.16} = 21$$

In order to get a good estimate of the average fruit size for raspberry growers in our area we should obtain samples from 21 growers.

Using SAS® as a sampling tool

SAS® provides some very powerful options for sample selection with PROC SURVEYSELECT. This procedure provides methods for probability-based sample selection ranging from simple to complex. In order to use PROC SURVEYSELECT you need to provide SAS® with a dataset that includes the list of individuals in your population from which you intend to sample. You also specify options pertaining to desired selection method, sample size or sampling rate, as well as other selection parameters. We will cover the options applicable to plant science surveys as we cover the various methods.

Sampling from a Population with No Imposed Experimental Design

Simple Random Sampling

Description

In simple random sampling, every individual in the population has an equal chance of being selected for any sample. In addition, each sample has an equal chance of being drawn. The population must be defined and each individual in the population must be unique and identifiable, such as a single farm, a plant or a single research plot. The population must also be homogeneous. If the population is not homogeneous, another type of random sampling should be employed.

SAS® Method

Define the population and relevant characteristics of interest. Create a dataset list of individuals in the population and identify each with a label for identification, such as a number from 1 to N, where N is the total number of individuals in the population. The following dataset example includes information about 66 members of a population of trees from small apple orchards we wish to sample. We include a number identifying the label of each individual (tree) followed by information regarding cultivar (cv), rootstock (rs), tree age (years) and state of production (state). Some of the information will be used in later sampling examples and is included here for completeness.

```
1   TITLE1 'Chapter 19 Example 1 Simple Random Sampling';
2   DATA one;
3   INPUT tree cv $ rs $ years state $ @@;
4   CARDS;
5   1 reddel emla7 4 pa 2 reddel emla7 4 pa
6   3 reddel emla7 4 pa 4 reddel emla7 4 pa
7   5 reddel bud9 8 pa 6 reddel bud9 8 pa
8   7 reddel bud9 8 pa 8 reddel m9 5 pa
9   9 reddel m9 5 pa 10 reddel m9 5 pa
10  11 reddel m9 5 pa 12 reddel emla7 3 nj
11  13 reddel emla7 3 nj 14 reddel emla7 3 nj
12  15 reddel emla7 3 nj 16 reddel bud9 7 nj
13  17 reddel bud9 7 nj 18 reddel bud9 7 nj
14  19 reddel m9 5 nj 20 reddel m9 5 nj
15  21 reddel m9 5 nj 22 reddel m9 5 nj
16  23 reddel emla7 3 nc 24 reddel emla7 3 nc
17  25 reddel emla7 3 nc 26 reddel emla7 3 nc
18  27 reddel bud9 7 nc 28 reddel bud9 7 nc
19  29 reddel bud9 7 nc 30 reddel m9 4 nc
20  31 reddel m9 4 nc 32 reddel m9 4 nc
21  33 reddel m9 4 nc 34 gdel emla7 4 pa
22  35 gdel emla7 4 pa 36 gdel emla7 4 pa
23  37 gdel emla7 4 pa 38 gdel bud9 8 pa
24  39 gdel bud9 8 pa 40 gdel bud9 8 pa
25  41 gdel m9 5 pa 42 gdel m9 5 pa
26  43 gdel m9 5 pa 44 gdel m9 5 pa
27  45 gdel emla7 3 nj 46 gdel emla7 3 nj
28  47 gdel emla7 3 nj 48 gdel emla7 3 nj
29  49 gdel bud9 7 nj 50 gdel bud9 7 nj
30  51 gdel bud9 7 nj 52 gdel m9 5 nj
31  53 gdel m9 5 nj 54 gdel m9 5 nj
32  55 gdel m9 5 nj 56 gdel emla7 3 nc
33  57 gdel emla7 3 nc 58 gdel emla7 3 nc
34  59 gdel emla7 3 nc 60 gdel bud9 7 nc
35  61 gdel bud9 7 nc 62 gdel bud9 7 nc
36  63 gdel m9 4 nc 64 gdel m9 4 nc
37  65 gdel m9 4 nc 66 gdel m9 4 nc
38  RUN;
39  PROC PRINT DATA = one;
40  RUN;
```

Suppose we wanted to randomly sample ten trees from this population. Simply include the following lines in the Example 1 program:

```
41  PROC SURVEYSELECT DATA=one METHOD=srs N=10 OUT=surveydata;
42  PROC PRINT DATA = surveydata;
43  RUN;
```

Let's examine PROC SURVEYSELECT. The 'DATA=one' option identifies the dataset to use for selecting the sample, 'METHOD=srs' indicates simple random sampling (srs), 'N' identifies how many individuals to include in the sample and 'OUT=surveydata' indicates what dataset should be used for the output of the procedure. PROC SURVEYSELECT does not print our list of individuals for sampling directly so we need to put the output in a dataset and then print the dataset. Note that the 'srs' option is simple random sampling without replacement. This means that an individual cannot be selected for inclusion in a sample more than once. While there are sampling methods which allow resampling, we don't use them in plant science research.

The results of our program (without the printout of DATA=one):

'Chapter 19 Example 1 Simple Random Sampling'

The SURVEYSELECT Procedure

Selection Method	Simple Random Sampling

Input Data Set	ONE
Random Number Seed	144587775
Sample Size	10
Selection Probability	0.151515
Sampling Weight	6.6
Output Data Set	SURVEYDATA

'Chapter 19 Example 1 Simple Random Sampling'

Obs	tree	cv	rs	years	state
1	3	reddel	emla7	4	pa
2	15	reddel	emla7	3	nj
3	32	reddel	m9	4	nc
4	33	reddel	m9	4	nc
5	38	gdel	bud9	8	pa
6	40	gdel	bud9	8	pa
7	48	gdel	emla7	3	nj
8	61	gdel	bud9	7	nc
9	62	gdel	bud9	7	nc
10	66	gdel	m9	4	nc

The top table in our printout indicates the dataset from which the sample was drawn (ONE), the random number seed (144587775) used by SAS® (obtained as the time and date on the computer clock) to begin random sample selection, the sample size (10), the selection probability for any individual in the defined population (*(sample size)/(population size)*), sampling weight (1/(*selection probability*)) and the output dataset name (SURVEYDATA). The bottom table produces a printout of the individuals in our population that have been randomly chosen for our sample.

Once we have our list of individuals in our sample, we would measure the characteristic(s) of interest and calculate whatever parameter estimates we need.

Benefits

There is no bias in the sample, since the individuals in the sample were randomly selected. The sample fairly and accurately represents the population of interest.

Problems

Simple random sampling requires planning and may require considerable time and expense to implement. It also might be somewhat difficult to prepare a list of all members of the population.

Systematic random sampling

Description

Systematic random sampling is a sampling method where the first individual sampled is randomly chosen from a population and all subsequent individuals in the sample are selected at a regular interval, for example, every 6th plant in a row. This type of sampling is used when the population is homogeneous. With systematic random sampling every individual in the population does not have an equal chance of being included in a sample and a list of all individuals in the population is desired but not required.

SAS® Method

Define the population and relevant characteristics of interest. Create the dataset of individuals in the population.

Decide on the sample size needed (S) and an interval for sampling (I). The interval for sampling (I) would be N/S. Select the first individual for sampling at random, then sample every Ith individual. If you can't calculate 'I' because you don't know 'N', then select a sampling interval. Even if you don't know the exact size of your population, you should be able to approximate its size. You should also know how many samples you'll need, thus you can determine an appropriate interval. One way to do this is to first determine an appropriate range of intervals you might consider, say every 5th to 15th plant, then randomly select one of these intervals.

We'll reuse DATA=one from Example 1 for our example of systematic random sampling.

```
1   TITLE1 'Chapter 19 Example 2 Systematic Random Sampling';
2   DATA one;
3   INPUT tree cv $ rs $ years state $ @@;
4   CARDS;
5   1 reddel emla7 4 pa 2 reddel emla7 4 pa
6   3 reddel emla7 4 pa 4 reddel emla7 4 pa
7   5 reddel bud9 8 pa 6 reddel bud9 8 pa
8   7 reddel bud9 8 pa 8 reddel m9 5 pa
9   9 reddel m9 5 pa 10 reddel m9 5 pa
10  11 reddel m9 5 pa 12 reddel emla7 3 nj
11  13 reddel emla7 3 nj 14 reddel emla7 3 nj
12  15 reddel emla7 3 nj 16 reddel bud9 7 nj
13  17 reddel bud9 7 nj 18 reddel bud9 7 nj
14  19 reddel m9 5 nj 20 reddel m9 5 nj
15  21 reddel m9 5 nj 22 reddel m9 5 nj
16  23 reddel emla7 3 nc 24 reddel emla7 3 nc
17  25 reddel emla7 3 nc 26 reddel emla7 3 nc
18  27 reddel bud9 7 nc 28 reddel bud9 7 nc
19  29 reddel bud9 7 nc 30 reddel m9 4 nc
20  31 reddel m9 4 nc 32 reddel m9 4 nc
21  33 reddel m9 4 nc 34 gdel emla7 4 pa
22  35 gdel emla7 4 pa 36 gdel emla7 4 pa
23  37 gdel emla7 4 pa 38 gdel bud9 8 pa
24  39 gdel bud9 8 pa 40 gdel bud9 8 pa
25  41 gdel m9 5 pa 42 gdel m9 5 pa
26  43 gdel m9 5 pa 44 gdel m9 5 pa
27  45 gdel emla7 3 nj 46 gdel emla7 3 nj
28  47 gdel emla7 3 nj 48 gdel emla7 3 nj
29  49 gdel bud9 7 nj 50 gdel bud9 7 nj
30  51 gdel bud9 7 nj 52 gdel m9 5 nj
31  53 gdel m9 5 nj 54 gdel m9 5 nj
32  55 gdel m9 5 nj 56 gdel emla7 3 nc
33  57 gdel emla7 3 nc 58 gdel emla7 3 nc
34  59 gdel emla7 3 nc 60 gdel bud9 7 nc
35  61 gdel bud9 7 nc 62 gdel bud9 7 nc
36  63 gdel m9 4 nc 64 gdel m9 4 nc
37  65 gdel m9 4 nc 66 gdel m9 4 nc
38  RUN;
39  PROC PRINT DATA = one;
40  RUN;
41  PROC SURVEYSELECT DATA=one METHOD=sys SAMPSIZE=10
    OUT=surveydata;
42  PROC PRINT DATA = surveydata;
43  RUN;
```

Systematic random sampling is requested in PROC SURVEYSELECT by using 'sys' as the selected option for 'METHOD' (line 41). PROC SURVEYSELECT will randomly select the first individual in the sample. How the rest of the members of the sample are chosen depends on options you select. To keep things simple and less confusing, I'll guide you in one direction for an appropriate selection protocol for systematic random sampling. Indicate the desired sample size via the 'SAMPSIZE=' option (line 41) and PROC SURVEYSELECT will select every (N/n)th individual after the initial randomly selected individual.

Results:

'Chapter 19 Example 2 Systematic Random Sampling'

The SURVEYSELECT Procedure

Selection Method	Systematic Random Sampling

Input Data Set	ONE
Random Number Seed	278702065
Sample Size	10
Selection Probability	0.151515
Sampling Weight	6.6
Output Data Set	SURVEYDATA

'Chapter 19 Example 2 Systematic Random Sampling'

Obs	tree	cv	rs	years	state
1	5	reddel	bud9	8	pa
2	11	reddel	m9	5	pa
3	18	reddel	bud9	7	nj
4	24	reddel	emla7	3	nc
5	31	reddel	m9	4	nc
6	38	gdel	bud9	8	pa
7	44	gdel	m9	5	pa
8	51	gdel	bud9	7	nj
9	57	gdel	emla7	3	nc
10	64	gdel	m9	4	nc

Note in the top table that the sampling weight (which is 1/(*selection probability*)) is 6.6. This is the sampling interval SAS® uses in selecting your systematic random sample after the initial selection of tree #5 as the first individual in the sample. Since 6.6 is not an integer, SAS® oscillates between 6 and 7 when selecting the *n*th sequential individual for inclusion in the sample. Note that the trees selected are 5, 11, 18, 24, 31, 38, 44, 51, 57 and 64.

You would sample the trees listed above to measure the characteristic(s) of interest, then estimate desired parameters from that data.

Benefits

One of the benefits of this type of sampling is that it provides a method for obtaining a random and representative sample from a population even when you can't list out all of the members of the population. Since the sample is random and every Ith member is sampled, it ensures that the sample represents the entire population.

Problems

While this type of sampling requires some planning and considerable effort, these two factors should not necessarily be considered problems. In order to get reliable data, planning and effort are always required. If the order of the population is systematic in any way, systematic error may enter the sampling. When there is a chance that the order is systematic, some other sampling procedure should be used to account for the systematic factor. This might include stratified or clustered random sampling.

AN EXAMPLE WHEN A LIST OF THE POPULATION IS NOT AVAILABLE   In the above example, we had a list of trees available for sampling. What if you didn't have a list of individuals in the population from which you wished to sample? Suppose you had advertised a farmer's market in your local paper and wanted to determine how effective it really was at bringing folks to your location. You have recently had about 1000 visitors to your market each time it was held and know that you simply can't ask them all whether or not they had seen the ad. Also you don't know how many people will show up on any given day, thus you really can't make a list of individuals in your population. In order to get a relatively good answer to the question, you could ask the 4th customer (randomly selected before the market opens) entering the market whether they saw the ad or not. You could then proceed to ask every 20th customer. You are systematically sampling your population.

ANOTHER EXAMPLE WHEN A LIST OF THE POPULATION IS AVAILABLE   Suppose you had a mailing list of all the folks who had visited your farmer's market last fall (your population) and want to determine how many of them plan on returning in the spring. You have a mailing list of 4000 individuals, and you don't want to send everyone on the list a survey since you are sending each survey participant a $10 voucher for participating. You decide you can send out 100 surveys. You could randomly start your list of potential participants with the 14th person on the list and then select every 4000/100 = 40th individual on the list to send the survey and voucher to.

As an aside, one problem with e-mail surveys is that most people will not respond. That is the reason for the enticement with a $10 voucher. I have found that very often you get a bimodal distribution of responses particularly when you are asking about product/service satisfaction issues. You often receive a significant number of responses from those who were not satisfied and another grouping of those who were satisfied. It is important to be aware of such possibilities when designing surveys.

Stratified random sampling

Description

Stratified random sampling is used when the population consists of individuals that can be grouped according to a specific characteristic. Samples are then drawn from each group so that a representative sample from the entire population is achieved. If samples from such a population are taken at random, estimates of variability will contain variation due to both between-individual variation and between-strata-group variation. When samples are stratified then sampled, variation among strata group can be accounted for separately. Think of this as similar to the situation when we use a randomized complete block design. Each sub-group is called a stratum (plural: strata). For example, you might need to sample from an apple orchard and might consider grouping according to cultivar, rootstock, tree age or any combination of these characteristics. The idea is that you would not want to risk the chance that all or most of the individuals in your sample had come from predominantly one group, such as all with the same rootstock. To ensure this doesn't happen, you will select individuals from each group.

Two types of stratified sampling are often considered. Proportional allocation is often used when a specific proportion of the individuals in each stratum are sampled, say 20%. A stratum with 100 individuals would have 20 individuals sampled from it while a stratum with 200 individuals would have 40 individuals sampled from it. Equal allocation is used to draw the same number of individuals from each stratum.

SAS® Method:

Define the population and the characteristic(s) of interest as well as groups for determining strata. Prepare a list of all individuals in each stratum and determine whether you want proportional or equal allocation. If you are not able to prepare a list of all members for each stratum, every *n*th individual from each stratum should be sampled, similar to systematic random sampling.

Reusing our DATASET=one:

```
1   TITLE1 'Chapter 19 Example 3 Stratified Random Sampling';
2   DATA one;
3   INPUT tree cv $ rs $ years state $ @@;
4   CARDS;
5   1 reddel emla7 4 pa 2 reddel emla7 4 pa
6   3 reddel emla7 4 pa 4 reddel emla7 4 pa
7   5 reddel bud9 8 pa 6 reddel bud9 8 pa
8   7 reddel bud9 8 pa 8 reddel m9 5 pa
9   9 reddel m9 5 pa 10 reddel m9 5 pa
10  11 reddel m9 5 pa 12 reddel emla7 3 nj
11  13 reddel emla7 3 nj 14 reddel emla7 3 nj
12  15 reddel emla7 3 nj 16 reddel bud9 7 nj
13  17 reddel bud9 7 nj 18 reddel bud9 7 nj
14  19 reddel m9 5 nj 20 reddel m9 5 nj
15  21 reddel m9 5 nj 22 reddel m9 5 nj
```

```
16  23 reddel emla7 3 nc 24 reddel emla7 3 nc
17  25 reddel emla7 3 nc 26 reddel emla7 3 nc
18  27 reddel bud9 7 nc 28 reddel bud9 7 nc
19  29 reddel bud9 7 nc 30 reddel m9 4 nc
20  31 reddel m9 4 nc 32 reddel m9 4 nc
21  33 reddel m9 4 nc 34 gdel emla7 4 pa
22  35 gdel emla7 4 pa 36 gdel emla7 4 pa
23  37 gdel emla7 4 pa 38 gdel bud9 8 pa
24  39 gdel bud9 8 pa 40 gdel bud9 8 pa
25  41 gdel m9 5 pa 42 gdel m9 5 pa
26  43 gdel m9 5 pa 44 gdel m9 5 pa
27  45 gdel emla7 3 nj 46 gdel emla7 3 nj
28  47 gdel emla7 3 nj 48 gdel emla7 3 nj
29  49 gdel bud9 7 nj 50 gdel bud9 7 nj
30  51 gdel bud9 7 nj 52 gdel m9 5 nj
31  53 gdel m9 5 nj 54 gdel m9 5 nj
32  55 gdel m9 5 nj 56 gdel emla7 3 nc
33  57 gdel emla7 3 nc 58 gdel emla7 3 nc
34  59 gdel emla7 3 nc 60 gdel bud9 7 nc
35  61 gdel bud9 7 nc 62 gdel bud9 7 nc
36  63 gdel m9 4 nc 64 gdel m9 4 nc
37  65 gdel m9 4 nc 66 gdel m9 4 nc
38 RUN;
39 PROC PRINT DATA = one;
40 RUN;
```

Suppose we wanted to make sure our sample contained a certain number of trees from each state. We would stratify our population by state, then randomly sample from each state. We could use the following statements to sample four trees from each of three states for a total sample size of 12. PROC SURVEYSELECT requires that the data is sorted by the variable indicating strata. We can accomplish this using PROC SORT (line 41):

```
41 PROC SORT DATA=one; BY state;
42 PROC SURVEYSELECT DATA=one METHOD=srs N=4 OUT=surveydata;
43 STRATA state;
44 PROC PRINT DATA = surveydata;
45 RUN;
```

We indicate we want to use simple random sampling as our method (line 42) and we want samples of size N = 4 (line 42) drawn from each stratum determined by state (line 43).

The results:

'Chapter 19 Example 3 Stratified Random Sampling'

The SURVEYSELECT Procedure

Selection Method	Simple Random Sampling
Strata Variable	state

Input Data Set	ONE
Random Number Seed	183248975
Stratum Sample Size	4
Number of Strata	3
Total Sample Size	12
Output Data Set	SURVEYDATA

'Chapter 19 Example 3 Stratified Random Sampling'

Obs	state	tree	cv	rs	years	SelectionProb	SamplingWeight
1	nc	30	reddel	m9	4	0.18182	5.5
2	nc	32	reddel	m9	4	0.18182	5.5
3	nc	33	reddel	m9	4	0.18182	5.5
4	nc	61	gdel	bud9	7	0.18182	5.5
5	nj	16	reddel	bud9	7	0.18182	5.5
6	nj	48	gdel	emla7	3	0.18182	5.5
7	nj	53	gdel	m9	5	0.18182	5.5
8	nj	55	gdel	m9	5	0.18182	5.5
9	pa	1	reddel	emla7	4	0.18182	5.5
10	pa	4	reddel	emla7	4	0.18182	5.5
11	pa	11	reddel	m9	5	0.18182	5.5
12	pa	40	gdel	bud9	8	0.18182	5.5

You would sample the trees listed above to measure the characteristic(s) of interest, then estimate desired parameters from that data.

Benefits

The main benefit of stratified random sampling is that it provides a representative sample of a heterogeneous population, ensuring that individuals with different identifiable characteristics are included in the sample. The heterogeneity in our population is the differences among states. It would be unwise to assume homogeneity among states, since soil types, climate and general production recommendations vary from region to region.

Problems

Stratified random sampling requires time and effort as well as knowledge of characteristics of the population that can be used for selecting strata. The selection of strata is important, because incorrect categorization of individuals will lead to unreliable information since it might be biased with an excessive number of individuals with a particular characteristic which was not accounted for.

Multistage sampling

Description

Sometimes a list of individuals within a population of interest is difficult (if not impossible) to create, because members are spread out over a large area or they are simply difficult to enumerate. This presents a problem for simple random sampling and stratified random sampling since, for both methods, all individuals in the population must be identified prior to sampling. It is sometimes easier to initially sample larger aggregate groups of individuals, followed by further sampling. This type of sampling is called multistage sampling.

Multistage sampling involves multiple levels of sampling, each selecting sub-groups of individuals within larger groups from which to sample further. The final stage of sampling involves selection of individuals from which data is gathered. Since the individuals selected for sampling are not chosen until the final stage, identification of specific members of the population occurs at this final stage.

Multistage sampling is flexible in that it can accommodate a wide variety of situations where sampling is needed. The number of stages sampled can vary as well as the number of members included for each stage. For example, you want to multistage sample farms to determine which ones grow bell peppers in the US and you want to consider three stages of sampling. The primary stage consists of selecting six of the 12 Northeastern states in the US with simple random sampling. The secondary stage selects three counties from each state using simple random sampling. The tertiary and final stage consists of randomly sampling five farms per county. Part of the researcher's job is to carefully define the population of interest at each stage of sampling. This makes multiple stage sampling extremely flexible. The entire sampling scheme for this example would consist of a total of 90 farms (6 states × 3 counties × 5 farms). The entire population consists of 2980 farms.

SAS® Method

Define the populations and the characteristic(s) of interest as well as groups for any stratification desired. Create a SAS® dataset from which to sample that includes the information necessary for sampling at each stage. The data is in a 'csv' file which consists of state, county and farm identifying variables. To simplify things, there were only ten farms per county. In reality, counties would likely have different numbers of farms and this number could vary widely.

```
1 TITLE1 'Chapter 19 Example 4 Multistage Random Sampling';
2 DATA one;
3 INFILE '/folders/myfolders/CABI Text/Chapter 19/Chapter 19
  Example 4 States and counties.csv' DLM =',' FIRSTOBS=2;
4 INPUT state $ county farm;
5 CARDS;
```

```
6   RUN;
7   PROC PRINT DATA = one;
8   RUN;
9   PROC SORT DATA=one; BY state;
10  PROC SURVEYSELECT DATA=one METHOD=srs SAMPSIZE=6
    OUT=selectedstates;
11  CLUSTER state;
12  PROC PRINT DATA = selectedstates;
13  RUN;
14  PROC SORT DATA=selectedstates; BY state;
15  PROC SURVEYSELECT DATA=selectedstates METHOD=srs SAMPSIZE=3
    OUT=selectedcounties;
16  STRATA state;
17  CLUSTER county;
18  PROC PRINT DATA = selectedcounties;
19  RUN;
20  PROC SORT DATA=selectedcounties; BY state county;
21  PROC SURVEYSELECT DATA=selectedcounties METHOD=srs SAMPSIZE=5
    OUT=selectedfarms;
22  STRATA state county;
23  DATA selectedfarms2; SET selectedfarms;
24  KEEP state county farm;
25  PROC PRINT DATA = selectedfarms2;
26  RUN;
```

Bear in mind that, with this approach to sampling, the number of counties in each state varies widely, for example DE has 3 while PA has 67. If DE and PA happen to be selected as states from which counties are sampled, all counties would be sampled in DE while only 3/67 = 1.5% of the counties in PA would be sampled. This may or may not present a problem, depending on the nature of the sampling in question. If it is a problem, I would recommend visiting a statistician to help determine the best approach.

It's worth examining the previous program to consider what each statement accomplishes in our multistage sampling so that you may modify it to suit your particular needs.

The first step is to create and print the dataset from which to sample. This is accomplished in lines 1 through 8. The data for creating the dataset are in the file 'Chapter 19 Example 4 States and counties.csv' listing 12 Northeastern states, counties within each state identified by number and farms within county number 1 to 10 (we limited the number of farms to ten to simplify the example).

```
1   TITLE1 'Chapter 19 Example 4 Multistage Random Sampling';
2   DATA one;
3   INFILE '/folders/myfolders/CABI Text/Chapter 19/Chapter 19
    Example 4 States and counties.csv' DLM =',' FIRSTOBS=2;
4   INPUT state $ county farm;
5   CARDS;
6   RUN;
7   PROC PRINT DATA = one;
8   RUN;
```

The first stage of sampling is accomplished with lines 9 through 13. We first sort the data in dataset 'one' by state so that we can sample 6 ('SAMPSIZE=6') states from our list of states ('CLUSTER state' (line 11)) by simple random sampling '(METHOD=srs') and create an output dataset we call 'selectedstates' ('OUT=selectedstates').

```
9   PROC SORT DATA=one; BY state;
10  PROC SURVEYSELECT DATA=one METHOD=srs SAMPSIZE=6
    OUT=selectedstates;
11  CLUSTER state;
12  PROC PRINT DATA = selectedstates;
13  RUN;
```

The second stage of sampling is accomplished with lines 14 through 19. We sort our dataset 'selectedstates' by state to select 3 ('SAMPSIZE=3') counties ('CLUSTER county') from each state ('STRATA state') to create the output dataset 'selectedcounties' ('OUT=selectedcounties').

```
14  PROC SORT DATA=selectedstates; BY state;
15  PROC SURVEYSELECT DATA=selectedstates METHOD=srs SAMPSIZE=3
    OUT=selectedcounties;
16  STRATA state;
17  CLUSTER county;
18  PROC PRINT DATA = selectedcounties;
19  RUN;
```

The final stage of sampling to select farms from each county is accomplished with lines 20 through 26. We sort our dataset 'selectedcounties' by state and county to be able to select 5 ('SAMPSIZE=5') farms from each county within state '(STRATA state county') to be included in the dataset 'selectedfarms' ('OUT=selectedfarms'). The 'selectedfarms' dataset has more information than we need, so we clean it up by creating a final dataset called 'selectedfarms2' ('DATA selectedfarms2') from the 'selectedfarms' dataset ('SET selectedfarms'), keeping only the state, county and farm information we need ('KEEP state county farm').

```
20  PROC SORT DATA=selectedcounties; BY state county;
21  PROC SURVEYSELECT DATA=selectedcounties METHOD=srs
    SAMPSIZE=5 OUT=selectedfarms;
22  STRATA state county;
23  DATA selectedfarms2; SET selectedfarms;
24  KEEP state county farm;
25  PROC PRINT DATA = selectedfarms2;
26  RUN;
```

The printout of our final sample looks like this (first 20 out of 90 observations only):

Obs	state	county	farm
1	ME	2	1
2	ME	2	2
3	ME	2	5
4	ME	2	6
5	ME	2	7
6	ME	5	1
7	ME	5	6
8	ME	5	8
9	ME	5	9
10	ME	5	10
11	ME	10	2
12	ME	10	6
13	ME	10	7
14	ME	10	8
15	ME	10	9
16	NH	2	1
17	NH	2	2
18	NH	2	3
19	NH	2	5
20	NH	2	9

Benefits

The benefits of multistage sampling are that it is flexible and fairly easy to plan, given adequate population definitions, while providing an acceptable and reliable estimate of the population in question.

Problems

The only real 'problem' with this type of sampling is in defining the population you wish to sample from and developing the criteria on which to base your stages of sampling. If poor planning occurs, the samples taken will not truly reflect the population from which they were drawn.

Sampling from Designed Experimental Plots

When designed experiments are conducted in the field, greenhouse or lab, sampling occurs as an intrinsic component of the research. During the planning stages of an experiment the experimental unit which is ultimately measured is precisely defined so that sampling units are easily identifiable. For example, if single plants are the experimental units, then single plants are usually measured. Sometimes the experimental units are too large to measure entirely. For example, suppose trees in an orchard are the experimental units and you decide you want to estimate the number of leaves per tree. You probably can't count all of the leaves on the tree. Another example might be if you wish to count fungal spores on a petri dish (the experimental unit). In these and many other cases, you may choose to sample only a portion of the experimental unit. This is called sub-sampling and there are specific methods for appropriate and statistically correct approaches for these circumstances.

The are many instances of plant science research where very large plots are used as the experimental units for practical reasons. Very large plots are usually used when the methodology of the research requires such plot size, for example when large field equipment must be used to apply treatments, such as sprayers, fertilizer applicators or combines for harvest. Most of the time, data do not need to be acquired from the entire plot and a smaller sample will suffice. In addition, data may be required for the entire plot for one characteristic (yield for example) but much smaller samples can be used for other data (nutrient content for example).

When sub-samples are taken from larger plots we are striving to obtain estimates that are as close to the value that the characteristic under consideration would have had if the entire plot would have been harvested. Any deviation from the actual value for the entire plot observed in the sample estimate is called the sampling error. We are always striving to achieve as small a sampling error as possible.

One way to think of sampling from a large experimental unit is to consider the experimental unit as a population of sorts where we will sample individuals to obtain our estimate. The key is to identify what constitutes an individual. In some cases, such as a plot with 50 strawberry plants, each plant would be an individual. In another setting, the individual might be a unit area of the plot. In the case of an entire orchard tree serving as the experimental unit or 'plot', the individual might be defined as a branch of a certain size, a single fruit or perhaps a single leaf. The researcher must effectively define the individuals within the experimental unit.

The individuals within the experimental unit must be easy to identify and measure, and provide an acceptable estimate with minimal variance while being relatively low cost (usually time and effort spent on measuring). The number of individuals we need to sample from a larger unit (sample size) depends on the variability among the individuals within the same experimental unit and the degree of precision needed from the estimate. Greater precision requirements and/or larger variability require larger sample sizes. Often a researcher can specify that they want to estimate the value of a characteristic ± a certain percentage, such as ± 10% (the margin of error). If an estimate of variance is available from the literature or from previous research and a margin of error is defined, a suitable sample size can be calculated using the formulas previously described in this chapter.

Once the experimental units are evaluated and the population(s) defined so that individual members of the population can be identified for sampling, one of the previously described methods for sampling can be implemented. The SAS® programs used to generate the lists of samples are exactly the same.

Most of the sampling in designed experiments is some variation of multistage random sampling. Suppose you had a CRD design with five treatments and four replications for a total of 20 experimental units (field plots). You want to take three samples from each plot for a total of 60 samples. The reason you don't want to simply take one measurement from each experimental unit is because the plots are large and you want to make sure you take a representative sample from each plot. Therefore you divide the plots into a grid consisting of a total of 15 sampling units per experimental plot. Thus you would have a total of 300 (5 × 4 × 15) subplots to sample from. You would randomly select three subplots from each experimental unit. Since you want to make sure you select three subplots from each experimental plot, you consider the sample a stratified random sample with the strata defined by plot. If you performed a simple random sample from among the 300 subplots, you would not be assured that three subplots would be selected from each experimental plot. The following SAS® program would provide the list of subplots to sample.

```
1   TITLE1 'Chapter 19 Example 5 Sampling from a CRD';
2   DATA one;
3   INFILE '/folders/myfolders/CABI Text/Chapter 19/Chapter 19
    Example 5 Sampling from a CRD.csv' DLM =',' FIRSTOBS=2;
4   INPUT plot rep treatment $ subplot;
5   CARDS;
6   RUN;
7   PROC PRINT DATA = one;
8   RUN;
9   PROC SORT DATA=one;
10  BY plot;
11  PROC SURVEYSELECT DATA=one METHOD=srs N=3 OUT=surveydata;
12  STRATA plot;
13  DATA surveydata2; SET surveydata;
14  KEEP plot rep treatment subplot;
15  PROC PRINT DATA = surveydata2;
16  RUN;
```

Here is a printout of the first 18 (out of 60 total) subplots to sample (edited to reduce size):

'Chapter 19 Example 5 Sampling from a CRD'

Obs	plot	rep	treatment	subplot
1	1	1	a	2
2	1	1	a	9
3	1	1	a	14
4	2	1	b	2
5	2	1	b	4
6	2	1	b	14
7	3	1	c	1
8	3	1	c	3
9	3	1	c	7
10	4	1	d	1
11	4	1	d	7
12	4	1	d	9
13	5	1	e	2
14	5	1	e	6
15	5	1	e	8
16	6	2	a	2
17	6	2	a	5
18	6	2	a	9

Similar approaches would be taken to randomly sub-sample experimental plots from other designed experiments, including those from randomized complete blocks, split-plots and variants and Latin squares.

Sometimes a slightly more sophisticated sampling routine utilizing multistage sampling is needed. For example, suppose you had an experiment in an RCBD utilizing whole trees in an apple orchard as the experimental unit. There are 20 treatments and ten blocks for a total of 200 trees. Suppose you wanted to test fruit at harvest for firmness. You don't want to risk simply taking a random sample of harvested fruits, since they may originate from the same area on a tree. You want to make sure you take samples from all areas on a tree. To facilitate such sampling, you partition the tree into lower and upper regions of the north, south, east and west quadrants so that each tree ultimately has 8 zones. You can't sample from each zone, due to the size of your experiment (sampling all zones would mean 1600 samples and you can only process 800) and the time you can devote to the project, yet at the same time you want to make sure you have two samples from the lower zone and two samples from the upper zone from each tree. You could use multistage random sampling to determine which fruit to measure. The following SAS® program would provide the list of subplots to sample.

```
1   TITLE1 'Chapter 19 Example 6 Multistage sampling from a
    RCBD';
2   DATA one;
3   INFILE '/folders/myfolders/CABI Text/Chapter 19/Chapter 19
    Example 6 Multistage sampling from a RCBD.csv' DLM =','
    FIRSTOBS=2;
4   INPUT tree block treatment height $ direction $ zone;
5   CARDS;
6   RUN;
7   PROC SORT DATA=one;
8   BY tree height;
9   PROC SURVEYSELECT DATA=one METHOD=srs SAMPSIZE=2
    OUT=selectedzones;
10  STRATA tree height;
11  DATA selectedzones2; SET selectedzones;
12  KEEP tree block treatment height direction zone;
13  PROC PRINT DATA = selectedzones2;
14  RUN;
```

Printout for the first two trees:

'Chapter 19 Example 6 Multistage sampling from a RCBD'

Obs	tree	height	block	treatment	direction	zone
1	1	bottom	1	1	n	5
2	1	bottom	1	1	s	6
3	1	top	1	1	n	1
4	1	top	1	1	w	4
5	2	bottom	1	2	n	5
6	2	bottom	1	2	s	6
7	2	top	1	2	s	2
8	2	top	1	2	e	3

Multistage random sampling can be as simple or complex as necessary to fulfill the research mission. Regardless of how simple or how complex a chosen sampling design is, remember that a sampling design simply describes which individuals in a population, i.e. which plants in a plot or branches on a tree, should be measured for inclusion in the sample. Also remember that a good plan minimizes both variance estimates and bias.

Sampling Over Time

Many plant science experiments require the measurement of some characteristic of interest over time, e.g. plant height at 1, 2 and 3 months after field establishment. The question often arises as to whether or not the same plants should be measured on each sampling date or whether different samples should be taken. The main consideration is that measurements made over time on the same plant are often correlated with each other and are not independent, thus the assumptions for parametric analysis is not necessarily valid. Another concern is that the mere act of handling a plant during measurement may cause it to perform differently than plants that are not measured; in other words, measurement causes some response in the plant. If different plants are sampled for each measurement, increased variability is likely since more plant-to-plant variability is introduced with a greater number of sampled plants.

Depending on the crop and research goals, careful handling of plants over the course of any experiment will often minimize any aberrant effects of handling. A simple solution to this problem is to plan ahead and make sure enough plants are available to sample new ones on each observation date. The observations would then be independent. Keep in mind that some experiments require re-measurement of the same plants to quantify treatment effects on development over time. Also, keep in mind that if the sampling over time is destructive, for example to measure tissue nutrient content over the season, adequate plant material must be established at the beginning of the experiment to ensure success.

20 When Data Doesn't Behave

Introduction

When we use many of the procedures outlined in this text, particularly those based on an ANOVA, we assume that our data meet three basic conditions:

1. The errors are independent.
2. The errors are normally distributed.
3. The errors have homogeneous variances.

While much of the data collected for plant science research possesses these qualities, there are times when it doesn't. If one or more of these assumptions are violated, the level of significance and the sensitivity of statistical tests are compromised. Before using parametric procedures such as ANOVAs and linear regression on such data, we must address these issues.

Most of the time these problems are addressed by transforming variables and analyzing the transformed data with parametric methods (as long as the transformed data meet the assumptions). Another approach is to utilize a non-parametric procedure as discussed in Chapter 18. In general, the order of impact on our inferences is as follows: independence is most important, normality is least important (but still important) and homogeneous variances is in between the two. Remember, we are concerned about the errors, not the data themselves.

Since we can never really know error values, we estimate them with residuals, the deviations of our observations from their expected values considering the model chosen for analysis. We check these assumptions regarding our residuals using various methods. By knowing the extent of these violations, we can have a better idea of how confident we can be regarding our inferences.

Data properties aren't necessarily clear cut. Nearly all of the data we encounter have at least some slight deviation from the above assumptions, i.e. there's always at least a little bit of non-normality, dependence, and heterogeneous variances. Thus it's not really whether or not our data violate the assumptions, but rather, by how much. If the amount of deviation from these assumptions is not excessive, then parametric tests will provide good tests since they are fairly robust. With larger deviations, confidence in our tests is reduced. We evaluate the extent of these violations to give us an idea of how valid our tests are.

The process of transforming data can be cumbersome, since a transformation to correct one problem with the data assumptions may inadvertently lead to another. While the process of finding the right transformation to correct the problem at hand without causing another can be challenging, there are some transformations that should be evaluated based on the type of data being processed. This chapter will present some of the more common transformations with a discussion of why, when and how to use them as well as how to implement them in SAS®.

What Do These Assumptions Mean?

Independence

Independence of errors and the residuals that estimate them implies that the value of an observation on one experimental unit in no way affects the value of an observation on another. When we follow randomization protocols for designed experiments, we are virtually guaranteeing that our observations, their errors and the residuals that estimate them are independent. One exception is when we take observations over time when values of later observations may be affected by the values of observations taken at an earlier time. Also, if there is any systematic pattern in the arrangement of treatments from one replication to another, errors may not be independent.

Normality

The normality assumption asserts that the data from which we estimate our errors via residuals are taken from a population with a normal distribution. Not all plant science data are normally distributed. Data might come from a population with a binomial, Poisson or chi-squared distribution.

Homogeneous variances

This assumption asserts that variance estimates derived within and among the different groups of an ANOVA are constant and are all estimating the same variance. They may differ slightly in value, but, within reason, they are all estimating the same thing. We can statistically test whether or not the differences we might observe are truly 'within reason'.

Outliers

This is a good time to bring up data points that are very different than others in the same treatment or replicate. They are known as outliers. Any data point that is suspected of being an outlier should be further investigated to determine if any action is needed before proceeding with the analysis. An outlier might simply be miscoded data or transposed numbers during data entry. These cases are easy to fix. With outliers that seem to be valid data points, further inquiry is justified. Was the experimental unit similar to others in appearance? Was it located on the edge of a plot or field? Was it the correct cultivar? There are many issues that should be considered. The most important consideration is that you do not simply discard a data point because it appears to be an outlier. A discussion of how to handle the situation is given below in the section describing remedies for problem data.

Tests for Validity of Assumptions

Independence

It is very difficult, if not impossible, to test for independence of errors. As previously stated, when good randomization procedures are followed as part of the experimental protocol, the likelihood of independence is enhanced. However, when observations are collected over time, the independence assumption is often violated. Plots of residuals against any time variable (such as observation order) or levels of a factor in an experiment can often provide a clue as to whether or not the residuals are independent.

Any pattern in a plot that is not random suggests a lack of independence. One potential lack of independence (serial dependence or autocorrelation) that might be observed with time-dependent observations is that observations close to each other in time tend to be too similar (positive dependence) or too dissimilar (negative dependence). This type of dependence can be seen in plots of residuals on the vertical axis against the time factor on the horizontal axis. Positive dependence appears as a drifting pattern across the plot, while negative dependence appears as residuals centered near zero but with rapidly alternating positive and negative residuals.

The Durbin–Watson test statistic can be used to check whether or not there is serial dependence (autocorrelation) over time. The residuals must be sorted in time order. The DW statistic should be around 2 if there is no serial correlation. Positive serial correlation produces DWs < 2 and negative serial correlation produces DWs > 2.

The following is an example of using the DW statistic to check for autocorrelation in a small dataset of yield data collected over time. Remember, the data must be input or sorted in time order for the DW statistic to be meaningful. The dataset was purposefully kept small to keep things simple.

```
1   TITLE1 'Chapter 20 Example 1.sas';
2   TITLE2 'Data Problems';
3   TITLE3 'DW Test for Independence (Autocorrelation)';
4   DATA one;
5   INPUT time yield;
6   CARDS;
7   1 20
8   2 30
9   3 15
10  4 35
11  5 20
12  6 40
13  7 30
14  8 25
15  9 30
16  10 20
17  RUN;
18  PROC PRINT;
19  RUN;
20  PROC REG DATA = one PLOTS = none;
21  MODEL yield = time / dwProb;
22  RUN;
```

The output:

Number of Observations Read	10
Number of Observations Used	10

Analysis of Variance					
Source	DF	Sum of Squares	Mean Square	F Value	Pr > F
Model	1	9.16667	9.16667	0.13	0.7229
Error	8	543.33333	67.91667		
Corrected Total	9	552.50000			

Root MSE	8.24116	R-Square	0.0166
Dependent Mean	26.50000	Adj R-Sq	-0.1063
Coeff Var	31.09871		

Parameter Estimates					
Variable	DF	Parameter Estimate	Standard Error	t Value	Pr > |t|
Intercept	1	24.66667	5.62978	4.38	0.0023
time	1	0.33333	0.90732	0.37	0.7229

'Chapter 20 Example 1.sas'
'Data Problems'
DW Test for Indepedence (Autocorrelation)

The REG Procedure
Model: MODEL1
Dependent Variable: yield

Durbin-Watson D	2.947
Pr < DW	0.8990
Pr > DW	0.1010
Number of Observations	10
1st Order Autocorrelation	-0.555

Note: Pr<DW is the p-value for testing positive autocorrelation, and Pr>DW is the p-value for testing negative autocorrelation.

The DW test indicates that there is no problem with the independence assumption, since neither the $Pr < DW$ nor $Pr > DW$ are significant at $\alpha = 0.05$.

Dependence may also arise from close proximity of experimental units. Plots called variograms can be used to detect this type of dependence. Each possible pair of units must be plotted. The plots consist of the distance between the pair on the horizontal axis and the squared difference between their residuals on the vertical axis. If any pattern is discernible, there is dependence and the assumption of independence does not hold. The process of creating and analyzing variogram plots is beyond the scope of this text. To give you an idea of how complicated the process is, 134 pages are dedicated to it in the SAS/ STAT® Guide.

Normality

'PROC UNIVARIATE normal' provides our test for normality of residuals. The test statistic which SAS® uses (W) is the Shapiro–Wilks test if $N \leq 2000$ and the Kolmogorov D test otherwise. We reject the assumption of normality if the α value for the relevant test is generally ≤ 0.05. Since we explored the test for normality in Chapter 3, there is no need to present an example here.

Homogeneous variance

We use Levene's test of variance homogeneity. The test is requested by using the '/ HOVTEST=LEVENE' option in the means statement of a one factor ANOVA (see Chapter 9). In addition, plots of residuals on the vertical axis against predicted values on the horizontal axis give a visual assessment of variance heterogeneity. The plot will appear as a series of vertical stripes, one for each treatment group. With constant variance, the spread in each stripe will be similar so that no real pattern is discernible as you look at the plot. Non-constant variance is observed as a pattern in the spread of the residuals. The pattern can be such that it resembles a megaphone opening towards the right of the plot (variances increasing with increasing means), a left-opening megaphone (variances decreasing with increasing means) or two megaphones opening at the center of the plot, one from each direction (usually observed when the data are proportions). Other patterns may be observed and if the pattern is anything but random, there is good reason to believe there is non-constant variance.

Let's reconsider a modified version of Example 3 from Chapter 12. The experiment evaluated lettuce cultivars and seed sources in an RCB design. There were five blocks and 20 treatments.

```
1   TITLE1 'Chapter 20 Example 2.sas';
2   TITLE2 'RCBD, 5 blocks 20 treatments';
3   TITLE3 'Lettuce cultivar and seed source evaluation';
4   DATA one;
5   INPUT block treatment cultivar $ ssource $ cases;
6   CARDS;
7   1 1 j a 740
8   2 1 j a 655
9   3 1 j a 1162
10  4 1 j a 621
11  5 1 j a 601
12  1 2 j g 869
```

```
13  2   2 j g  759
14  3   2 j g  969
15  4   2 j g  630
16  5   2 j g  714
17  1   3 j j  972
18  2   3 j j  755
19  3   3 j j  545
20  4   3 j j  724
21  5   3 j j  807
22  1   4 j t  1075
23  2   4 j t  662
24  3   4 j t  547
25  4   4 j t  838
26  5   4 j t  911
27  1   5 j w  1179
28  2   5 j w  644
29  3   5 j w  652
30  4   5 j w  931
31  5   5 j w  489
32  1  16 n a  712
33  2  16 n a  570
34  3  16 n a  474
35  4  16 n a  1028
36  5  16 n a  615
37  1  17 n g  810
38  2  17 n g  756
39  3  17 n g  674
40  4  17 n g  1118
41  5  17 n g  611
42  1  18 n j  817
43  2  18 n j  805
44  3  18 n j  783
45  4  18 n j  918
46  5  18 n j  623
47  1  19 n t  354
48  2  19 n t  959
49  3  19 n t  773
50  4  19 n t  599
51  5  19 n t  713
52  1  20 n w  455
53  2  20 n w  1071
54  3  20 n w  685
55  4  20 n w  506
56  5  20 n w  826
57  1   6 p a  980
58  2   6 p a  655
59  3   6 p a  753
60  4   6 p a  1031
```

```
 61  5   6 p a  523
 62  1   7 p g  521
 63  2   7 p g  748
 64  3   7 p g  851
 65  4   7 p g  202
 66  5   7 p g  692
 67  1   8 p j  527
 68  2   8 p j  860
 69  3   8 p j  951
 70  4   8 p j  510
 71  5   8 p j  709
 72  1   9 p t  629
 73  2   9 p t  951
 74  3   9 p t  455
 75  4   9 p t  502
 76  5   9 p t  895
 77  1  10 p w  734
 78  2  10 p w 1066
 79  3  10 p w  563
 80  4  10 p w  502
 81  5  10 p w  806
 82  1  11 r a  829
 83  2  11 r a  257
 84  3  11 r a  656
 85  4  11 r a  505
 86  5  11 r a  399
 87  1  12 r g  934
 88  2  12 r g  566
 89  3  12 r g  766
 90  4  12 r g  515
 91  5  12 r g  403
 92  1  13 r j  402
 93  2  13 r j  557
 94  3  13 r j  863
 95  4  13 r j  711
 96  5  13 r j  600
 97  1  14 r t  509
 98  2  14 r t  568
 99  3  14 r t  865
100  4  14 r t  816
101  5  14 r t  704
102  1  15 r w  608
103  2  15 r w  560
104  3  15 r w  368
105  4  15 r w  919
106  5  15 r w  703
107 RUN;
108 PROC PRINT DATA = one;
```

```
109  RUN;
110  PROC UNIVARIATE NORMAL DATA = one;
111  VAR cases;
112  PROC ANOVA;
113  CLASSES block treatment;
114  MODEL cases = treatment;
115  MEANS treatment / HOVTEST = LEVENE;
116  RUN;
```
The results of the normality test indicate that the Shapiro–Wilk test (since $N < 2000$), $W = 0.992503$, $\alpha = 0.8564$, indicates that we fail to reject the null hypothesis of normality. The results for Levene's test for homogeneity of variance:

'Chapter 20 Example 2.sas'
'RCBD, 5 blocks 20 treatments'
'Lettuce cultivar and seed source evaluation'

The ANOVA Procedure

Levene's Test for Homogeneity of cases Variance ANOVA of Squared Deviations from Group Means					
Source	DF	Sum of Squares	Mean Square	F Value	Pr > F
treatment	19	1.558E10	8.1983E8	0.49	0.9597
Error	80	1.339E11	1.674E9		

Since $\alpha = 0.9597$, we fail to reject the hypothesis of homogeneous variances. Some references suggest that tests such as Levene's are not reliable for ascertaining homogeneity of variance, since they are also very sensitive to normality issues. In other words, if the data are not normal, Levene's test, when significant, may be indicating significant variance heterogeneity and/or non-normality of the data. In this example, the data appear to be normal, thus Levene's test is reliable.

When you're not sure how reliable Levene's test is or when you want another mode of evaluation for variance homogeneity, residual plots can help. Here are the SAS® statements for generating residual plots to evaluate variance homogeneity for the same experiment.

```
1   TITLE1 'Chapter 20 Example 3.sas';
2   PROC GLM DATA = one PLOTS = none;
3   CLASSES block treatment;
4   MODEL cases = block treatment;
5   OUTPUT OUT = diagnostics RESIDUAL = res PREDICTED = pred
    COOKD = CooksD;
6   PROC PLOT;
7   PLOT res*treatment = '+';
8   PLOT res*pred = '*';
9   PLOT CooksD*cases = cases / VREF = 0.04;
10  RUN;
11  PROC PRINT DATA = diagnostics;
12  RUN;
```
The output for residual × treatment plot:

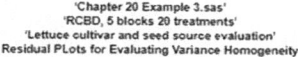

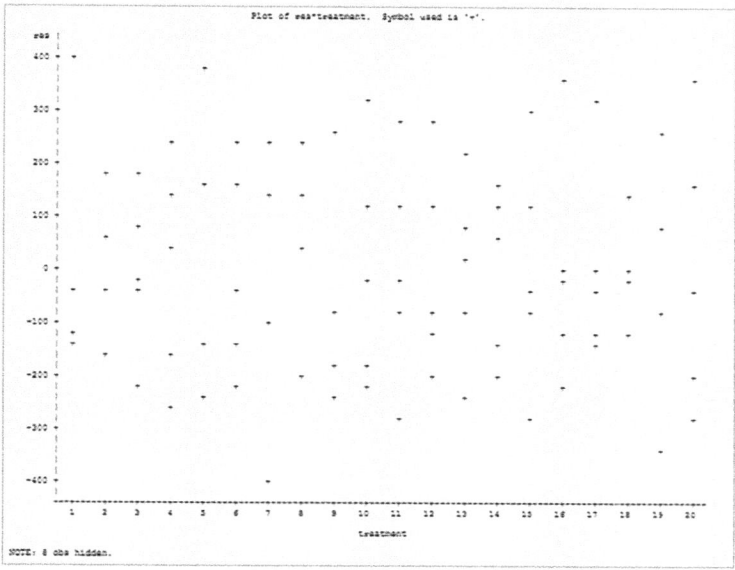

This plot in general does not reveal any significant pattern to the plots of residuals versus treatment. This is in agreement with Levene's test that there is homogeneity of variances among the treatments.

The output for residual × predicted plot:

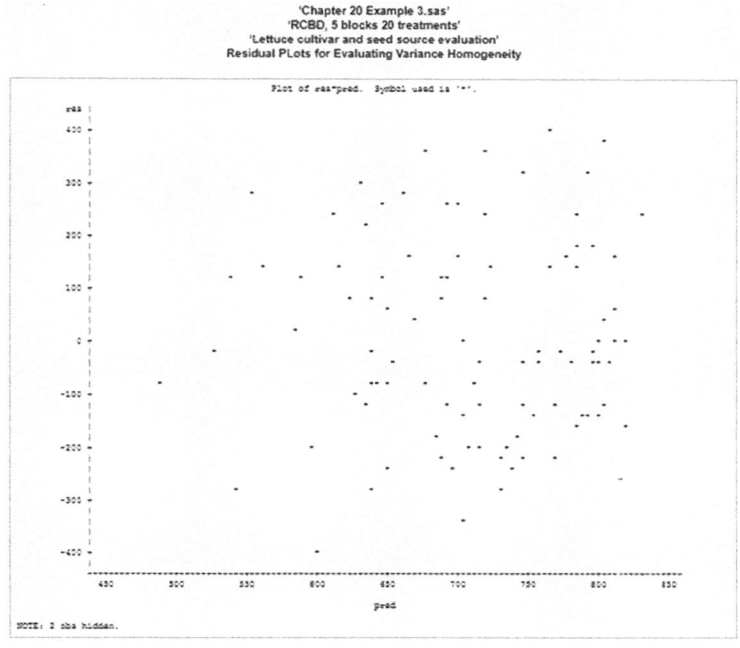

In this plot you should again look for patterns such as the left- or right-opening megaphones to provide an idea of whether or not variances are homogeneous. The above plot appears fairly random, agreeing with the previous plot of residual × treatment and Levene's test. If you look hard enough you might see a very slightly right-opening megaphone, suggesting increasing variances with increasing means. However, based on the Levene's test and the other plot, I would still fail to reject the assumption of variance homogeneity.

Outliers

Outliers are identified by data plots or by Cook's D statistic. In the above residual plots, look carefully for observations that are plotted way out of range from other observations. This may indicate the need to evaluate the observation as an outlier. It doesn't mean that there is necessarily anything wrong with the observation. It is simply a flag for you to evaluate the observation and to make sure it is legitimate. We will discuss other options for possible outliers shortly.

Cook's D was requested in line 5 above. A plot of Cook's D values × cases was requested in line 9. Influential data points (outliers) are indicated by values of Cook's $D \geq 4/n$, in our case, Cook's $D \geq 0.04$. The plot simply makes it easier to spot potentially influential data rather than looking through pages of dataset printouts with the request for a horizontal line at the Cook's D value of 0.04 (using '/ VREF = 0.04;' at the end of line 9).

Here is the plot:

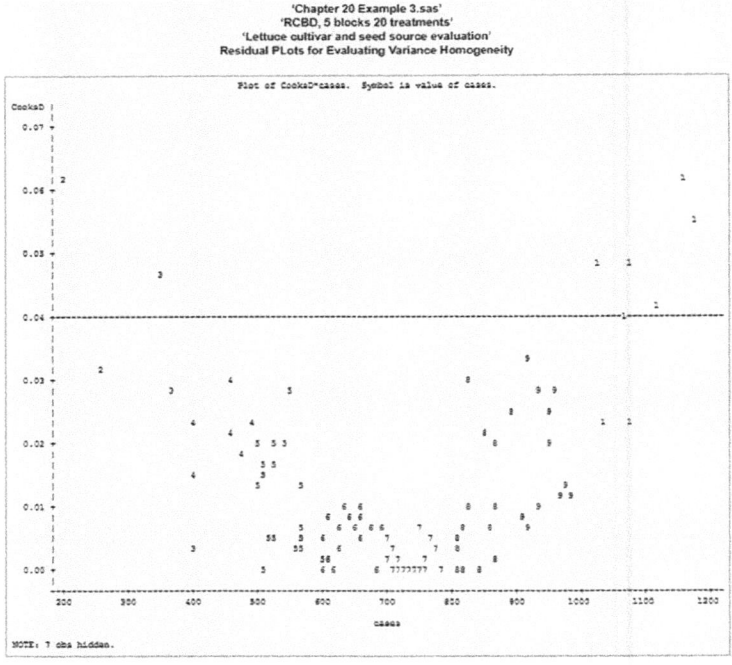

There are *eight* potentially influential data points (outliers) that need to be investigated and verified as legitimate data values.

Effects of Assumption Violations

Understanding the consequences of using standard parametric tests on data that do not meet the above-listed assumptions helps you understand why you need to pursue a remedy.

Lack of independent errors

If residuals are not independent, treatment means are unbiased (good); the overall estimate of variance (MSE) is fairly unbiased (with larger sample sizes) (good); however, the variance of a mean is no longer σ^2/n, where n is the sample size (bad). Why is that bad? Because now the estimates of standard errors used when computing differences among means or in contrasts are biased and the claimed α levels are not valid, thus Type I and II error rates are not valid. The ANOVA F-test will also suffer, since the α level may not be valid. The ANOVA F-test is robust when there is not independence, but this is only true when the α level is averaged over all possible randomization schemes for any given experiment. For a specific randomization, the α level may be significantly impacted, therefore the Type I and II error rates may be significantly impacted.

Non-normality of errors

When considering whether or not a distribution is normal it helps to have some measure of just how normal it is. Two measures which help in this regard are skewness (measure of asymmetry) and kurtosis (measure of spread). A normal distribution is symmetric about the mean with equal spread above and below it (remember the bell-shaped curve?), thus the values of skewness and kurtosis for a normal distribution are 0. Positively skewed distributions have a longer right tail, while negatively skewed distributions have a longer left tail. Long tails have positive kurtosis, while short tails have negative kurtosis. Estimating skewness and kurtosis values for different samples has limited use for diagnosing problems, but understanding what the two terms mean helps one understand how non-normal data will impact inferences drawn from them.

Treatment means and the common variance estimate (MSE) are still relatively unbiased (good) and F-test p-values are relatively unaffected by slightly non-normal data (good). Long tails (positive kurtosis) tend to make F-tests too conservative, thus there is a tendency for Type II errors; and short tails (negative kurtosis) tend to make F-tests too liberal, thus a tendency for Type I errors. Skewness (asymmetry) has less of an effect on p-values. Also, these effects of non-normality decrease quickly with increasing sample size.

Confidence intervals are more affected by skewness than by kurtosis. With normal data, two-sided confidence intervals are evenly split above and below the mean. With non-zero skewness, the range of the confidence interval is correct, but it tends to have too much of it above or below the mean rather than equally distributed to both sides. One-sided confidence intervals can be seriously incorrect, depending on the nature of the skewness.

Non-constant variances

Non-constant variances can adversely affect the Type I error rate if there are more than two groups in the comparison or test or if sample sizes are unequal. If larger sample sizes have larger variances, then our test is too conservative (lower Type I error rate). On the other hand, if large sample sizes are associated with smaller variances, our tests are too liberal (higher Type I error rate). Confidence intervals will be too large when comparing groups with smaller variances and too small when comparing groups with larger variances.

Outliers

Outliers can severely influence estimates of treatment means and the variance. It is important to correctly identify outliers and evaluate their validity and take any steps necessary to remedy the problem.

Transformations to 'Remedy' Violations

Transformations are tools we use to bring our data more in line with the above listed assumptions. When the data satisfy our assumptions more closely, we can have greater confidence in the inferences we make from our analyses. Most of the time we are transforming data due to problems with non-normality or heterogeneous variances.

Remedies for lack of independence

There are no simple solutions for dealing with dependent (correlated) data. If the problem is of concern, some sort of time series or spatial analysis should be pursued as well as the help of a statistician, since both approaches are beyond the scope of this text.

Remedies for non-normality

The ART transformation was covered in depth in Chapter 18 and it is useful in remedying nearly all plant science instances of non-normal data.

Remedies for heterogeneous variances

There are generally two situations where we might have to deal with non-constant variances: (1) when the variances are proportional to the means; and (2) when there is no discernible relationship between the means and the variances. The first type of variance heterogeneity usually occurs when the data are not normal. This often happens with count data, such as number of flowers per plant, insects per leaf or weeds per plot. These types of count data often follow a Poisson distribution. Another type of data with heterogeneous variances is data describing a proportion where an outcome is one of two values, such as proportion of living plants, infected leaves, pollinated flowers, etc. These data often follow a binomial distribution. The second type of variance heterogeneity, when the variance and the mean are not related, occurs when some of the treatments have larger variances compared with others due to the nature of the treatments. For example, in breeding programs differences in homogeneity within genetic lines being compared may be substantial. Another example is in trials testing yield responses to some applied substance, such as fertilizer or pest control, where the application of the treatment may not be as uniform as desired. The variability in treated plots may be significantly different than the variability in control plots, due to poor techniques.

Many statistics texts will present a number of options for transforming variables to make them normal or to promote constant variances. That approach consists of: 'if X is the situation' then 'this Z is the transformation' followed by 'if Z doesn't work (render the variances homogeneous) then try a different transformation'. This approach can be frustrating and tedious. Rather than presenting all the options out there, this text will take the approach of utilizing what is called the Box–Cox Transformation.

Box–Cox transformation

The most common type of transformation is one where the dependent variable (Y) is transformed into a new variable Y_t and the transformed Y_t is evaluated for normality and constant variance (actually the residuals are evaluated, but we normally just say the transformed values are evaluated). The transformations normally used are called power transformations, and the most common power transformations are the Box–Cox power transformations.

Using the Box–Cox transformation allows you to harness the power of SAS® while saving a lot of time and preventing the aggravation often associated with searching for an appropriate transformation. Essentially, the Box–Cox transformation will use complex mathematical techniques to determine the most appropriate transformation for your data. The mathematics involved are beyond the scope of this text, but the methodology's usefulness is not.

You must make sure that all data values are non-zero and positive. If you have some values of data that are 0 or negative numbers, you must add a constant to all data values to produce all non-zero, non-negative values. For example, if you are only dealing with data which are 0 or positive values, simply add 1 to each value. If you are dealing with some negative values, determine the most-negative value and add the absolute value of it + 1 to all data values. For example, if a data value is –38, add 39 to all data points. Once this pre-transformation 'data clean-up' is complete, you can start the Box–Cox process.

The Box–Cox process produces a value called lambda (λ) which will be presented in the output. Different values of λ correspond to different transformations in the power transformation family. The value of λ calculated with Box–Cox is that value of λ that minimizes the sums of squares error. The following table presents the transformations and associated λ values.

λ	Transformation	y - value
1	None	y
0.5	Square root	$\sqrt{y}$
0	Log	$\ln(y)$
−0.5	Reciprocal square root	$\dfrac{1}{\sqrt{y}}$
−1	Reciprocal	$\dfrac{1}{y}$

The following code will find the Box–Cox transformation for our data from the previous example (Example 3).

```
1    TITLE1 'Chapter 20 Example 4.sas';
2    TITLE2 'RCBD, 5 blocks 20 treatments';
3    TITLE3 'Lettuce cultivar and seed source evaluation';
4    TITLE4 'Box-Cox Transformation';
5    DATA one;
6    INPUT block treatment cultivar $ ssource $ cases @@;
7    CARDS;
8    1 1 j a 740 2 1 j a 655 3 1 j a 1162 4 1 j a 621 5 1 j a 601
9    1 2 j g 869 2 2 j g 759 3 2 j g 969 4 2 j g 630 5 2 j g 714
10   1 3 j j 972 2 3 j j 755 3 3 j j 545 4 3 j j 724 5 3 j j 807
11   1 4 j t 1075 2 4 j t 662 3 4 j t 547 4 4 j t 838 5 4 j t 911
12   1 5 j w 1179 2 5 j w 644 3 5 j w 652 4 5 j w 931 5 5 j w 489
13   1 16 n a 712 2 16 n a 570 3 16 n a 474 4 16 n a 1028 5 16 n a 615
14   1 17 n g 810 2 17 n g 756 3 17 n g 674 4 17 n g 1118 5 17 n g 611
15   1 18 n j 817 2 18 n j 805 3 18 n j 783 4 18 n j 918 5 18 n j 623
16   1 19 n t 354 2 19 n t 959 3 19 n t 773 4 19 n t 599 5 19 n t 713
17   1 20 n w 455 2 20 n w 1071 3 20 n w 685 4 20 n w 506 5 20 n w 826
18   1 6 p a 980 2 6 p a 655 3 6 p a 753 4 6 p a 1031 5 6 p a 523
19   1 7 p g 521 2 7 p g 748 3 7 p g 851 4 7 p g 202 5 7 p g 692
20   1 8 p j 527 2 8 p j 860 3 8 p j 951 4 8 p j 510 5 8 p j 709
21   1 9 p t 629 2 9 p t 951 3 9 p t 455 4 9 p t 502 5 9 p t 895
22   1 10 p w 734 2 10 p w 1066 3 10 p w 563 4 10 p w 502 5 10 p w 806
23   1 11 r a 829 2 11 r a 257 3 11 r a 656 4 11 r a 505 5 11 r a 399
24   1 12 r g 934 2 12 r g 566 3 12 r g 766 4 12 r g 515 5 12 r g 403
25   1 13 r j 402 2 13 r j 557 3 13 r j 863 4 13 r j 711 5 13 r j 600
26   1 14 r t 509 2 14 r t 568 3 14 r t 865 4 14 r t 816 5 14 r t 704
27   1 15 r w 608 2 15 r w 560 3 15 r w 368 4 15 r w 919 5 15 r w 703
28   RUN;
29   PROC TRANSREG MAXITER=0 NOZEROCONSTANT DATA = one;
30   MODEL BoxCox(cases) = identity(treatment);
31   ODS GRAPHICS off;
32   RUN;
```

For your specific dataset, simply change line 29 to reflect the name of your dataset and change line 30 so that your dependent variable replaces 'cases' and your treatment variable replaces 'treatment'. The rest of lines 28 through 32 remains as it is.

The output:

'Chapter 20 Example 4.sas'
'RCBD, 5 blocks 20 treatments'
'Lettuce cultivar and seed source evaluation'
Box-Cox Transformation

The TRANSREG Procedure

Box-Cox Transformation Information for cases			
Lambda	R-Square	Log Like	
-3.00	0.00	-686.260	
-2.75	0.00	-666.142	
-2.50	0.00	-647.026	
-2.25	0.00	-629.053	
-2.00	0.00	-612.371	
-1.75	0.00	-597.122	
-1.50	0.00	-583.427	
-1.25	0.00	-571.375	
-1.00	0.01	-561.008	
-0.75	0.01	-552.313	
-0.50	0.01	-545.233	
-0.25	0.01	-539.670	
0.00	0.01	-535.500	
0.25	0.01	-532.589	
0.50	0.01	-530.804	*
0.75	0.01	-530.022	<
1.00	+	0.01	-530.132	*
1.25	0.01	-531.038	*
1.50	0.01	-532.659	
1.75	0.01	-534.925	
2.00	0.01	-537.777	
2.25	0.01	-541.167	
2.50	0.01	-545.051	
2.75	0.01	-549.393	
3.00	0.01	-554.161	

< - Best Lambda
* - 95% Confidence Interval
+ - Convenient Lambda

The λ associated with the best power transformation is indicated in the output, and in this case λ = 0.75.

(Remember that our data does not appear to be in violation of the major assumptions to begin with, thus this Box–Cox transformation is for illustrating the methodology.)

Now you would transform your original data with the statement:

```
DATA two;
SET one;
ty = POW(y,λ);
RUN;
```

where 'y' is the dependent variable you wish to transform and 'λ' is the value of λ generated by the Box–Cox method.

For our example above:

```
DATA two;
SET one;
tcases = POW(cases,0.75);
RUN;
```

You would then re-test the data to examine assumptions regarding the residuals just like you did for the original data, i.e. evaluate normality, homogeneity of variances and influential data points (outliers).

So that you can see how the Box–Cox transformation works, let's examine a fictitious dataset that is purposefully generated so that it is not normal. The dataset is a set of 5000 observations randomly generated from an exponential distribution.

```
 1  TITLE1 'Chapter 20 Example 5.sas';
 2  TITLE2 'Generated non-normal data';
 3  TITLE3 'Box-Cox Transformation';
 4  DATA one;
 5  z = 0;
 6  DO i = 1 to 5000;
 7  y = RANEXP(12345);
 8  OUTPUT;
 9  END;
10  RUN;
11  PROC TRANSREG MAXITER=0 NOZEROCONSTANT DATA = one;
12  MODEL BoxCox(y) = identity(z);
13  OUTPUT;
14  RUN;
15  PROC UNIVARIATE noprint;
16  HISTOGRAM y ty;
17  RUN;
```

The output from this example for the Box–Cox process is graphical, since we did not turn off the graphics like we did in the previous example (Example 4, line 31). The best λ is 0.25.

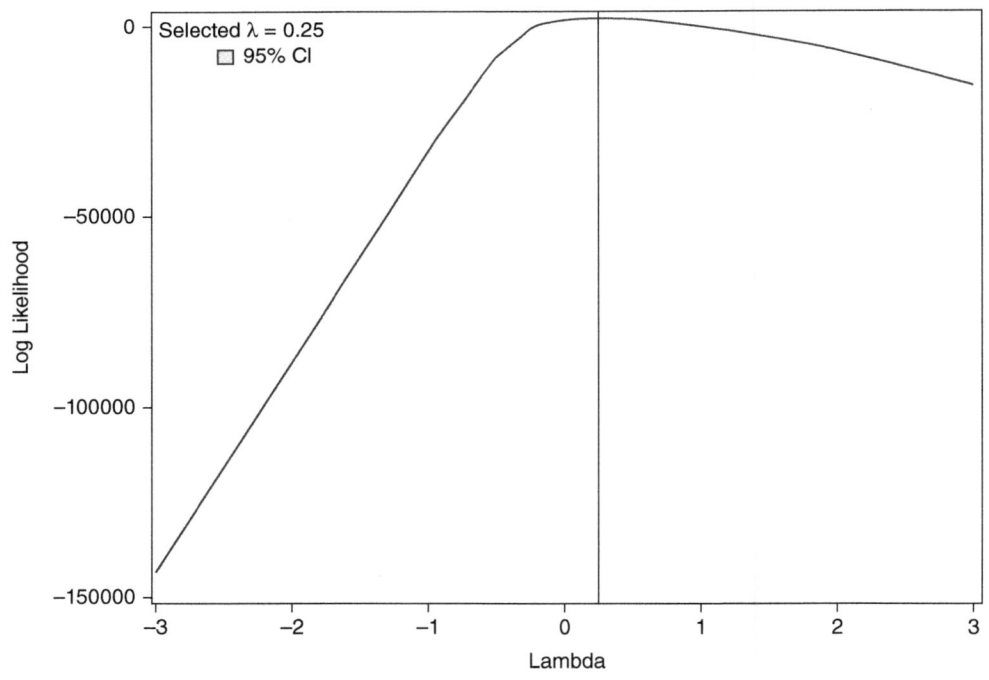

The best way to see how well the transformation did in making the data more normal is to look at plots of the data before transformation (not very normal!):

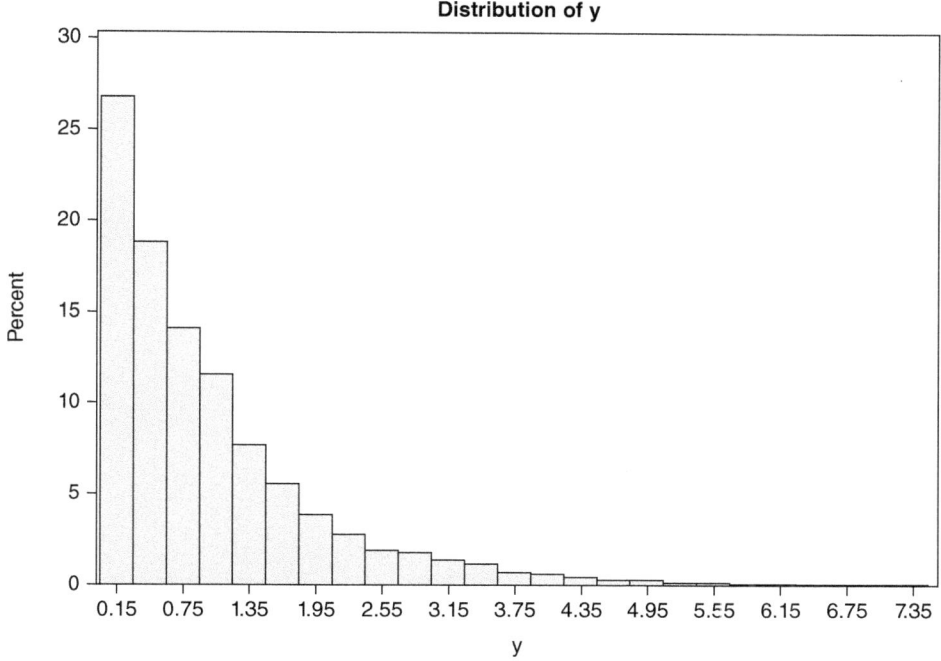

'Chapter 20 Example 5.sas'
'Generated non-normal data'
Box-Cox Transformation

The UNIVARIATE Procedure

Distribution of y

and after transformation (looks pretty normal):

'Chapter 20 Example 5.sas'
'Generated non-normal data'
Box-Cox Transformation

The UNIVARIATE Procedure

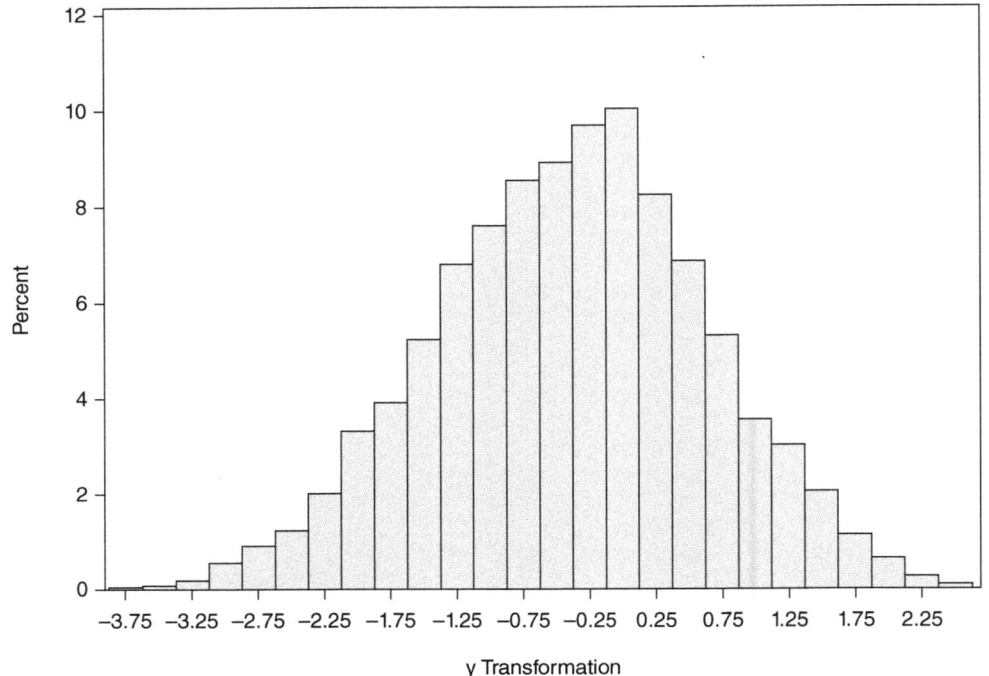

Distribution of Ty

y Transformation

Of course, you would transform the data and evaluate whether or not the transformed data satisfy the assumptions of normality, equal variances and no outliers.

Remedies for Outliers

Outliers are very common in most research work. While we can use Cook's D statistic along with residual plots to identify influential data points, it is up to the researcher to determine what, if any, remedy should be administered. Just because an observation lies outside the general range of the rest of the data does not mean that it is incorrect. Once a data point is identified as potentially influential, several steps should be taken to validate the data.

The first thing to do is to check to make sure the data was entered into the dataset correctly by making sure digits were not transposed, additional digits weren't added by mistake and that decimal points are in the right place. In addition, verify that the measurement recorded is a possible value for the observation; for example, if a variable can have a value between 20 and 50, and data point 10 has a value of 12, then it's simply incorrect. You should try to determine why it is incorrect if possible. Data may have been incorrectly recorded.

Once outliers are identified, their influence on the analysis is often examined by conducting the complete analysis with influential data included, followed by re-analysis of the data without the influential data included. The inferences from the two analyses are then compared. If multiple outliers are detected, the researcher should investigate why they occurred, not only that they occurred. Perhaps sloppy technique or poor data transcription is the cause. The source of outlier generation must be identified and rectified, but keep in mind that outliers cannot be corrected by simply discarding the data.

When no explanation for the outlier is revealed, the data should be included in the analysis. There is nothing wrong with exposing the data point for what it is in any discussion of results. In addition, presentation of results from both sets of analyses (with and without the outlier) is acceptable as long as this is made clear to the audience.

Presentation of Research Results

21

Introduction

The purpose of this chapter is to describe some general principles for the summarization and presentation of research data in a clear, coherent and concise manner. The first step is usually a statistical analysis (the preceding 20 chapters) followed by a second, equally important step: summarization and presentation of the results as text, tables or graphs.

Before deciding on a presentation style, it is imperative to determine who your audience is (lay or scientific) and what form of presentation you are preparing (oral or written). If the presentation is oral, familiarize yourself with the accepted style of presentation for your intended audience. The talk you present to a group of scientists is different than the one you would present to a garden club even though the topic might be the same for both. Scientific audiences tolerate more in-depth and detailed presentations, while lay audiences want 'just the facts', or in other words, 'What do I need to know and why do I need to know it?' If you are writing it is important to distinguish among styles for an article in the popular press, a thesis or dissertation or an article in a refereed scientific journal. Each outlet normally has a recommended style and depth of content and it's important to follow it.

In-text Presentation

Sometimes research results are important enough to be included in a written presentation but there is simply not enough data to warrant another form of presentation, such as a table or a graph. In these cases, insertion of relevant results directly into the text is needed. Be clear and succinct, making sure you present all the information needed in the discussion. This might include means, sample sizes, standard errors and α levels. This process can sometimes be difficult, since you want to be clear and concise without cluttering the text yet you must provide enough information to convince the reader of your assertions. The time spent on careful crafting of sentences is worth it.

Tables

Tables are an effective vehicle for summarizing many types of data and are often included in both written and oral presentations. Tables for oral presentations should follow the same general guidelines as those for written presentations. However, oral presentation tables are often smaller (in content) than written versions. I create my tables as if they were for a written presentation, then edit it into an oral version with the realization that I will be conveying much of the information verbally and can therefore have a simpler table. Tables for an oral presentation must be simple enough for clear presentation. There's nothing worse than seeing a table of 100 numbers and hearing the speaker say

DOI: 10.1079/9781789249927.0021

'I want you to look at this number' while trying to find the mean they are looking for among the hundred! Make sure you distill tables into manageable sizes to include only presentation-relevant information in them.

Table Tips

1. For either oral or written presentations, follow the format generally used for the intended audience.

2. For a written presentation, follow the specific format required for the intended publication whether it's a lay publication, a scientific journal article or a thesis or dissertation. Many journals provide downloadable templates for authors with explicit instructions for creating tables. Also, most colleges and universities have specific guidelines for thesis and dissertation formats.

3. Each column should have a heading with units of measurement. Don't separate the columns with vertical lines, they are distracting and most publishers don't want them.

4. Align decimals in all columns containing numbers.

5. Use a suitable number of digits in your number presentations, that is, one more place to the right than measured or calculated if reporting proportions.

6. Tables should present new information, not repeat information already presented in the text or graphs. Also make sure tables are numbered consecutively as they are presented in the text, and present them after mentioning them in the text.

7. Make sure that readers can interpret your tables without having to reference the text.

8. Most publishers want editable text-based tables. This means text created in one of the popular word-processing or spreadsheet programs.

9. If your table appears too cramped or cluttered, or if it doesn't appear that your table will fit onto one published page in your intended publication, consider reformatting the table or breaking it into two or perhaps three smaller tables.

10. Be consistent across tables in the same article with respect to font, font size, etc. Most templates will intrinsically set these to the default of the publication for which they were downloaded.

11. Notes used in tables are usually identified via superscripted letters of the alphabet from the upper left to lower right section of the table, as needed. Superscripts are usually identified in footnotes. Footnotes are also used to define acronyms or abbreviations.

12. Check the spelling in your tables. Double-check the data in your tables for accuracy.

Producing Tables with SAS®

Most tables for inclusion in oral or written presentations are produced using word-processing or spreadsheet software. Data for inclusion in the table are usually retrieved from statistical analyses that have provided the summary information such as means, sample size, etc. Rough sketches of tables are created from such output and final versions formatted using templates from the publisher. Sometimes it is convenient to use the table generation facilities of SAS® to review various configurations of the same data in order to determine which presentation is best suited for your needs. This is especially true for large datasets. Data summaries can be produced using SAS® rather than by hand, with the resulting tables inserted into a formatted table.

SAS® has a number of procedures available which can be used for generating tables and the same table may be generated using different techniques. For this chapter, we'll examine the SAS® procedure PROC TABULATE for generating tables from datasets. Be aware that books have been written on how to use the features of PROC TABULATE and we will only scratch the surface. It will, however, be enough of an introduction so that you can use the procedure to investigate different table formats for your presentations.

We will use the dataset from Chapter 20 in our examples.

```
1   TITLE1 'Chapter 21 Example 1.sas';
2   TITLE2 'RCBD, 5 blocks 20 treatments';
3   TITLE3 'Lettuce cultivar and seed source evaluation';
4   TITLE4 'Table Generation Using PROC TABULATE';
5   DATA one;
6   INPUT block treatment cultivar $ ssource $ cases @@;
7   CARDS;
8   1 1 j a 740 2 1 j a 655 3 1 j a 1162 4 1 j a 621 5 1 j a 601
9   1 2 j g 869 2 2 j g 759 3 2 j g 969 4 2 j g 630 5 2 j g 714
10  1 3 j j 972 2 3 j j 755 3 3 j j 545 4 3 j j 724 5 3 j j 807
11  1 4 j t 1075 2 4 j t 662 3 4 j t 547 4 4 j t 838 5 4 j t 911
12  1 5 j w 1179 2 5 j w 644 3 5 j w 652 4 5 j w 931 5 5 j w 489
13  1 16 n a 712 2 16 n a 570 3 16 n a 474 4 16 n a 1028 5 16 n a 615
14  1 17 n g 810 2 17 n g 756 3 17 n g 674 4 17 n g 1118 5 17 n g 611
15  1 18 n j 817 2 18 n j 805 3 18 n j 783 4 18 n j 918 5 18 n j 623
16  1 19 n t 354 2 19 n t 959 3 19 n t 773 4 19 n t 599 5 19 n t 713
17  1 20 n w 455 2 20 n w 1071 3 20 n w 685 4 20 n w 506 5 20 n w 826
18  1 6 p a 980 2 6 p a 655 3 6 p a 753 4 6 p a 1031 5 6 p a 523
19  1 7 p g 521 2 7 p g 748 3 7 p g 851 4 7 p g 202 5 7 p g 692
20  1 8 p j 527 2 8 p j 860 3 8 p j 951 4 8 p j 510 5 8 p j 709
21  1 9 p t 629 2 9 p t 951 3 9 p t 455 4 9 p t 502 5 9 p t 895
22  1 10 p w 734 2 10 p w 1066 3 10 p w 563 4 10 p w 502 5 10 p w 806
23  1 11 r a 829 2 11 r a 257 3 11 r a 656 4 11 r a 505 5 11 r a 399
24  1 12 r g 934 2 12 r g 566 3 12 r g 766 4 12 r g 515 5 12 r g 403
25  1 13 r j 402 2 13 r j 557 3 13 r j 863 4 13 r j 711 5 13 r j 600
26  1 14 r t 509 2 14 r t 568 3 14 r t 865 4 14 r t 816 5 14 r t 704
27  1 15 r w 608 2 15 r w 560 3 15 r w 368 4 15 r w 919 5 15 r w 703
28  RUN;
```

PROC TABULATE

A very simple program requesting a table of treatment means for the variable 'cases' can be generated with the following:

```
01  PROC TABULATE data = one;
02  CLASS treatment;
03  VAR cases;
```

```
04 TABLE cases*mean*treatment;
05 RUN;
```
Line 1 invokes the TABULATE procedure using dataset 'one'. To indicate the categories upon which we want our table built, we use the statement 'CLASS treatment' in line 2 and indicate the dependent variable 'cases' in line 3 as the variable for presentation. The TABLE keyword of line 4 instructs SAS® how to present the table: a table of cases means for the treatments in the dataset.

The output:

'Chapter 21 Example 1.sas'
'RCBD, 5 blocks 20 treatments'
'Lettuce cultivar and seed source evaluation'
'Table Generation Using PROC TABULATE'

										cases									
										Mean									
										treatment									
1	2	3	4	5	6	7	8	9	10	11	12	13	14	15	16	17	18	19	20
755.80	788.20	760.60	806.60	779.00	788.40	602.80	711.40	686.40	734.20	529.20	636.80	626.60	692.40	631.60	679.80	793.80	789.20	679.60	708.60

By changing the TABLE statement in line 4, we can change the look of our table to vertical rather than horizontal:

```
01 TITLE1 'Chapter 21 Example 2.sas';
02 PROC TABULATE data = one;
03 CLASS treatment;
04 VAR cases;
05 TABLE treatment, cases*mean;
06 RUN;
```
which produces:

treatment	cases Mean
1	755.80
2	788.20
3	760.60
4	806.60
5	779.00
6	788.40
7	602.80
8	711.40
9	686.40
10	734.20
11	529.20
12	636.80
13	626.60
14	692.40
15	631.60
16	679.80
17	793.80
18	789.20
19	679.60
20	708.60

Now let's consider cultivar and seed source in generating our table:

```
01 TITLE1 'Chapter 21 Example 3.sas';
02 PROC TABULATE data = one;
03 CLASS cultivar ssource;
04 VAR cases;
05 TABLE cultivar*ssource*cases*mean;
06 RUN;
```

Note the changes in lines 3 and 5 to include 'cultivar' and 'ssource' rather than treatment in our table. This produces:

'Chapter 21 Example 3.sas'

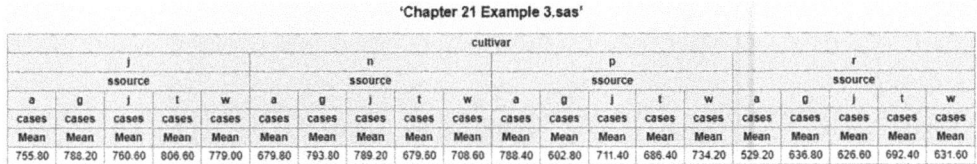

cultivar																			
j					n					p					r				
ssource					ssource					ssource					ssource				
a	g	j	t	w	a	g	j	t	w	a	g	j	t	w	a	g	j	t	w
cases	cases	cases	cases	cases	cases	cases	cases	cases	cases	cases	cases	cases	cases	cases	cases	cases	cases	cases	cases
Mean	Mean	Mean	Mean	Mean	Mean	Mean	Mean	Mean	Mean	Mean	Mean	Mean	Mean	Mean	Mean	Mean	Mean	Mean	Mean
755.80	788.20	760.60	806.60	779.00	679.80	793.80	789.20	679.60	708.60	788.40	602.80	711.40	686.40	734.20	529.20	636.80	626.60	692.40	631.60

However, we would probably want either 'cultivar' or 'ssource' as the first column and the remaining category as the header for the remaining columns.

The code for 'cultivar' as the first column:

```
01 TITLE1 'Chapter 21 Example 4.sas';
02 PROC TABULATE data = one;
03 CLASS cultivar ssource;
04 VAR cases;
05 TABLE cultivar, ssource*cases*mean;
06 RUN;
```

which produces:

'Chapter 21 Example 4.sas'

	ssource				
	a	g	j	t	w
	cases	cases	cases	cases	cases
	Mean	Mean	Mean	Mean	Mean
cultivar					
j	755.80	788.20	760.60	806.60	779.00
n	679.80	793.80	789.20	679.60	708.60
p	788.40	602.80	711.40	686.40	734.20
r	529.20	636.80	626.60	692.40	631.60

The code for 'ssource' as the first column:

```
01 TITLE1 'Chapter 21 Example 5.sas';
02 PROC TABULATE data = one;
03 CLASS cultivar ssource;
04 VAR cases;
05 TABLE ssource, cultivar*cases*mean;
06 RUN;
```

which produces:

'Chapter 21 Example 5.sas'

	cultivar			
	j	n	p	r
	cases	cases	cases	cases
	Mean	Mean	Mean	Mean
ssource				
a	755.80	679.80	788.40	529.20
g	788.20	793.80	602.80	636.80
j	760.60	789.20	711.40	626.60
t	806.60	679.60	686.40	692.40
w	779.00	708.60	734.20	631.60

By changing line 5, many different table configurations can be evaluated to determine which looks best for your purposes. Note that PROC TABULATE calculates a number of different statistics which you can request for inclusion in your table by replacing the 'means' in line 5 with the statistic of interest. You can request more than one statistic at a time for inclusion in a table by changing line 5 to read (for example):

```
05 TABLE ssource, cultivar*cases*(mean n);
```

which produces:

'Chapter 21 Example 6.sas'

	cultivar							
	j		n		p		r	
	cases		cases		cases		cases	
	Mean	N	Mean	N	Mean	N	Mean	N
ssource								
a	755.80	5	679.80	5	788.40	5	529.20	5
g	788.20	5	793.80	5	602.80	5	636.80	5
j	760.60	5	789.20	5	711.40	5	626.60	5
t	806.60	5	679.60	5	686.40	5	692.40	5
w	779.00	5	708.60	5	734.20	5	631.60	5

The statistics available include, but are not limited to: MEDIAN, MAX, MIN, MEAN, N, STDDEV, STDERR and SUM. For a complete list go to: https://documentation.sas.com/?docsetId=proc&docsetTarget=n00yutbvvckjwrn1ldg5xkvjy1pu.htm&docsetVersion=9.4&locale=en

There are many formatting options available in SAS® that would make the tables more attractive, but we can't possibly cover them here. In addition, the present context of using PROC TABULATE is as a tool to get a general idea of row and column arrangements for table production in word-processing or spreadsheet software. An excellent starting point for using some of the fancier features of PROC TABULATE can be found by Eberhart (2010) at:

https://www.researchgate.net/publication/259147230_PROC_TABULATE_-_Building_Tables_With_Style

SAS® Graphs

Many different types of graphs are available for representing data from research, but a limited number of graph styles are normally used for most plant science presentations. We will discuss the appropriate use of each type of graph for different situations as well as the methodology of producing them with SAS® using a general format. Publication-specific alterations to the generalized format are usually fairly easy to accommodate. An excellent reference for producing graphs with SAS® is *Statistical Graphics Procedures by Example: Effective Graphs Using SAS®* by Matange and Heath (2011).

The five graph styles we'll examine include scatter graphs (also known as an X-Y graph), line graphs, regression lines, response surfaces and bar charts.

Before we discuss particular types and examples, let's review a few tips regarding graph use in plant science presentations.

Graph Tips

1. For either oral or written presentations, follow formats generally used for the intended audience.

2. For a written presentation, follow the specific format required for the intended publication.

3. Graphs should present new information, not repeat information already presented in the text or tables. Also make sure graphs are numbered consecutively as they are presented in the text, and present them after mentioning them in the text.

4. Make sure that readers can interpret your tables without having to reference the text.

5. Include all necessary information such as labels for both axes, measurement units, symbol identification in a clear legend, error bar labels, mean separation labels, etc. so that your audience does not have to guess about any aspect of your graph.

6. Minimize the amount of eye movement around the graph that is needed for correct interpretation of the information in the graph. For example, excessive eye movement may occur if one has to keep referring to a legend rather than simply reading a label placed effectively on the graph.

7. Keep all graphs for publication black and white, even if your publisher offers color printing options. Even though your color graphs may look great, remember that many folks might be looking at photocopies or black and white printouts of your work. Additionally, if you are using bars in a graph don't use shades of grey, but rather use different, easily discernible patterns. Shades of grey are often hard to discern when photocopied or printed out on a lower-resolution printer.

8. For oral presentations use color tastefully. Too much color and poorly chosen color combinations are distracting and annoying. Most software used for graphic generation has default well-planned, color combinations. Use them. Don't use red and blue close together on the same graph, since humans have a hard time focusing on both colors at the same time. Also, don't use red and green for bar charts or symbols, since up to 10% of the males in your audience might have red–green colorblindness.

9. Be consistent across graphs in the same article with respect to font, font size, etc.

10. Check the spelling in your graphs and double-check the data used in generating them.

11. Don't clutter up your graphs with grid lines, background patterns, 3-D effects, unnecessary legends, arrows, extra text, etc. Keep your graph as clean and concise as possible while adhering to all the rules above.

The Output Delivery System (ODS) Graphics system of SAS®

SAS® utilizes what is called the ODS Graphics system to generate and present graphics for many of its procedures. All you have to do to have the graphics associated with any SAS® procedure generated is to make sure that the system is turned on. You do this by adding the following statement at the beginning of your SAS® program:

```
ODS GRAPHICS ON;
```

The graphs generated by different SAS® procedures automatically are created internally by SAS® using the Graph Template Language (GTL) which is part of SAS®. The GTL is also used in many of the Statistical Graphics (SG) procedures which we will shortly cover. These procedures are part of the ODS Graphics system and produce graphs very similar to the automated graphs produced by SAS®.

The approach we will take in this text is similar to that of the reference cited above. Rather than presenting all possible options for each graph we cover, we will present an example that produces a graph with all of the essential elements that you can modify to suit your needs. If you need more information or options, see the reference or other SAS® resources.

The Basics of PROC SGPLOT

The SAS® procedure we most often use for graphs is PROC SGPLOT. It consists of five main statements that are used to create and modify basic graphs. All five may or may not be required for your specific graph. The only statement which is absolutely required within PROC SGPLOT is the PLOT statement. The other statements (REFLINE, INSET, AXIS and KEYLEGEND) may or may not be needed for your graph.

The general syntax of a SAS® program for creating a graph (Example 5) is:

```
01 PROC SGPLOT DATA = data-set < options >;
02 PLOT required-parameters < / options >;
03 REFLINE statement(s);
04 INSET statement(s);
05 AXIS statement(s);
06 KEYLEGEND statement(s);
07 RUN;
```

Remember, only lines 1, 2 and 7 are required, while lines 3 through 6 are optional.

Graphs Styles

Scatter plot (also known as an X-Y graph)

This is your basic plot for pairs of numbers with often inferred causality of X on Y. The horizontal axis is X and the vertical axis is Y.

```
01 TITLE1 'Chapter 21, Example 7.sas';
02 TITLE2 'Basic Scatter Plot';
03 DATA one;
04 INPUT cultivar $ x y @@;
05 CARDS;
06 a 0 13.8 a 0 13.5 b 0 13.2 b 0 14.0 c 0 10 c 0 9.5
07 a 10 15.5 a 10 15 b 10 15.2 b 10 14.9 c 10 12.0 c 10 14.7
08 a 20 21 a 20 22.7 b 20 24.3 b 20 21.7 c 20 19.7 c 20 19.1
09 a 30 18.9 a 30 18.3 b 30 19.6 b 30 18.7 c 30 17.2 c 30 18.0
10 RUN;
```

```
11 PROC SGPLOT DATA=one;
12 SCATTER x=x y=y;
13 RUN;
```
Output:

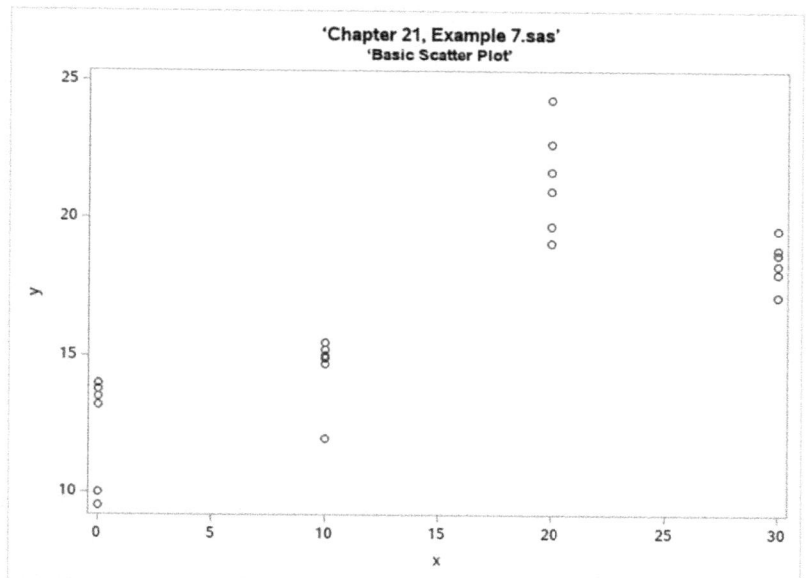

We can add some code to have different symbols for the three different cultivars. We must add the code of line 1 to change a characteristic of the ODS system so that we can specify the symbols to use for the different cultivars. We specify the symbols we want to use for the cultivars in line 13. In this line we set the data symbol style attributes to the values in parentheses (x y z). If you wanted to use a, b and c to indicate the cultivars, you could change the values in parentheses to (a b c).

```
01 ODS GRAPHICS ON / ATTRPRIORITY=none;
02 TITLE1 'Chapter 21, Example 8.sas';
03 TITLE2 'Basic Scatter Plot';
04 DATA one;
05 INPUT cultivar $ x y @@;
06 CARDS;
07 a 0 13.8 a 0 13.5 b 0 13.2 b 0 14.0 c 0 10 c C 9.5
08 a 10 15.5 a 10 15 b 10 15.2 b 10 14.9 c 10 12.0 c 10 14.7
09 a 20 21 a 20 22.7 b 20 24.3 b 20 21.7 c 20 19.7 c 20 19.1
10 a 30 18.9 a 30 18.3 b 30 19.6 b 30 18.7 c 30 17.2 c 30 18.0
11 RUN;
12 PROC SGPLOT DATA = one;
13 STYLEATTRS DATASYMBOLS = (x y z);
14 SCATTER x=x y=y / GROUP = cultivar;
15 RUN;
```
The output:

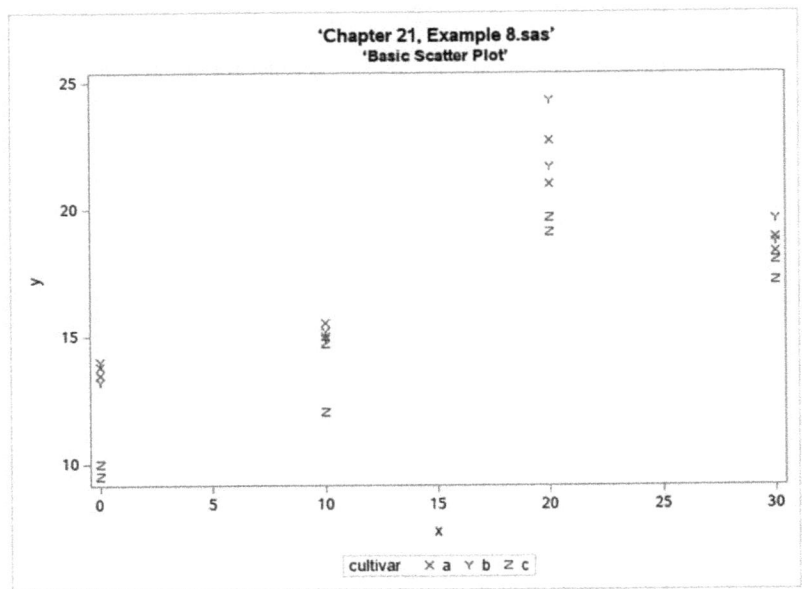

Now, let's move the legend (changing the label from cultivar with a lower case 'c' to Cultivar with an uppercase 'C' (TITLE='Cultivar: ') to the upper right position (POSITION=topright) inside the graph frame (LOCATION=inside) and change it from a horizontally presented legend to a vertically presented legend (ACROSS=1), all options of the KEYLEGEND statement, by inserting line 15, and give the X and Y axes new names with lines 16 and 17, respectively.

```
01 ODS GRAPHICS ON / ATTRPRIORITY=none;
02 TITLE1 'Chapter 21, Example 9.sas';
03 TITLE2 'Basic Scatter Plot';
04 DATA one;
05 INPUT cultivar $ x y @@;
06 CARDS;
07 a 0 13.8 a 0 13.5 b 0 13.2 b 0 14.0 c 0 10 c 0 9.5
08 a 10 15.5 a 10 15 b 10 15.2 b 10 14.9 c 10 12.0 c 10 14.7
09 a 20 21 a 20 22.7 b 20 24.3 b 20 21.7 c 20 19.7 c 20 19.1
10 a 30 18.9 a 30 18.3 b 30 19.6 b 30 18.7 c 30 17.2 c 30 18.0
11 RUN;
12 PROC SGPLOT DATA = one;
13 STYLEATTRS DATASYMBOLS = (x y z);
14 SCATTER x=x y=y / GROUP = cultivar;
15 KEYLEGEND / TITLE='Cultivar: ' LOCATION=inside POSITION=topright
   ACROSS=1;
16 XAXIS LABEL="Independent Variable";
17 YAXIS LABEL="Dependent Variable";
18 RUN;
```

Output:

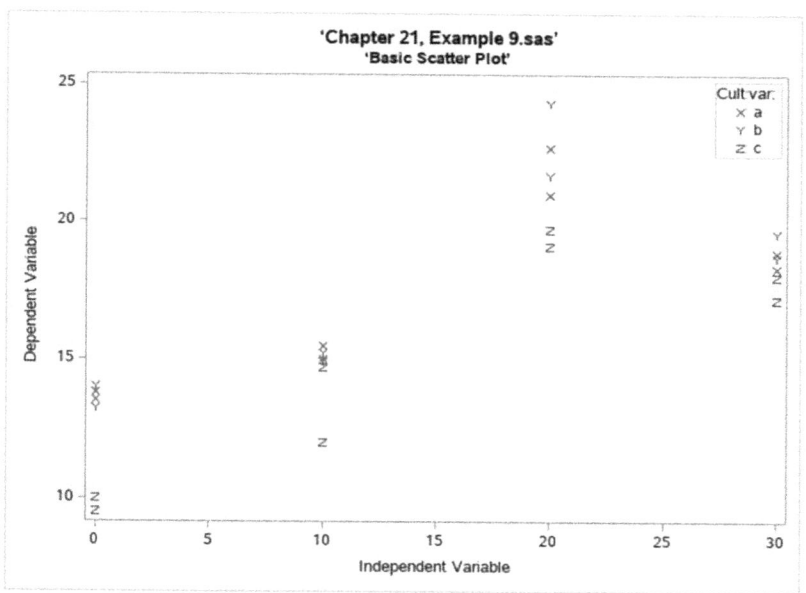

'Chapter 21, Example 9.sas'
'Basic Scatter Plot'

With only a few lines of code, we have been able to create a nice scatter plot of our data. There are many other options available to modify the appearance of your graph as well as how the image created is output. You can request graphics output in many different formats but probably the most useful format is a pdf file. Once you have a pdf file of your graph, it can be imported into many word-processing, spreadsheet and presentation software packages.

To create a pdf of your graph, use the following code, paying attention to lines 2, 20 and 21. Line 2 creates the pdf file and sends it to your 'myfolders' folder you set up when you installed SAS®. You must put the pdf file in this folder or you will get an error message saying that you have insufficient authorization to access the file. Don't worry about it, just make sure you indicate the correct folder. You can change the filename 'Graphpdfsample' to one of your choosing if you wish. Line 20 ends any SAS® procedures that are still open (this just prevents problems later in the same SAS® session) and line 21 turns off the pdf output option of the ODS.

```
01 ODS GRAPHICS ON / ATTRPRIORITY=none;
02 ODS pdf file='/folders/myfolders/Graphpdfsample.pdf';
03 TITLE1 'Chapter 21, Example 10.sas';
04 TITLE2 'Basic Scatter Plot';
05 DATA one;
06 INPUT cultivar $ x y @@;
07 CARDS;
08 a 0 13.8 a 0 13.5 b 0 13.2 b 0 14.0 c 0 10 c C 9.5
09 a 10 15.5 a 10 15 b 10 15.2 b 10 14.9 c 10 12.0 c 10 14.7
10 a 20 21 a 20 22.7 b 20 24.3 b 20 21.7 c 20 19.7 c 20 19.1
11 a 30 18.9 a 30 18.3 b 30 19.6 b 30 18.7 c 30 17.2 c 30 18.0
12 RUN;
13 PROC SGPLOT DATA=one;
```

```
14 STYLEATTRS DATASYMBOLS=(x y z);
15 SCATTER x=x y=y / GROUP=cultivar;
16 KEYLEGEND / TITLE='Cultivar: ' LOCATION=inside POSITION=topright
   ACROSS=1;
17 XAXIS LABEL="Independent Variable";
18 YAXIS LABEL="Dependent Variable";
19 RUN;
20 QUIT;
21 ODS pdf close;
```
Now that we are familiar with how to create a simple scatter graph, let's move on to some other graphs that are often used for presenting results of plant science research.

Line graphs

Line graphs are very similar to scatter plots with the basic difference that related data points are connected via a line. Suppose that in our previous example we wanted to connect the values of Y across the values of X for each cultivar. This would create a series of three plots all on the same graph. These types of graphs are appropriately called series plots.

Using the previous example's data, creating a series of three plots would be appropriate if a plot of the relationship between the mean response of Y at each level of X was desired for each cultivar. In order to plot the means, we would generate their values in lines 12 through 17 in the following code, then proceed with the plot requests in lines 18 through 24.

In lines 12 and 13 we request the means for variable 'y' in dataset one. Line 14 identifies two class variables, cultivar and x, so that we can generate the means of all cultivar*x combinations in line 15 using the TYPES keyword. We want the means saved to an output dataset we are calling 'two' and we are labeling the means 'y' with the (MEAN=y) instruction. We request the series plots in line 15, one for each cultivar.

```
01 ODS GRAPHICS ON / ATTRPRIORITY=none;
02 TITLE1 'Chapter 21, Example 11.sas';
03 TITLE2 'Basic Series Plot';
04 DATA one;
05 INPUT cultivar $ x y @@;
06 CARDS;
07 a 0 13.8 a 0 13.5 b 0 13.2 b 0 14.0 c 0 10 c 0 9.5
08 a 10 15.5 a 10 15 b 10 15.2 b 10 14.9 c 10 12.0 c 10 14.7
09 a 20 21 a 20 22.7 b 20 24.3 b 20 21.7 c 20 19.7 c 20 19.1
10 a 30 18.9 a 30 18.3 b 30 19.6 b 30 18.7 c 30 17.2 c 30 18.0
11 RUN;
12 PROC MEANS data = one;
13 VAR y;
14 CLASS cultivar x;
15 TYPES cultivar*x;
16 OUTPUT OUT = two MEAN=y;
17 RUN;
```

```
18 PROC SGPLOT DATA = two;
19 STYLEATTRS DATASYMBOLS=(x y z);
20 SERIES x=x y=y / GROUP=cultivar;
21 KEYLEGEND / TITLE='Cultivar: ' LOCATION=inside POSITION=topright
   ACROSS=1;
22 XAXIS LABEL="Independent Variable";
23 YAXIS LABEL="Dependent Variable";
24 RUN;
```

Output:

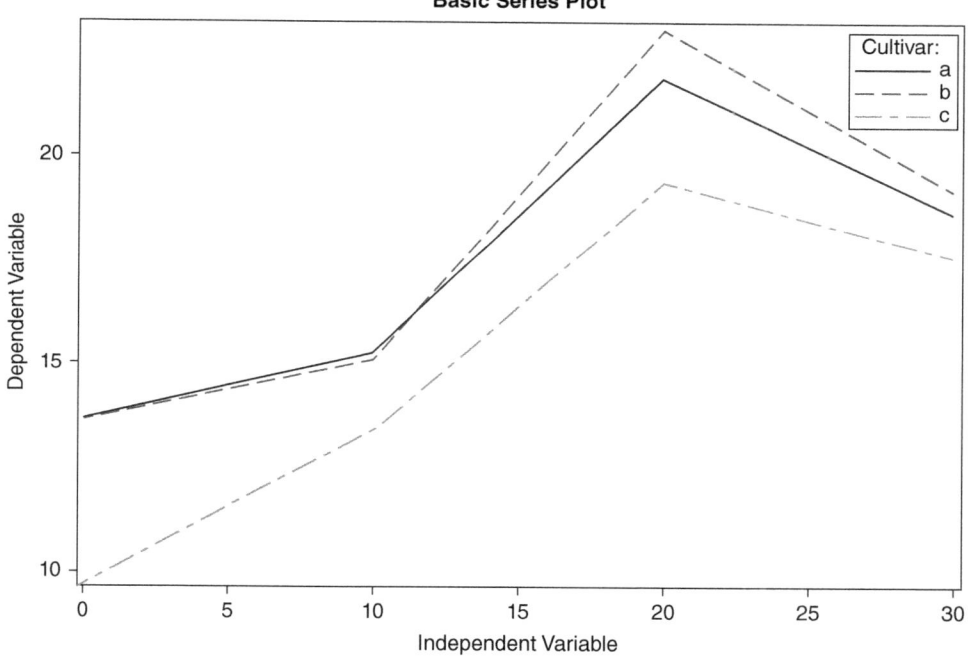

For a simple line plot of the mean response combining all three cultivars, you could use the following code. We request the means at each level of x with line 15 and simplify the plotting instructions in line 19.

```
01 ODS GRAPHICS ON / ATTRPRIORITY=none;
02 TITLE1 'Chapter 21, Example 12.sas';
03 TITLE2 'Basic Series Plot';
04 DATA one;
05 INPUT cultivar $ x y @@;
06 CARDS;
07 a 0 13.8 a 0 13.5 b 0 13.2 b 0 14.0 c 0 10 c C 9.5
08 a 10 15.5 a 10 15 b 10 15.2 b 10 14.9 c 10 12.0 c 10 14.7
09 a 20 21 a 20 22.7 b 20 24.3 b 20 21.7 c 20 19.7 c 20 19.1
10 a 30 18.9 a 30 18.3 b 30 19.6 b 30 18.7 c 30 17.2 c 30 18.0
11 RUN;
```

```
12 PROC MEANS data = one;
13 VAR y;
14 CLASS cultivar x;
15 types x;
16 OUTPUT OUT = two MEAN=y;
17 RUN;
18 PROC SGPLOT DATA = two;
19 SERIES x=x y=y ;
20 XAXIS LABEL="Independent Variable";
21 YAXIS LABEL="Dependent Variable";
22 RUN;
```

Output:

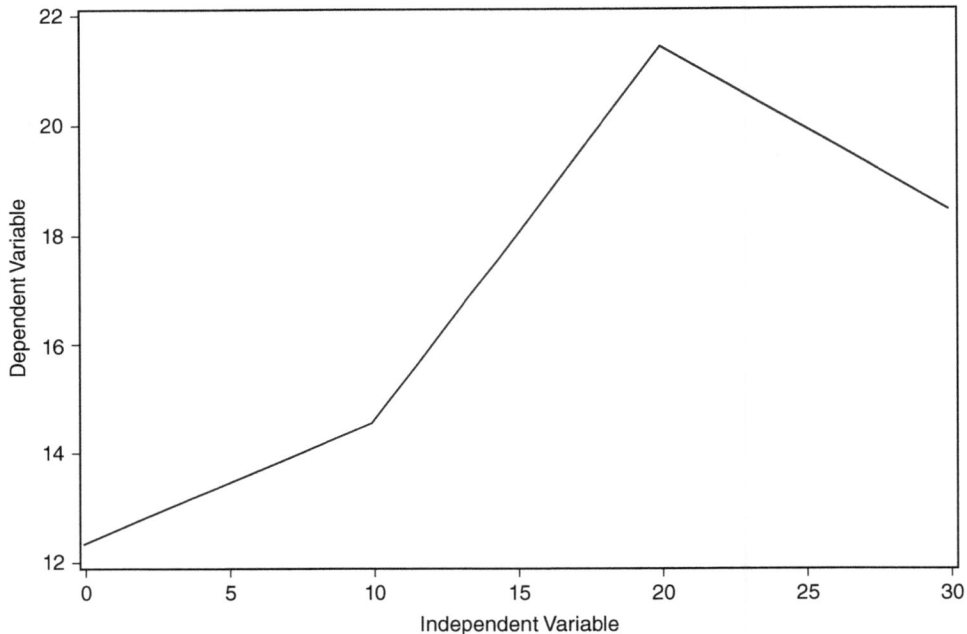

'Chapter 21, Example 12.sas'
'Basic Series Plot'

Regression lines

In Chapter 9 we discussed regression analysis and some of the plots associated with it. The following discussion will focus on the plotting of regression analyses, not the methodology of the analysis itself. Review Chapter 9 for that discussion.

The following is Example 10 from Chapter 9. We have renamed it to fit this chapter.

```
01 ODS GRAPHICS ON / ATTRPRIORITY=none;
02 TITLE1 'Chapter 21 Example 13.sas';
03 TITLE2 'Regression and PROC SGPLOT';
```

```
04 DATA one;
05 INPUT x y @@;
06 LABEL x = pH;
07 LABEL y = Biomass (kg);
08 CARDS;
09 4.5 23 4.9 41 4.9 40 4.8 38 4.8 34 4.8 37
10 5.1 47 5.2 52 5.2 50 5.3 58 5.4 61 5.5 63
11 RUN;
12 PROC SGPLOT DATA = one;
13 REG x = x y = y / CLM NOLEGCLM NOLEGFIT;
14 RUN;
```

which produces:

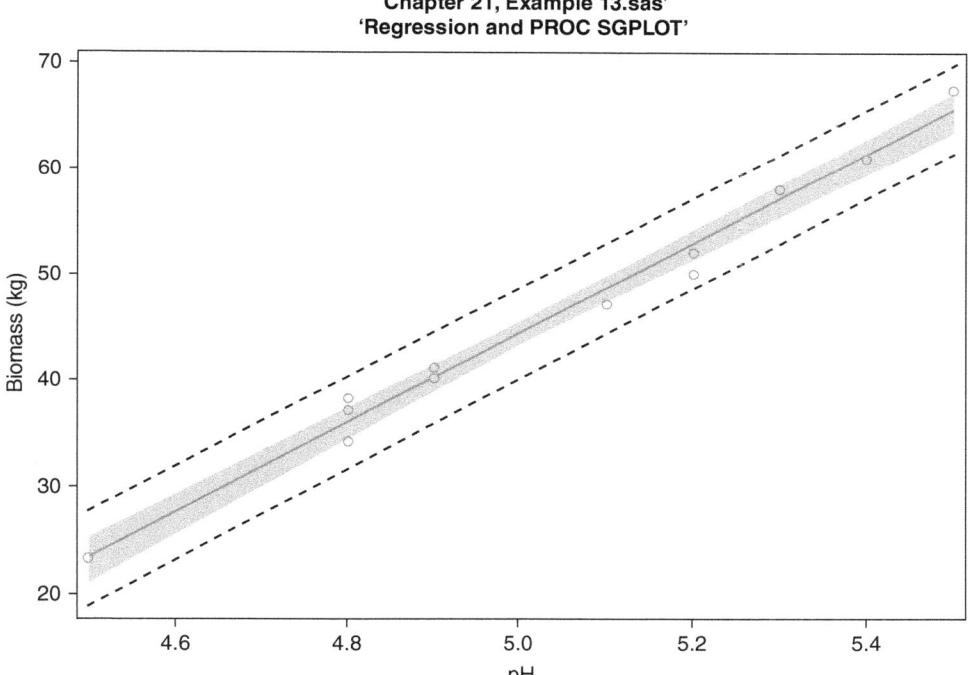

'Chapter 21, Example 13.sas'
'Regression and PROC SGPLOT'

All that is needed for enhancement is a legend indicating the line equation with a description of the confidence bands for the estimated line. The automatic legends produced by PROC SGPLOT are not sufficient for our needs, as they are too rudimentary. We will create our own 'legend' and place it on our graph using INSET statements within PROC SGPLOT. It takes a small effort, but in the end you will have a very nice graph ready for publication.

There are four main steps to producing this graph. The first step is completing a full regression analysis with diagnostics. If you haven't, refer to Chapter 9 and complete this step.

```
01 TITLE1 'Chapter 21, Example 14.sas';
02 TITLE2 'Regression and PROC SGPLOT';
03 DATA one;
04 INPUT x y @@;
05 LABEL x = pH;
06 LABEL y = Biomass (kg);
07 CARDS;
08 4.5 23 4.9 41 4.9 40 4.8 38 4.8 34 4.8 37
09 5.1 47 5.2 52 5.2 50 5.3 58 5.4 61 5.5 68
10 RUN;
```

The second step is to save the parameter estimates as well as the R^2 value (since this is often presented with the regression line equation) for your regression line in SAS® datasets. This is accomplished with the ODS OUPUT statement of line 14 below. As a side note, be aware that there are many different OUTPUT datasets created by SAS® during the execution of a procedure such as REG or ANOVA. In order to know what they contain and how to access them, refer to online SAS® help for the procedure of interest.

```
11. ODS GRAPHICS OFF;
12. PROC REG DATA = one;
13. MODEL y = x;
14. ODS OUTPUT ParameterEstimates=ParmEst FitStatistics=FitStat;
15. RUN;
```

Step 3 consists of saving the parameter estimates as variables which can be used in a SAS® Macro statement (a convenient way to reuse SAS® code).

```
16 DATA tempo;
17 SET ParmEst;
18 IF _n_ = 1 THEN CALL symput('Int', PUT(estimate, BEST6.));
19 ELSE CALL symput('Slope', PUT(estimate, BEST6.));
20 RUN;
21 DATA tempo2;
22 SET FitStat;
23 IF _n_ = 1 THEN CALL symput('R2', PUT(nValue2, BEST6.));
24 RUN;
```

Step 4 uses the INSET option in PROC SGPLOT to place a text box on the graph using the following code:

```
25 ODS GRAPHICS ON / ATTRPRIORITY=none;
26 PROC SGPLOT DATA = one;
27 REG y=y x=x / CLM NOLEGFIT;
28 KEYLEGEND / NOBORDER POSITION=bottomright LOCATION=inside
   ACROSS=1;
29 INSET "Y = &Int + &Slope" "R2 = &R2" / POSITION =topleft;
30 RUN;
```

The &Int, &Slope and &R2 values in line 29 are using the values assigned to these MACRO variables to print the text box at the upper left of the graph using the INSET statement. If this is confusing, don't worry about it. As long as you follow the steps above, replacing your values in the appropriate places, you'll get the results you want.

Here is the output:

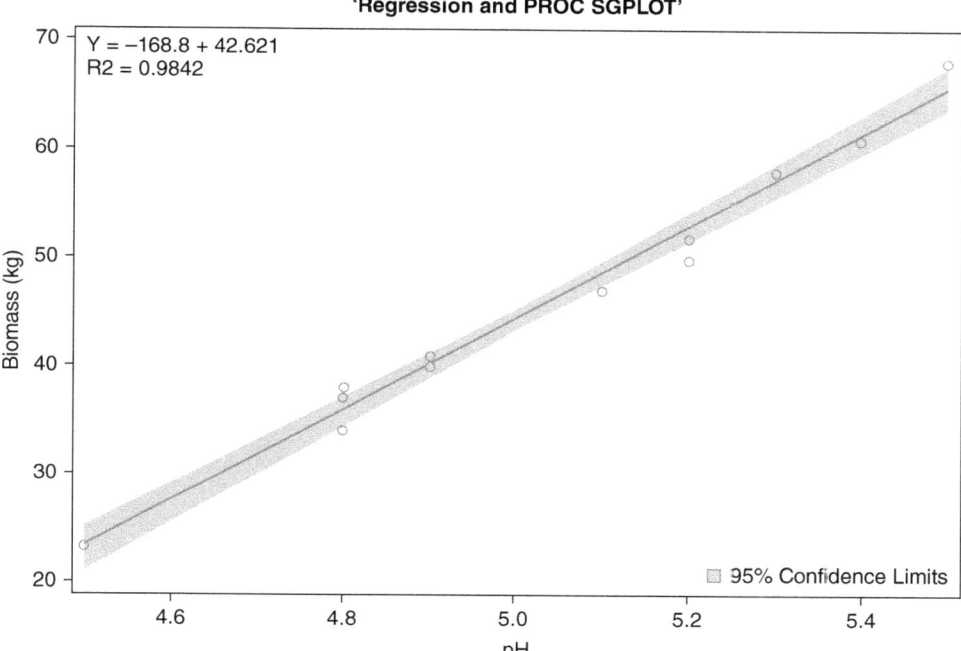

'Chapter 21, Example 14.sas'
'Regression and PROC SGPLOT'

Response surface graphs

Even though we did not cover response surface regression methodology in Chapter 9, it is fitting to present a short discussion of producing 3-D response surface graphs using SAS. A response surface regression develops an equation for a dependent variable using two independent, quantitative variables (such as temperature, light level, fertility rate, seeding density, etc.). The model most often used is the full quadratic response surface model which considers the linear and quadratic components of each independent variable as well as the interactions of the two.

Suppose we have data which include the response variable Y and two independent variables, X_1 and X_2. In order to perform a response surface regression, we would create the quadratic components (X_1^2 and X_2^2) and the interactions of the two independent variables ($X_1 X_2$ and X_1^2 X_2^2) in our DATA statement in the program below.

The equation of a complete quadratic response surface is:

$$Y = \beta_0 + \beta_1 X_1 + \beta_2 X_2 + \beta_3 X_1^2 + \beta_4 X_2^2 + \beta_5 X_1 X_2 + \beta_6 X_1^2 X_2^2 + \varepsilon_i$$

which we would model with the SAS® statement:

MODEL y = x1 x2 x1x1 x2x2 x1x2 x1x1x2x2;

Ok, so how do we perform a response surface regression analysis on this data? We would run the complete model:

MODEL y = x1 x2 x1x1 x2x2 x1x2 x1x1x2x2;

Evaluate the P>F for the quadratic interaction component (the last component in the model statement, *x1x1x2x2*). If it is significant, your modeling task is over and you have your equation. If it is not significant at the desired level (most often 0.05), remove it from the model statement and re-run the regression with the new reduced model statement:

MODEL y = x1 x2 x1x1 x2x2 x1x2;

Now evaluate the P>F for the quadratic components and the linear interaction component (*x1x1 x2x2 x1x2*). If all three are significant, your modeling is complete. If any of them are not significant, remove the least significant one from the model statement and re-run the new model. Evaluate the significance of the two components remaining from the original three (*x1x1 x2x2 x1x2*). If both are significant, you are done. If not, delete the least significant component and re-run the model. Evaluate the remaining component, keep it if it is significant, delete it if it is not. Now you should only have to evaluate the significance of (*x1 x2*). Remember that linear component must remain in model (even if non-significant) if quadratic component for that variable or the linear interaction is significant.

This rule in polynomial regression (response surface regression is a form of multiple polynomial regression) is very important. You must include lower-order components in models with significant higher-order components. If you run the model in a stepwise fashion using the model with components in the order listed, you won't have a problem with this. Just remember these rules:

- If the $X_1^2 X_2^2$ component is significant, stop. All other components must remain in the model.
- If the X_1^2, X_2^2 or $X_1 X_2$ components are all significant, stop. All other components must remain in the model.
- If the X_1^2 component is significant, X_1 must remain whether it is significant or not.
- If the X_2^2 component is significant, X_2 must remain whether it is significant or not.
- If the $X_1 X_2$ component is significant, both X_1 and X_2 must remain whether or not either are significant.
- If only X_1 and X_2 remain in the model, evaluate each separately for significance.

This process takes time and effort but is not difficult. It is presented here in greatly abbreviated form. If greater detail is needed, the reader should consider a good regression text such as *Applied Regression Analysis* by Draper and Smith (1998) or *Response Surface Methodology: Process and Product Optimization Using Designed Experiments* by Myers *et al.* (2016).

```
01  TITLE1 'Chapter 21, Example 15.sas';
02  TITLE2 'Response Surface Regression and PROC SGPLOT';
03  DATA one;
04  INPUT x1 x2 y @@;
05  LABEL x1 = 'Temperature';
06  LABEL x2 = 'Nitrogen Rate';
07  LABEL y = 'Yield';
08  x1x1 = x1*x1;
09  x2x2 = x2*x2;
10  x1x2 = x1*x2;
11  x1x1x2x2 = x1x1*x2x2;
12  CARDS;
13  100 0 11.1 200 0 4.1 300 0 10.1 400 0 5.1 500 0 14.1 600 0 21.1
14  100 50 16.6 200 50 9.6 300 50 9.6 400 50 10.6 500 50 13.6 600 50 16.6
15  100 100 22.1 200 100 14.1 300 100 10.1 400 100 14.1 500 100 7.1
    600 100 14.1
16  100 150 20.6 200 150 16.6 300 150 9.6 400 150 13.6 500 150 11.6
    600 150 16.6
17  100 0 11.1 200 0 8.1 300 0 5.1 400 0 10.1 500 0 15.1 600 0 19.1
18  100 50 13.6 200 50 6.6 300 50 13.6 400 50 8.6 500 50 11.6 600
    50 14.6
19  100 100 21.1 200 100 15.1 300 100 12.1 400 100 13.1 500 100 12.1
    600 100 12.1
20  100 150 22.6 200 150 15.6 300 150 11.6 400 150 14.6 500 150 13.6
    600 150 8.6
21  100 0 9.1 200 0 10.1 300 0 5.1 400 0 6.1 500 0 15.1 600 0 21.1
22  100 50 10.6 200 50 12.6 300 50 13.6 400 50 5.6 500 50 6.6 600 50
    13.6
23  100 100 16.1 200 100 9.1 300 100 12.1 400 100 12.1 500 100 10.1 600
    100 13.1
24  100 150 22.6 200 150 18.6 300 150 7.6 400 150 5.6 500 150 11.6 600
    150 11.6
25  100 0 15.1 200 0 13.1 300 0 4.1 400 0 13.1 500 0 7.1 600 0 14.1
26  100 50 12.6 200 50 9.6 300 50 6.6 400 50 12.6 500 50 14.6 600 50
    18.6
27  100 100 14.1 200 100 11.1 300 100 8.1 400 100 10.1 500 100 15.1 600
    100 12.1
28  100 150 22.6 200 150 11.6 300 150 13.6 400 150 9.6 500 150 7.6 600
    150 16.6
29  100 0 9.1 200 0 7.1 300 0 4.1 400 0 4.1 500 0 13.1 600 0 16.1
30  100 50 12.6 200 50 9.6 300 50 13.6 400 50 7.6 500 50 10.6 600 50
    18.6
31  100 100 14.1 200 100 15.1 300 100 14.1 400 100 7.1 500 100 12.1 600
    100 12.1
32  100 150 21.6 200 150 17.6 300 150 13.6 400 150 9.6 500 150 9.6 600
    150 16.6
```

```
33  100 0 16.1 200 0 12.1 300 0 9.1 400 0 9.1 500 0 9.1 600 0 12.1
34  100 50 12.6 200 50 15.6 300 50 10.6 400 50 12.6 500 50 6.6 600
    50 13.6
35  100 100 23.1 200 100 9.1 300 100 10.1 400 100 13.1 500 100 14.1
    600 100 12.1
36  100 150 17.6 200 150 14.6 300 150 9.6 400 150 6.6 500 150 14.6
    600 150 15.6
37  100 0 15.1 200 0 10.1 300 0 9.1 400 0 6.1 500 0 7.1 600 0 19.1
38  100 50 13.6 200 50 7.6 300 50 5.6 400 50 9.6 500 50 11.6 600 50 15.6
39  100 100 19.1 200 100 13.1 300 100 9.1 400 100 13.1 500 100 13.1
    600 100 17.1
40  100 150 19.6 200 150 17.6 300 150 16.6 400 150 9.6 500 150 14.6
    600 150 10.6
41  100 0 7.1 200 0 11.1 300 0 7.1 400 0 4.1 500 0 11.1 600 0 18.1
42  100 50 11.6 200 50 11.6 300 50 10.6 400 50 8.6 500 50 15.6 600 50
    19.6
43  100 100 15.1 200 100 10.1 300 100 14.1 400 100 11.1 500 100 7.1
    600 100 16.1
44  100 150 23.6 200 150 18.6 300 150 12.6 400 150 14.6 500 150 5.6
    600 150 10.6
45  RUN;
46  PROC REG DATA = one PLOTS=none;
47  MODEL Y = x1 x2 x1x1 x2x2 x1x2 ;
48  MODEL Y = x1 x2 x1x1 x1x2 ;
49  OUTPUT OUT = two p = z;
50  DATA three; SET two;
51  LABEL z = 'Predicted Yield';
52  RUN;
```

We label our variables in lines 5 through 7. In addition, we generate the components we need for our model statement in lines 8 through 11. These statements generate our quadratic components (X_1^2 and X_2^2) and the interaction between X_1 and X_2 (X_1X_2) and X_1^2 and X_2^2 ($X_1^2X_2^2$). In lines 46 through 52 we perform a response surface regression analysis following the rules set out above for eliminating variables from the model. Two model statements have been included to illustrate the process. In practice, the first model statement would be run to provide the following output:

'Chapter 21, Example 15.sas'
'Response Surface Regression and PROC SGPLOT'

The REG Procedure
Model: MODEL1
Dependent Variable: y Yield

Number of Observations Read	192
Number of Observations Used	192

Analysis of Variance					
Source	DF	Sum of Squares	Mean Square	F Value	Pr > F
Model	5	2024.31964	404.86393	49.77	<.0001
Error	186	1513.15952	8.13527		
Corrected Total	191	3537.47917			

Root MSE	2.85224	R-Square	0.5722
Dependent Mean	12.40208	Adj R-Sq	0.5608
Coeff Var	22.99806		

Parameter Estimates						
Variable	Label	DF	Parameter Estimate	Standard Error	t Value	Pr > |t|
Intercept	Intercept	1	15.50417	1.11915	13.85	<.0001
x1	Temperature	1	-0.05543	0.00612	-9.06	<.0001
x2	Nitrogen Rate	1	0.08825	0.01493	5.91	<.0001
x1x1		1	0.00009621	0.00000825	11.66	<.0001
x2x2		1	-0.00000833	0.00008234	-0.10	0.9195
x1x2		1	-0.00018429	0.00002156	-8.55	<.0001

Since the 'x2x2' $\left(X_2^2 \right)$ component is not significant, it would be eliminated from the model, and the regression re-run using the model statement in line 48 to produce the following output:

'Response Surface Regression and PROC SGPLOT'

The REG Procedure
Model: MODEL2
Dependent Variable: y Yield

Number of Observations Read	192
Number of Observations Used	192

Analysis of Variance					
Source	DF	Sum of Squares	Mean Square	F Value	Pr > F
Model	4	2024.23631	506.05908	62.54	<.0001
Error	187	1513.24286	8.09221		
Corrected Total	191	3537.47917			

Root MSE	2.84468	R-Square	0.5722
Dependent Mean	12.40208	Adj R-Sq	0.5631
Coeff Var	22.93712		

Parameter Estimates						
Variable	Label	DF	Parameter Estimate	Standard Error	t Value	Pr > |t|
Intercept	Intercept	1	15.52500	1.09714	14.15	<.0001
x1	Temperature	1	-0.05543	0.00610	-9.08	<.0001
x2	Nitrogen Rate	1	0.08700	0.00837	10.39	<.0001
x1x1		1	0.00009621	0.00000823	11.69	<.0001
x1x2		1	-0.00018429	0.00002150	-8.57	<.0001

All components are significant, thus our modeling is complete (of course, all regression diagnostics as described in Chapter 9 were utilized to verify that we have no problems with our data such as non-normality, heterogeneous variances or outliers). The equation for our response surface is:

$$\hat{Y} = 15.525 - 0.05543 * X_1 + 0.087 * X_2 + 0.00009621 * X_1^2 - 0.00018429 * X_1 X_2$$

But how do we get a plot of our response surface graph? Unfortunately, PROC SGPLOT does not have an option for producing a 3-dimensional surface plot. We first generate a template for our graph using the PROC TEMPLATE procedure in SAS® (line 53). We define a template we call 'surface' in lines 54 through 62. In line 56 we provide a title for our plot ('Response Surface Plot') which you can change to suit your needs. In line 58, we provide the variable names for plotting, in this case x, y and z where x = X_1, y = X_2 and z = the predicted value generated by our regression equation and output to dataset two as variable 'z' which was labeled 'Predicted Yield' in dataset three. We then use the template for plotting in lines 63 and 64.

```
53 PROC TEMPLATE;
54 DEFINE STATGRAPH surface;
```

```
55 BEGINGRAPH;
56 ENTRYTITLE 'Response Surface Plot';
57 LAYOUT OVERLAY3D / cube=false;
58 SURFACEPLOTPARM x=x1 y=x2 z=z / SURFACETYPE=fillgrid;
59 ENDLAYOUT;
60 ENDGRAPH;
61 END;
62 RUN;
63 PROC SGRENDER DATA=three TEMPLATE=surface;
64 RUN;
```

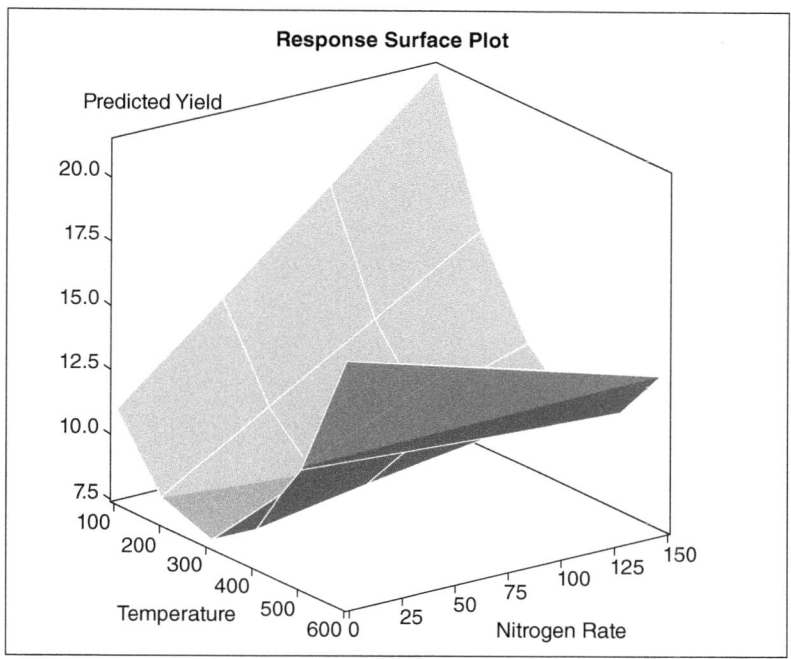

If you need a fancier graph, you can delve into the SAS® coding required for desired modifications.

Bar graphs

One final type of graph which is often used in plant science research is the bar graph. This type of graph is often used to display values for different qualitative treatments such as cultivar, fertilizer source or pesticide used. We will again use PROC SGPLOT for producing graphs for this type of data.

Let's generate a dataset of yield values for five cultivars.

```
01 ODS GRAPHICS ON / ATTRPRIORITY=none;
02 TITLE1 'Chapter 21, Example 16.sas';
03 TITLE2 'Basic Bar Graph';
04 DATA one;
05 INPUT cultivar $ yield @@;
06 LABEL cultivar = 'Cultivar';
07 LABEL yield = 'Yield (Tons/Ha)';
08 CARDS;
09 a 19 a 17 a 27 a 29 a 20
10 b 19 b 32 b 25 b 19 b 31
11 c 23 c 24 c 34 c 32 c 33
12 d 44 d 35 d 32 d 50 d 25
13 e 8 e 14 e 18 e 9 e 9
14 RUN;
```

Note that we have given our variables LABELS in lines 6 and 7. We can create a nice bar graph with the following three lines:

```
15 PROC SGPLOT DATA=one;
16 VBAR cultivar / RESPONSE=yield STAT=mean LIMITS=both;
17 RUN;
```

With line 16 you are requesting a vertical bar graph for the variable 'cultivar', graphing the mean (STAT=mean) yield (RESPONSE = yield) with the 95% confidence limits (LIMITS=both) for each cultivar.

The graph:

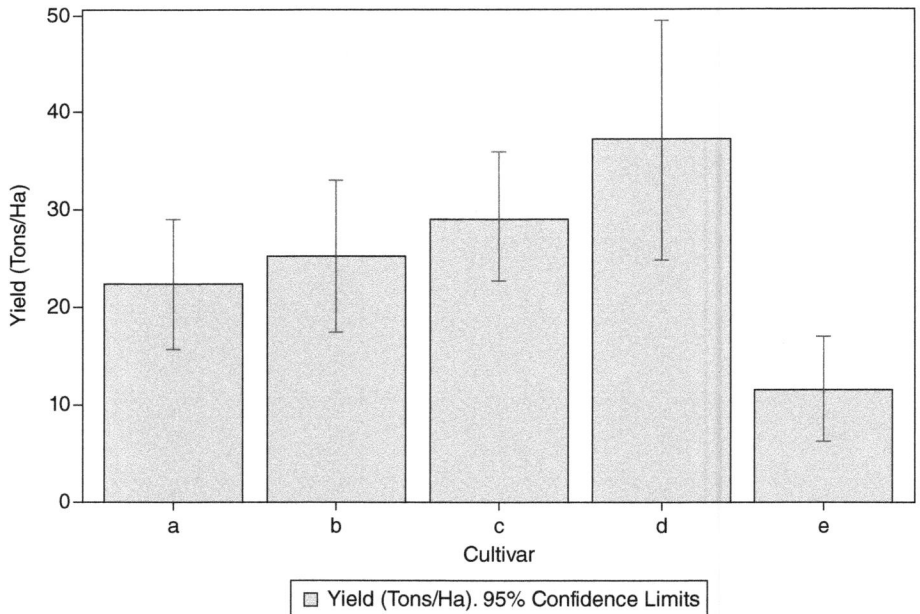

Many variations of this basic graph can be produced rather easily using the many options available in PROC SGPLOT. Many examples are presented in the text by Matange and Heath (2011).

References

Draper, N.R. and Smith, H. (1998) *Applied Regression Analysis*. Wiley Series in Probability and Statistics, John Wiley and Sons, New York.

Eberhart, M. (2010) PROC TABULATE – Building Tables With Style. Available at: https://www.researchgate.net/publication/259147230_PROC_TABULATE_-_Building_Tables_With_Style (accessed 9 April 2020)

Matange, S. and Heath, D. (2011) *Statistical Graphics Procedures by Example: Effective Graphs Using SAS®*. SAS Institute Inc., Cary, North Carolina.

Myers, R. H., Montgomery, D.C. and Anderson-Cook, C.M. (2016) *Response Surface Methodology: Process and Product Optimization Using Designed Experiments*, 4th edn. Wiley Series in Probability and Statistics, John Wiley and Sons, New York.

Index